Surface Phenomena and Fine Particles in Water-Based Coatings and Printing Technology

Surface Phenomena and Fine Particles in Water-Based Coatings and Printing Technology

Edited by

Mahendra K. Sharma

Eastman Kodak Company
Kingsport, Tennessee

and

F. J. Micale

Lehigh University
Bethlehem, Pennsylvania

Springer Science+Business Media, LLC

Library of Congress Cataloging-in-Publication Data

Fine Particle Society Symposium on Surface Phenomena and Fine
  Particles in Water-based Coatings and Printing Technology (1989 :
  Boston, Mass.)
    Surface phenomena and fine particles in water-based coatings and
  printing technology / edited by Mahendra K. Sharma and F.J. Micale.
        p.   cm.
    "Proceedings of Fine Particle Society Symposium on Surface
  Phenomena and Fine Particles in Water-based Coatings and Printing
  Technology, held August 22-26, 1989, in Boston, Massachusetts"--T.p.
  verso.
    Includes bibliographical references and index.
    ISBN 978-1-4613-6700-0   ISBN 978-1-4615-3812-7 (eBook)
    DOI 10.1007/978-1-4615-3812-7
    1. Paper coatings--Congresses.   2. Printing-ink--Congresses.
  3. Surface chemistry--Congresses.   I. Sharma, Mahendra K.
  II. Micale, F. J.   III. Fine Particle Society.   IV. Title.
  TS1118.F5F54 1989
  620'.44--dc20
                                                              91-11802
                                                                   CIP

Proceedings of Fine Particle Society Symposium on Surface Phenomena
and Fine Particles in Water-Based Coatings and Printing Technology,
held August 22-26, 1989, in Boston, Massachusetts

ISBN 978-1-4613-6700-0

© 1991 Springer Science+Business Media New York
Originally published by Plenum Press in 1991
Softcover reprint of the hardcover 1st edition 1991

PREFACE

THE CURRENT STATE OF THE ART of several aspects of water-based
coatings and printing processes is presented in this volume. It
documents the proceedings of the Internationl Symposium on Surface
Phenomena and Fine Particles in Water-Based Coatings and Printing
Technology sponsored by the Fine Particle Society (FPS). This meeting
was held in Boston, Massachusetts, August 21-25, 1989. The symposium
upon which this volume is based was organized in six sessions
emphasizing various basic and applied areas of research on
water-based technology. Major topics discussed involve surface
phenomena in coatings, printing defects and their remedies, surface
tension effects in water-based coatings and printing inks, surface
energies of polymer substrates, wettability, aqueous polymeric film
coating of pharmaceuticals, flexographic and gravure printing
processes, characterization of coating materials, pigment dispersion,
wax emulsions for surface modifications, and the role of polymer in
particle/surface deposition.

This edition includes the twenty four selected papers presented
in the symposium. These papers are divided in three broad categories:
(1) Water-Based Inks and Coatings, (2) Emulsions and Adhesion in
Coatings, and (3) Characterization of Coating and Printing Materials.
Several types of coating and printing on different substrates using
water-based formulations with special reference to surface phenomena
and particle technology are described in these sections.

This proceedings volume includes discussions of various processes
occuring at molecular, microscopic, and macroscopic levels in
water-based coatings and printing processes. We hope that this volume
will serve its intended objective of reflecting our current
understanding of formulation and process problems related to
water-based coating and ink systems. In addition, it will be a
valuable reference source for both novices as well as experts in the
field of water-based technology. It will also help the readers to
understand underlying surface phenomena and will enhance the reader's
potential for solving critical formulation and process problems.

We would like to convey our sincere thanks and appreciation to
the chairmen of the sessions: Dr. D. Barr, Dr. P. D. Berger,
Dr. G. J. O'Neill, Dr. R. M. Podhajny, Professor S. N. Srivastava and
Pofessor M. W. Urban. We wish to convey our thanks to the Fine
Particle Society for the generous support that allowed me to invite
many researchers from several countries to participate in the
symposium. We also would like to express our thanks and appreciation
to Ms. Patricia M. Vann and to the Editorial Staff of the Plenum
Publishing Corporation for their continued interest in this project.

We are grateful to reviewers of the manuscripts for their time and efforts. We wish to convey our sincere thanks and appreciation to all authors and coauthors for their contributions, enthusiasm and patience. One of us (MKS) would like to express his thanks to the appropriate management of the Eastman Chemical Company (ECC) for allowing him to participate in the organization of the symposium and to edit this proceedings volume. His special thanks are due to Mr. J. C. Martin (ECC) for his cooperation and understanding during the tenure of editing this proceedings volume.

Finally, one of us (MKS) is grateful to his colleagues and friends for their assistance and encouragement throughout this project. Also MKS would like to acknowledge the assistance and cooperation of his wife, Rama, in more ways than one, and extends his appreciation to his children (Amol and Anuj) for allowing him to spend many evenings and weekends working on this volume.

M. K. Sharma
Research Laboratories
Eastman Chemical Company
Kingsport, TN 37660

F. J. Micale
Zettlemoyer Center for Surface Studies
Lehigh University
Bethlehem, PA 18015

CONTENTS

CHARACTERIZATION OF COATING AND PRINTING MATERIALS

SURFACE PHENOMENA IN COATINGS AND PRINTING TECHNOLOGY

Mahendra K. Sharma

Eastman Chemical Company
Kingsport, TN 37662

This paper describes various aspects of water-based coatings and printing processes with special emphasis on the surface characteristics of coating/printing films. The film formation depends significantly on the surface properties of formulated coating/ink, and their interactions with substrates. Several surface parameters in relation to coating defects are briefly described. The mechanisms of printing processes and coating/ink film formation by water-based systems are presented. It has been shown that the formation of surface tension gradient during film curing determines the quality of the coating and printing films. Results demonstrate that the incorporation of suitable additives in the formulation can considerably minimize the crater formation. The hydrophilic-lipophilic balance (HLB) concept and the effect of surfactant concentration on pigment dispersion in an aqueous medium are discussed. An attempt was made to correlate the performance parameters of ink and coated film with the properties of the formulated coatings or inks containing various ingredients in order to obtain desired properties of the coating/printing films.

INTRODUCTION

Several possible methods such as solvent recovery, incineration, high solids coatings, reactive coating systems (e.g. UV curable, EV curable) and water-based coating and printing systems are available to converters to comply with federal, state and local Environmental Protection Agency (EPA) standards. Among these alternatives, it was difficult to estimate which compliance method would be most effective and economical for operating each plant. Each of these alternatives has various advantages and disadvantages, as well as needs firm commitment from converters in terms of time, manpower and capital for equipment modification. As a long-term compliance strategy, many converters have decided to evaluate the water-based coating and printing systems.

Surface Phenomena and Fine Particles in Water-Based Coatings and Printing Technology
Edited by M.K. Sharma and F.J. Micale, Plenum Press, New York, 1991

1

Water-based coating and printing systems have several advantages: cost effective, reduce volatile organic compound (VOC) emissions, involve minimum use of and exposure to hazardous organic solvents, easily handled and performs well in certain applications. On the other hand, the main disadvantages include: slow drying, foaming, machinability problem, stability of formulated coatings and printing inks and poor wetting of and/or adhesion to low surface energy substrates. During the past years, considerable progress has been made to overcome these problems employing water-based systems.[1-20]

Several water-based coating/printing formulations involve binder in the form of fine particles dispersed in the system. These are often known as pseudolatex (e.g. colloidal dispersion). If binder employed in coating/printing formulations is present in the dispersion form, the mechanism of film formation is considerably different as compared to film formation from binder in solution form. The pseudolatex particles must coalesce in order to form a continuous and smooth film. The mechanism of film formation from colloidal dispersion is schematically illustrated in Figure 1.

Most colloidal dispersions are milky in color due to large particle size compared to molecular solution (e.g. clear). In order to achieve desired coating/printing properties, the size of these particles must be reduced to a significant extent, which can be obtained by adding plasticizer, suitable solvent as well as by curing at high temperature. One must consider these alternatives for coating/printing from water-based systems containing binder in the form of colloidal dispersion form.

The studies published so far on water-based coating/printing systems often have little appreciation of the surface and interfacial aspects of

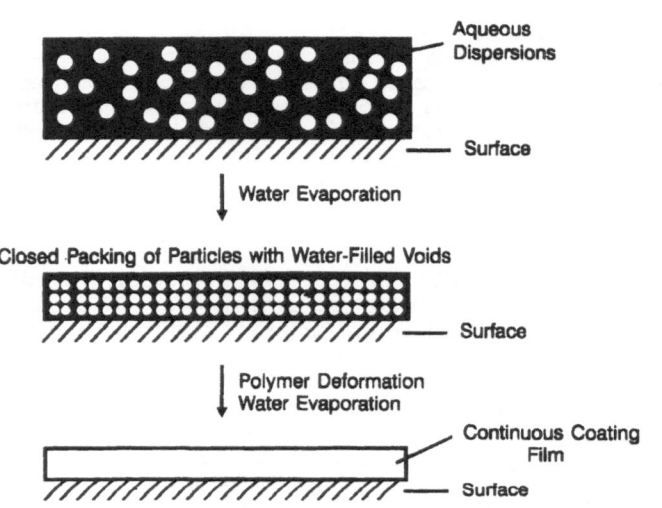

Fig. 1.   Mechanism of Film Formation from Aqueous Dispersions of Polymer.

the phenomena involved during these processes. It is, therefore, the purpose of this article to describe these mechanisms in terms of surface chemistry, which hopefully will lead to more meaningful and successful coating/ink formulations and coating/printing processes. An attempt was also made to describe various surface and colloidal parameters and their role in the stability of formulated coating/ink, coating/printing processes and performance of coated/printed films.

EXPERIMENTAL

Surface Tension (γ)

The surface tension of several liquids used as a solvent in combination with water in coating/ink formulations was measured with a Wilhelmy Plate/DuNouy ring method[21,22] at an ambient temperature. The platinum plate and/or platinum ring were cleaned by exposing to the flame of a Bunsen burner before use. Each measurement was repeated until three consistent values were obtained.

Contact Angle (θ) and Wettability

The wettability of a coating/printing substrate was examined by measuring contact angle of liquids or formulated coating/ink. A constant volume of liquid was drawn using a micrometer syringe for the formation of a drop on the solid coating/printing substrate, and the contact angle was measured with a contact angle goniometer[23] from Rame-Hart, Inc., Mountain Lake, New Jersey, U.S.A. The apparatus was calibrated before use with distilled water, which has a contact angle of 108 degrees on a clean and dry Teflon (tetrafluoroethylene) surface. A drop of the solvent/ink was placed on the desired coating/printing surface. Horizontal and vertical cross wires were adjusted according to the drop size, and the contact angle was determined directly from a circular graduated dial. The contact angle was also recorded from the opposite side of the drop, and the average of both readings is reported. The polar, nonpolar and total surface energies of the substrates were evaluated by measuring contact angle of water and methylene iodide as described previously.[14]

Surface Characterization

Electron spectroscopy for chemical analysis (ESCA) was used to analyze and determine the surface composition of the corona treated and untreated polymer substrates. These studies were conducted by using Perkin-Elmer instrument. The ESCA analyses provide the types of carbon-oxygen functionalities present at the surface of the polymer substrates. It can also determine the atomic percent of the elements on the surface, as well as the concentration of these elements as a function of depth into the surface of the coating/printing polymer substrates.

Evaluation of Coating/Printing Performance

Several selected techniques employed to characterize the coating/ink films formed on various substrates are described as follows:

Microscopic Studies. In order to examine the coating/printing defects, various photomicrographs were taken with the help of a camera attached to the microscope. The pictures were taken at different magnification for in-depth understanding and analysis of coating/printing films. As needed, SEM pictures and x-ray analyses were also conducted to examine various printed/coated films.

Block Resistance. The ink drawdowns were made on coating/printing substrates (e.g. coated paper) with number 3 Meyer rod. These drawdowns were allowed to dry in an oven at 100°C for 3-5 seconds or at an ambient temperature for 24 hours. These samples were evaluated for blocking temperature using the PI Sentinel Heat Sealer at 40 psi for 5 sec. The printed surface of the samples were folded face-to-face and placed under sealer at different temperatures. The test was repeated until the blocking occurred. The integrity of the printed film was visually assessed, and blocking resistance was rated as follows:

1. Poor: Picked and complete film removed
2. Fair: Picked, but partial film removed
3. Good: Slightly picked, but no film removed
4. Excellent: No picking and no film removed

Blocking temperature is defined as the highest temperature where the coatings and/or printed ink retains a blocking resistance rating of greater than 3. The visually assessed data were recorded as a function of temperature.

Water Resistance. Several water-based coatings or inks were employed to examine the water-resistance. The coatings or inks were applied on aluminum foil, polymer films and coated papers using number 3 Meyer rod. These coated samples were allowed to dry for 24 hours at an ambient temperature or dried in an oven at 100°C for 3-5 seconds. The water resistance of the coated/printed film was studied by a water spot test. The drops of distilled water were left for 5, 10, 15 and 20 minutes, and then wiped off gently with a facial tissue. The integrity of the coated film was visually assessed. The water spot test was rated as follows:

1. Poor: Total film removed
2. Fair: Partial film removed
3. Good: Dull and discolored film, but no removal
4. Excellent: The film is substantially unchanged

The visually assessed data for 20 minute water spot test are recorded for various coating/ink formulations, and discussed in the appropriate sections of this article.

RESULTS AND DISCUSSION

Coating/Printing Defects

Several coating and printing defects commonly observed are listed in Table 1. These defects are caused by inadequate formulation, process conditions and interactions with substrates. The possible causes and remedies for improving coating/printing defects are also described in Table 1. Among various coating defects, cratering, dewetting, Bernard cells, and mottling, are shown in Figures 2-5. These defects occur due to the flow of material in the coating or printing film after it is applied on the substrate. The major cause of material flow is the formation of surface tension gradient. The mechanism of the crater formation is shown in Figure 6. The differences in the surface tension in the same parts of the total surface cause cratering. During the drying process, rapid

4

Table 1. Various Coating and Printing Defects and Their Possible Remedies

Defects	Causes	Remedies
Cratering (Volcanoes)	Surface Tension Gradient, Fast Solvent Evaporation	Surfactant, Slow Drying
Mottling (Crawling)	Poor Wettability, Low Viscosity Surface Tension Gradient, Flocculation	Improve Wettability, Increase Viscosity
Pinholing	Poor Wettability, Trapped Bubbles	Improve Wettability
Be'nard Cells	Surface Tension Gradient, Flocculation, Low Viscosity	Surfactants, Increase Viscosity
Bumps and Sinks	Surface Tension Gradient, Low Viscosity	Surfactants, Increase Viscosity
Fogging (Scumming)	High Wettability, Inadequate Ink Wiping	Reduced Wettability, Fast Drying Solvent, Proper Ink Wiping
Fat Edges and Picture Framing	Surface Tension Gradient, Low Viscosity	Surfactants, Increase Viscosity
Orange Peel	High Surface Tension Flow, Poor Leveling	Surface Active Agents, Adjust Viscosity
Silking	Flocculation, Surface Tension Gradient	Surfactants, Increase Viscosity
Rub Up	Pigment Flocculation, Poor Binding	Surfactants, Increase Binder
Floating	Surface Tension Gradient, Flocculation	Surfactants, Increased Viscosity
Picking	Poor Paper Coating, Too Much Wetting, Low Temperature	Improve Adhesion, Reduce Wetting, Increase Temperature
Skipping	Trapped Air Bubbles, Insufficient Ink Supply, Contaminated Water-Ink Immiscibility	Increase Ink Supply, Anti-foaming Agents, Reduce Viscosity
Flooding	Floccuclation, Pigment Density, Size Differences	Improved Pigment Matching, Surfactants
Haze	Pigment Fines, Flocculation, Exudation	Surfactants, etc.

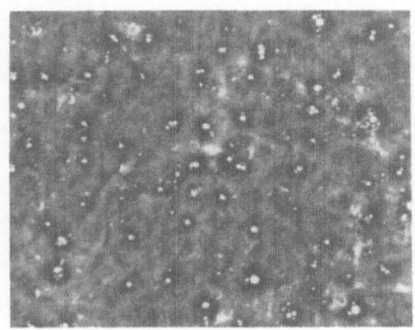 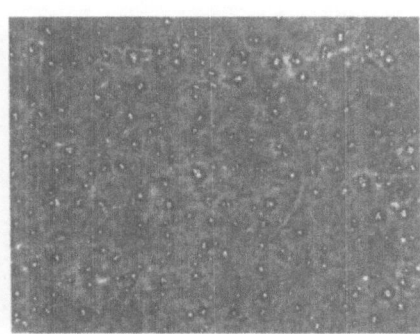

Fig. 2. Photomicrographs of Crater at Different Magnifications

 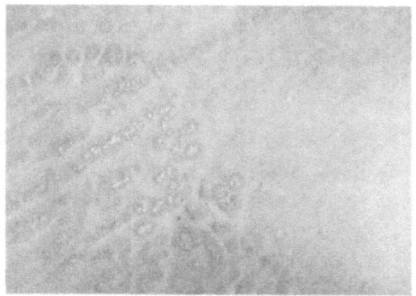

Fig. 3. Photomicrographs Show Dewetting Phenomena at Different Magnifications

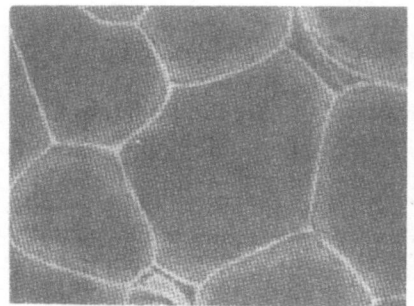 

Fig. 4. Photomicrographs of Benard Cells at Different Magnifications

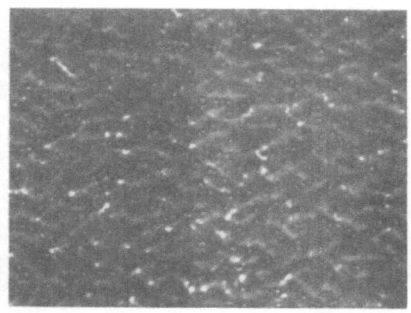 

Fig. 5. Photomicrographs Show Mottling/Crawling Phenomena at Different Magnifications

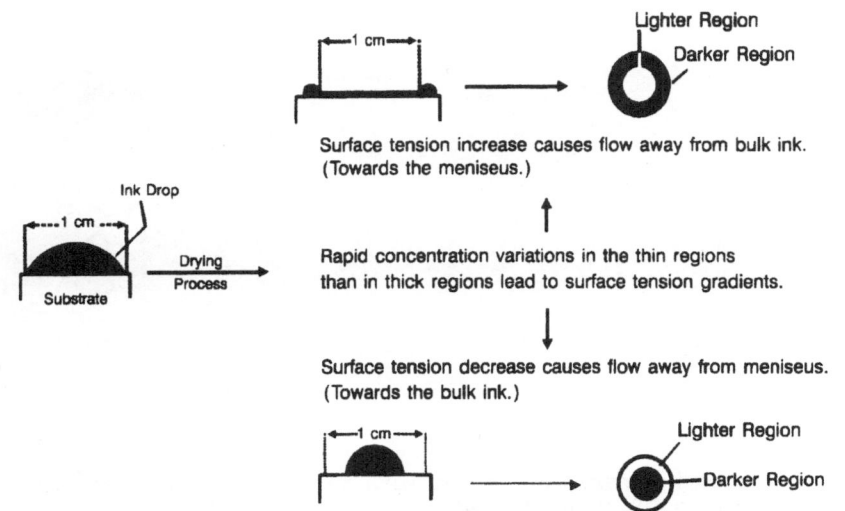

Fig. 6. A Schematic Illustration of the Mechanism of Crater Formation

concentration variations in the thin regions than thick regions lead to the formation of surface tension gradients. If the surface tension increases, the flow of ink/coating occurs away from bulk material (e.g. towards the meniseus). This leads to the formation of a darker region around the lighter region (e.g. crater) as shown in Figure 6. If the situation occurs in which surface tension is decreased, the coating/ink flows away from meniseus (e.g. towards the bulk coating/ink) resulting in the formation of a lighter region around the darker region. In order to overcome this problem, one should minimize the formation of surface tension gradient.

Figure 7 shows the photomicrograph of an aqueous printing on aluminum foil with and without additive. The aqueous printing ink contains Eastek polymer (e.g. EASTEK 1100 from Eastman Chemical Company) as a binder. In the absence of a suitable additive, the crater formation occurs. The incorporation of surface tension reducing additive in the ink formulation can eliminate the crater formation. In order to further understand the formation of craters, the SEM pictures of aqueous printing in the absence of additive were taken at various magnifications with and without fracturing the ink film on aluminum foil (Figure 8). The craters were also examined by x-ray dot-map analysis. As Eastek polymer used as a binder in water-based ink contains sulfur, the sulfur map is taken which indicates that an area in which the crater formed has almost no polymer. These results confirm that there is no polymer film in the area where crater formation occurred in the absence of a suitable additive due to difference in surface tension during film drying process. As mentioned earlier, the crater formation can be minimized or eliminated by adding suitable additive to the ink/coating which reduced the surface tension of the coating/ink formulation (Figure 7).

Coating/Printing

The coating/printing involve: (1) coating/printing processes; (2) substrate to coat/paint, and (3) formulated coating/ink products. In order to minimize the defects described in the previous section, as well as to understand the process problems which occur during coating/printing, the knowledge of ingredients employed in the formulated coating/ink products, substrates and processes are essential. Therefore, these parameters are described in the following sections:

(Craters)   (No Craters)
Without Additive   With Additive

Fig. 7.   Photomicrographs of Craters With and Without Additives in an
          Aqueous Ink Printed on Aluminum Foil

SEM

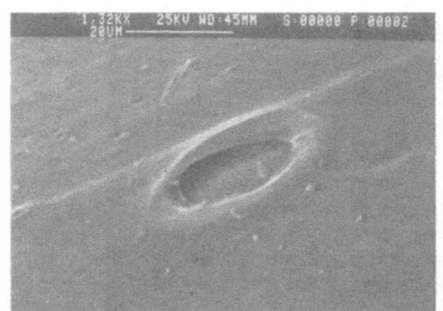 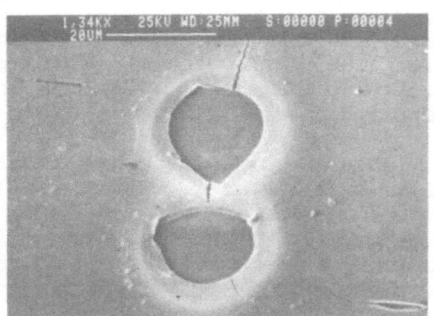

X-RAY DOT-MAP ANALYSIS

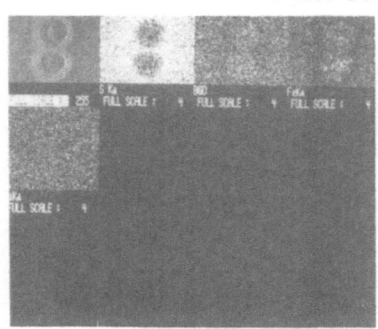 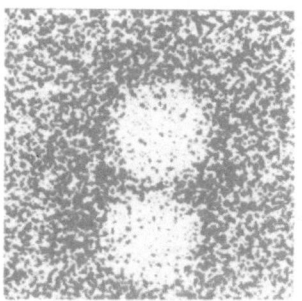

Fig. 8.   SEM Pictures and X-Ray Dot Map Analysis of Printed Aluminum Foil
          from Water-Based Ink

(1) Processes

Among the methods employed for coating/printing, the discussion is restricted to the gravure and flexographic printing processes because of their wide use in water-based printing as compared to other processes for printing a variety of substrates. These two processes are quite different in transferring the ink from roller to the printing substrates.

Gravure Printing Process. The gravure process is also known as the "intaglio" process meaning that the print image is sunk below the surface of the printing plate. The image to be printed is engraved into the plate. If the engravings in the plate are deep, the high viscosity inks are required for printing the substrates. As the engraved cells are shallow, the gravure inks are usually low viscosity (e.g. 20 - 50 cps) as compared to inks required for other printing processes.

Figure 9 schematically illustrates the gravure printing process. The etched cylinder is in contact with the ink fountain. The engraved cells fill with ink as the cylinder rotates, and ink is transferred to the printing substrate which touches the rotating cylinder. The mechanism of ink transfer from metal printing plate to substrate is schematically shown in Figure 10. In order to remove excess ink from the printing plate, a doctor blade is used before the cylinder touches the printing substrate. When the printing cylinder touches the substrate, a combined force due to hydraulic and penetration results in the deposition of ink from the engraved area onto the substrate. As the gravure cylinder is composed of metal which is a high energy surface, it may be difficult to transfer ink to the low energy printing substrates using this process.

Flexographic Printing Process. In the flexographic printing process, the printing plate is composed of rubber. An image area on the flexographic cylinder is raised from the printing plate or cylinder. Figure 11 schematically illustrates the flexographic printing process. Similar to the gravure process, a cylinder transfers ink to the substrate. However, the flexography involves a series of cylinders. A thin film of wet ink is first deposited onto the raised printing plate through a series of rollers. This raised printing plate then transfers ink to the substrate. The mechanism of ink transfer from raised flexographic plate onto substrate is schematically shown in Figure 12.

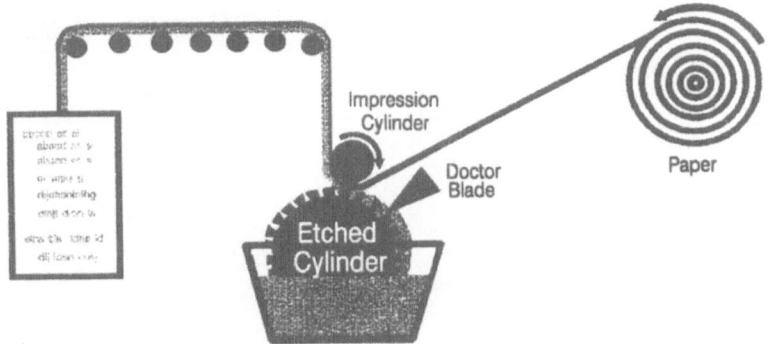

Fig. 9. A Schematic Illustration of the Gravure Printing Process

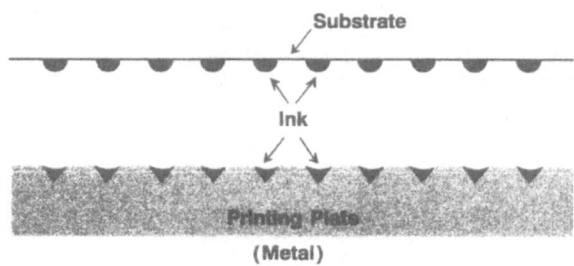

Fig. 10.   Mechanism of an Ink Transfer from Engraved Plate to Substrate
During Gravure Printing Process

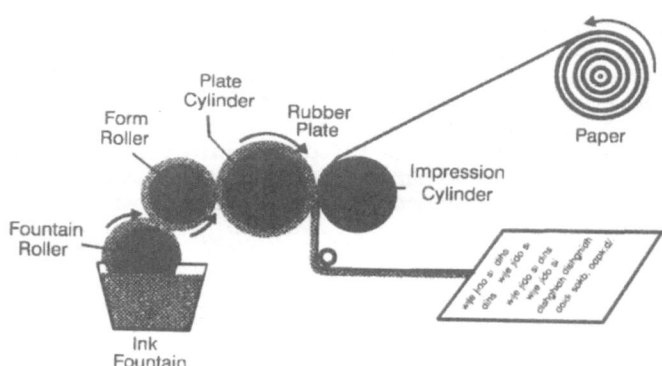

Fig. 11.   A Schematic Illustration of the Flexographic Process

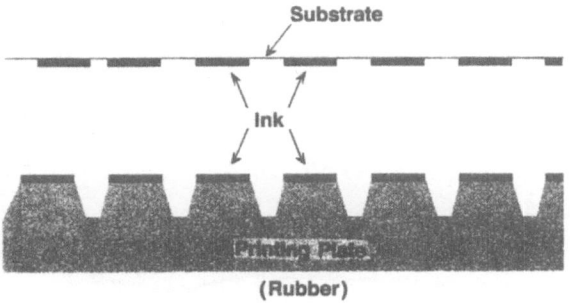

Fig. 12. Mechanism of an Ink Transfer from Flexographic Plate onto
Printing Substrate

This process is ideal for printing flexible substrates such as
polymer films as the raised image area on the rubber printing plate can
easily deform slightly when it touches the substrate without causing the
deformation or damage to the substrate. The viscosity of the
flexographic inks is in the range of 25-200 cps. In general, the dried
film thickness of ink is smaller for flexographic printing as compared to
that of gravure printing process.

Contrary to a gravure cylinder, the surface energy of flexographic
printing plate is low because usually it is composed of synthetic rubber.
Therefore, it may be advantageous to use flexographic process for
printing low energy surfaces like polymer films.

(2) Printing Substrates

Several types of substrate have been employed for printing by
different processes in order to achieve the desired end-use properties.
A wide variety of materials used for coating/printing are listed in Table
2. In general, these materials can be categorized in three broad
classes: (1) paper (coated and uncoated), (2) metallic substrates and (3)
polymer films. Among these three broad classes of printing substrates,
the printing of polymer films from water-based systems is usually
difficult due to their low surface energies. Therefore, these low energy
polymer films were examined for their surface composition, surface
energies, wettability and printability. An attempt was also made to
correlate the surface properties of polymer substrates with printability.

Surface Composition of Non-Porous Substrates (Polymer Films). It is
suggested that the surface energy of the printing substrates should be
higher (e.g. at least 10 dyne/cm) than that of the formulated inks. The
surface energy of water used as a major part of solvent in formulating
water-based coatings/inks is 72.8 dyne/cm, whereas the surface energy of
polymer substrates is in the range of 25-35 dynes/cm, which can be
increased to about 35-50 dynes/cm by corona discharge. It is common
practice to treat polymer films by corona in order to enhance wetting due
to increase in surface energy.

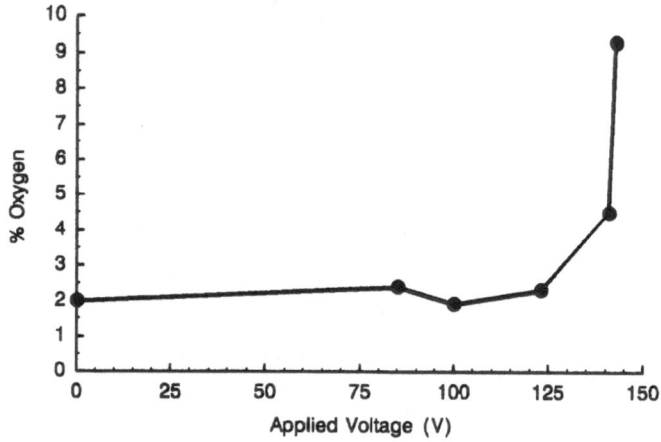

Fig. 13.   Effect of the Applied Voltage on Oxygen Content During Corona
Treatment of Polymer Film

The effect of the applied voltage during the corona treatment
process as a function of oxygen content in the PE film is shown in Figure
13.   Results indicate that the oxygen content in the PE film remains the
same (e.g. about 2.0%) in the range of 0-125 volts.   After 125 volts, the
oxygen content in the PE film increased sharply, and approached to 10.0%
oxygen content at about 150 volt applied voltage.   These data suggest
that for an oxidation of PE film, the applied voltage for the corona
treatment process should be above 125 volts.

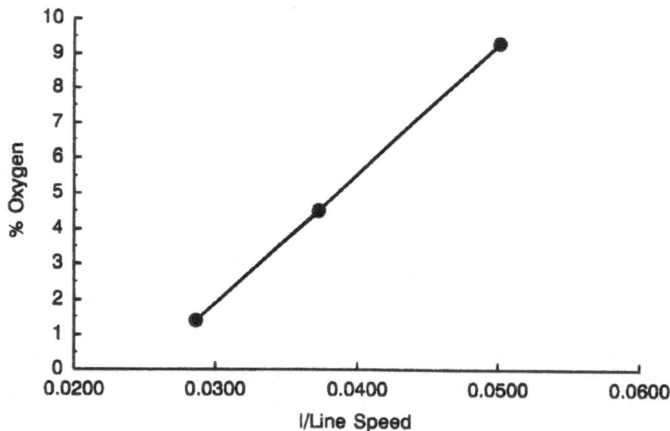

Fig. 14.   Effect of the Corona Treatment Time on Oxygen Uptake by
Polyethylene Film

Table 2.   Several Types of Coating/Printing Substrates

Uncoated paper	End papers for books
Clay coated paper	Onionskin papers
Cellulose nitrate coated paper	Parchment papers mottled and unmottled
Cellulose acetate coated paper	Ground wood paper
Polyethylene and polypropylene coated paper	Kraft paper
Polyethylene paper (Tyvek®)	Polyethylene film and sheeting
Polypropylene paper (Kimdura®)	Poly(vinyl chloride) films and sheets
Rubber latex polystyrene cellulose compositions	Polyester films & sheets
Ledger papers	Cellulose acetate films and sheets
Railroad boards	Polyvinyl acetate films
Bristol boards	Nylon films and sheeting
NCR papers (carbonless forms)	Polyurethane films and sheeting
Carbon collated papers	Wood laminate paneling
Flocked papers	Aluminum and steel sheets and foils
Book cover papers	Metal foils

Figure 14 shows the variation of oxygen uptake in the PE film as a function of the corona treatment time. A linear relationship is observed between the oxygen uptake and reverse of the line speed (e.g. treatment time). As the corona treatment time increases (e.g. 1/Line speed decreases), the oxygen uptake by the PE film increases. These results demonstrate that the treatment time significantly influences the oxygen uptake by the PE film.

The chemical composition of the corona treated surface of polyethylene (PE) film is recorded in Table 3. ESCA analysis indicates that the oxygen content for untreated PE film is less than 2.0%. The chemical composition listed in Table 3 is for different levels of corona treated PE film measured up to a depth of 30 A from the surface. It was observed that the oxygen content in the PE surface increases with increasing corona treatment time. The distribution of carbon-oxygen functionalities produced by the corona treatment process at 4.5% and 9.3% oxygen content in the PE film and their per cent composition are shown in Table 3. These PE films were made without additives. The chemical composition of the PE films remained unchanged for an aging period of six months. This may be due to the fact that no additive was incorporated in the PE film. It is known that the oxygen content in the PE film decreases with aging considerably in the presence of additives such as slip or lubricating agents used in the process of preparing the PE film.

Table 3.   Chemical Composition of the Corona Treated Polyethylene Film

Film Contained 4.5% Oxygen

95.3 %	-C-C-C-	(carbon chain)
2.4%	-C-O	(ethers)
1.8%	-C=O	(carbonyl; keto/aldehydo groups)
0.5%	$-CO_2-$	(acid-ester)

Film Contained 9.3% Oxygen

90.8%	-C-C-C-	(carbon chain)
6.8%	-C-O-	(ethers)
1.8%	-C=O	(carbonyl; keto/aldehydo groups)
0.6%	$-CO_2-$	(acid-ester)

Table 4. Oxygen Uptake by Polyethylene (PE) Film During Corona Treatment Versus Distance Between Discharge Electrode and PE Film

Treatment Distance, Inches	Voltage Applied, V	% Oxygen Uptake
0.0625	98	1.8
	123	2.3
	141	4.5
0.1250	98	2.0
	123	1.9
	141	2.0

The effect of the corona treatment (e.g. distance of PE film from discharge electrode) distance and oxygen uptake by PE film is recorded in Table 4. Results show that the oxygen content is high for corona treated PE films at shorter distances and at higher applied voltages. However, the oxygen content remains the same for PE film treated at 98 volts and different distances. By decreasing treatment distance to 0.05 inches and increasing applied voltage from 98 volts, a significant increase in the oxygen uptake can be achieved.

Surface Energy of the Non-Porous Printing Substrates. Several commercially available polymer films were characterized by measuring contact angle and critical surface tension (CST) or wetting tension. The CST was measured by standard ASTM method.[24] Results obtained are listed in Table 5. The critical surface tension (e.g. wetting tension) of these corona treated and untreated polymer films falls in the range of 30-72 dynes/cm. Results show that the CST of the corona treated films is higher as compared to that of untreated surface due to oxidation in the film by corona treatment process. The high CST film can wet easily with water. These films were also evaluated for printability. The printability was assigned based on the ease of printing the polymer substrates. Results show that there is no correlation between CST and printability.

Table 5. Contact Angle (θ) and Critical Surface Tension (CST) for Various Commercial Polymer Films

Film	Source	CST (dyne/cm)	Contact Angle (θ) Degree H_2O	CH_2I_2
Polyethylene High Density Treated	Deerfield	42 (1)	34 (1)	50 (1)
Polyethylene High Density Untreated	Deerfield	37	76	48
Polyethylene Low Density Treated	Deerfield	40 (4)	92 (4)	63 (4)
Polyethylene Low Density Untreated	Deerfield	31	92	62
Polypropylene	Mobil Oil Co.	45 (2)	56 (2)	42 (2)
Never-Tear (Coated)	Xerox	72	60	31
Polyethylene Terephthalate (PET)	DuPont	56 (3)	58 (3)	25 (3)

Printability = (1) > (2) > (3) > (4)

The contact angle of water and methylene iodide was measured (Table 5) for evaluation of surface energies. The contact angle of water on these films varies widely (e.g. 34 - 90 degree), whereas methylene iodide contact angle is in the range of 31-63 degrees. Results show a correlation between printability and contact angle of water. As the contact angle of water decreases, the film is easy to print. In addition, contact angle of methylene iodide has no correlation with printability.

The surface energies of various polymer films evaluated by using water and methylene iodide contact angle data are presented in Table 6. The polar energy of the corona treated films is higher as compared to untreated film for high density PE film. However, no significant difference in the surface energies was observed for treated and untreated films of low density polypropylene. The ratio of polar to nonpolar (P_E/P_D) was correlated with the printability of these films. Results show that as the P_E/P_D ratio increases, the films can be printed easily.

(3) Coating/Ink Formulations

The coating/ink formulated products mainly contain binder, solvent, pigment and additives in order to achieve printing at desired process conditions and end-use properties. The various properties and functions of these chemicals are described in this section.

Binder. The main function of binder is to form a smooth coating film in the presence of different ingredients used in the coatings or ink formulations. Various types of polymers are used as a binder. Several polymers/resins employed in the coating/printing formulations are listed in Table 7. The surface tension of these polymers varies widely from 19-58 dynes/cm. Based on the availability, end-use, cost and ease in processing, a suitable polymer can be selected as a binder for formulating coating/ink.

Table 6. Surface Energies of Various Commercial Polymer Films

Film	Source	Polar Energy (E_p)	Dispersive Energy (E_D)	Total Energy ($E_T = E_p + E_D$)	E_p/E_D Ratio
Polyethylene High Density Treated	Deerfield	37.1	24.4	61.5	1.517 (1)
Polyethylene High Density Untreated	Deerfield	7.1	31.3	38.4	0.228
Polyethylene Low Density Treated	Deerfield	2.8	24.8	27.6	0.112 (4)
Polyethylene Low Density Untreated	Deerfield	2.6	25.4	28.1	0.104
Polypropylene	Mobil Oil Co.	18.4	31.3	49.7	0.586 (2)
Never-Tear (Coated)	Xerox	13.3	37.4	50.7	0.356
Polyethylene Terephthalate (PET)	DuPont	13.6	39.5	53.1	0.345 (3)

Printability = (1) > (2) > (3) > (4)

Table 7. Surface Tension of Various Polymers/Resins Employed in
Coating/Printing Formulations as a Binder

Polymers/Resins	γ(dynes/cm)
Melamine Resin	57.6
Polyvinyl Butyral	53.6
Benzoguanamine Resin	52.0
Urea Resin	45.0
Polystyrene	42.5
Polyvinyl Chloride	41.9
Polymethyl Methacrylate (PMMA)	41.0
Polyvinyl Acetate	36.5
Polybutyl Methacrylate	34.6
Polyester (Melted)	23.0
Polydimethyl Siloxane	19.8

Table 8. Surface Tension of Various Solvents Used in Water-Based
Coating/Printing Formulations

Solvents	γ (dynes/cm)
Water	72.8
Ethylene Glycol	48.4
Propylene Glycol	36.0
Toluene	27.2
n-Butyl Acetate	25.3
n-Butyl Alcohol	24.6
Ethyl Acetate	23.9
n-Propyl Alcohol	23.9
Methyl Isobutyl Ketone	23.6
Methyl Alcohol	22.6
Ethyl Alcohol	22.0
Isopropyl Alcohol	21.4

Solvent. Several solvents are used for optimizing the properties of coating/printing ink formulations. The main function of solvent is to provide a suitable medium to the ingredients employed in the formulations. In general, the solvent for present water-based systems involves a mixture of water/volatile solvent (usually alcohols) ratios in the range of 95/5 - 80/20. It is desirable to reduce or eliminate completely volatile solvents from the water-based system. The surface tension of the solvents used in coating/ink formulations is also recorded in Table 8. The surface tension for most solvents except water is in the range of 20-50 dynes/cm.

As mentioned earlier, the common solvent for present water-based coating/printing ink formulations is water/alcohol and the surface tension of various water/alcohol mixtures is shown in Figure 15. A sharp decrease in surface tension was observed up to 10% alcohol content. Beyond 10% alcohol content, surface tension is either gradually decreased or almost constant. It is evident that the surface tension of water from 72 dynes/cm to about 30-45 dynes/cm can be achieved by incorporating alcohols, which provides desired formulation properties for water-based coating/ink formulations.

The escape of solvents from coating/printing ink film determines the quality of the coated/printed film. Therefore, the evaporation of alcohols from water/alcohol mixtures was studied at 60°C. It is evident from Figure 16 that the 2-propanol retained in the water/alcohol mixture is less as compared to 1-propanol. In other words, 2-propanol evaporates fast from the water/alcohol mixture resulting in the formation of high surface tension gradient during drying in the coated/printed film. The development of the surface tension gradient due to fast escape of alcohol can cause coating/printing defects. In order to minimize these defects, all solvents should evaporate at the same rate as water providing the same medium to the ingredients dissolved/dispersed in the solvent system.

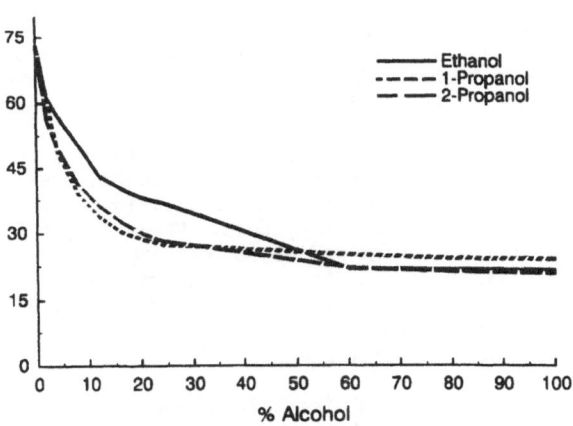

Fig. 15. Surface Tension as a Function of Various Ratios of Water/Alcohol Mixtures at 25°C

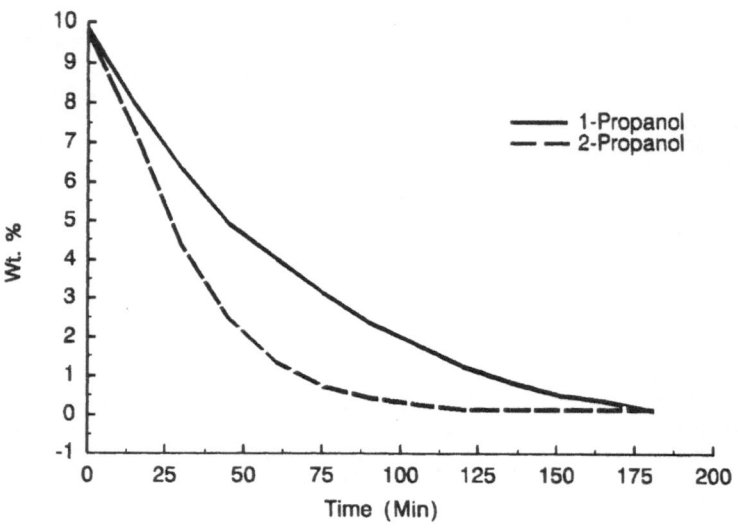

Fig. 16. Evaporation of Alcohols from Alcohol/Water (10/90 Wt/Wt)
Mixture at 60°C

The coating/ink formulations involve binder to form a film.
Therefore, the evaporation of alcohols from water/alcohol mixtures in the
presence of binder was studied, and data are presented in Table 9. The
binder used was a polyester commercially known as EASTEK® 1100 supplied
by the Eastman Chemical Company. It is interesting to note that a
different evaporation behavior of alcohols was observed. In the presence
of a water-dispersible polyester commercially known as EASTEK® 1100,
1-propanol evaporates faster as compared to 2-propanol from 90/10
water/alcohol mixtures in the presence of 30% polymer dispersions. These
data demonstrate that the presence of binder can significantly influence
the evaporation of volatile solvent mixtures. Based on these findings,
use of 2-propanol/water mixture used as a solvent in coating/ink
formulation can minimize the coating/printing defects as well as provide
better solvent medium to EASTEK® 1100 binder and other ingredients as
compared to 1-propanol/water mixture.

Pigments. The main function of pigments in the coating/printing
formulations is to provide desired color. In general, most ingredients
employed to provide desired properties and color are not soluble in the
solvent medium used for water-based systems. Therefore, the dispersion
of these pigments in the solvent medium is most important, which can be
achieved by incorporation of a suitable surfactant. The pigment surface
can be modified by surfactant adsorption. The surfactant structure and
behavior of a water drop on the substrate in the presence of an adsorbed
surfactant layer is schematically shown in Figure 17.

The surfactant molecule contains hydrophilic and lipophilic parts.
These opposite parts (forces) determine the surfactant solubility in
water and oil as well as orientation of surfactant molecules at the
interfaces. If the surfactant molecule is adsorbed at the pigment
surface in such a way that the hydrophilic group is facing away from the

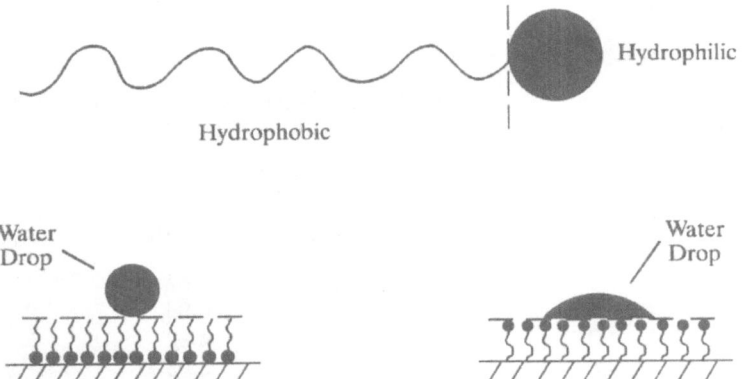

Fig. 17. Surfactant Structure and Behavior of a Water Drop on the
Substrate in the Presence of Adsorbed Monolayers Formed Due to
Different Orientations of Surfactant Molecules

pigment, the water-drop spreads, and the surface becomes water wettable.
If the lipophilic part is facing away from the pigment surface, water
will not spread and the surface remains water nonwettable (Figure 17).

Hydrophilic-lipophilic balance (HLB) values listed in Table 10 were
assigned, based on the water-oil solubility of surfactants. For oil
soluble surfactant, a lower HLB number was assigned, while a higher HLB
number was assigned for water soluble surfactants. The incorporation of
high HLB value surfactants is usually beneficial for water-based
formulations. In certain situations, a combination of low and high HLB
surfactants is also useful for pigment dispersion in an aqueous medium.
Based on the stability of dispersed pigments in an aqueous medium in the
presence of surfactants, the HLB range for various pigments is assigned
and listed in Table 11. These HLB data can assist in selecting a
surfactant for pigment dispersion. The surfactant with the same HLB
value as pigment may provide stable aqueous dispersions. In addition to
the proper surfactant HLB selection, pigment-surfactant interaction,
surfactant structure, surfactant-solvent interaction and surfactant
concentration also play a significant role in the dispersion stability.

Figure 18 schematically illustrates the HLB concept and optimum
surfactant concentration for pigment dispersion. If no surfactant is
present, pigment molecules, because of their hydrophobic nature, will not
disperse in an aqueous medium. The presence of an excess amount of a
surfactant with optimum HLB can also flocculate the pigments due to the
formation of bi-layer of surfactant molecules at the pigment surface
(Figure 18). If the HLB and concentration of surfactant are optimum,
stable pigment dispersions can be formed, and no flocculation occurs for
extended period of aging. The optimum concentration of surfactant in the
dispersed medium can also lead to minimum viscosity, maximum stability,
maximum compatibility and maximum color strength (Figure 18).

Additives. It is generally accepted that additives play an immense
role in optimizing the properties of water-based coating and printing ink
formulations. Several types of additives are employed in the water-based

Table 9. Alcohol Evaporation from Water/Alcohol (90/10 Wt/Wt) Mixture in the Presence of Binder (e.g. 30 wt. % EASTEK®1100) at 60°C

Weight, %

Time, Min	1-Propanol	2-Propanol
0	9.63	9.67
15	3.73	6.88
30	4.83	4.92
45	2.96	3.05
60	2.06	2.29
75	1.71	1.92
90	1.53	1.80
105	1.36	1.71
120	1.17	1.62
135	1.09	1.56
150	1.04	1.52
165	1.00	1.49

EASTEK® 1100 = 30% wt.; Alcohol = 10% wt.; Temperature = 60°C

Table 10. Hydrophilic – Lipophilic Balance (HLB) and Water-Oil Solubility of Surfactant Systems

Water Solubility	HLB
Insoluble	1-4
Poor	3-6
Milky Solution	6-10
Translucent	10-14
Clear Solution	14-20

Table 11. HLB Range for Various Pigments

Pigments	HLB Range
Toluidine Red	8-10
Lampblack	10-12
Quinacridones	11-14
Phthalo Green	12-14
Azo Yellows	13-14
Phthalo Blue	14-16
Titanium Dioxide	16-20

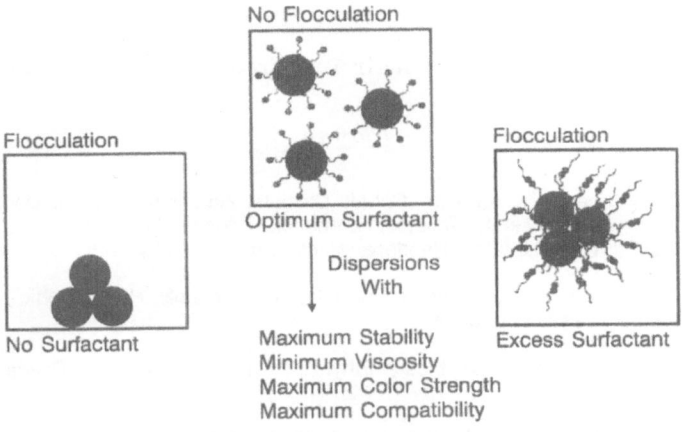

No Flocculation

Optimum Surfactant

Dispersions
With

Maximum Stability
Minimum Viscosity
Maximum Color Strength
Maximum Compatibility

Flocculation

No Surfactant

Flocculation

Excess Surfactant

Fig. 18.   A Schematic Illustration of HLB Concept and Optimum Surfactant
Concentration for Pigment Dispersion

formulations to obtain desired properties either during the
coating/printing processes or in the end-use of the coating/printing
materials. The additives and their functions are listed in Table 12.   It
is suggested that these additives require optimization for each
coating/printing ink formulations because of the presence of various
ingredients in the coatings and inks.

Performance of Water-Based Inks

The water-based inks were prepared for evaluation of printing
performance. Figure 19 schematically illustrates the pigment dispersion
in an aqueous ink with and without oil. The ink containing oil can be
referred to as an emulsion ink because hydrophobic oil is dispersed in
the form of oil globules in an aqueous medium. The pigment particle can
be adsorbed at the oil/water interface. The ink formulated without

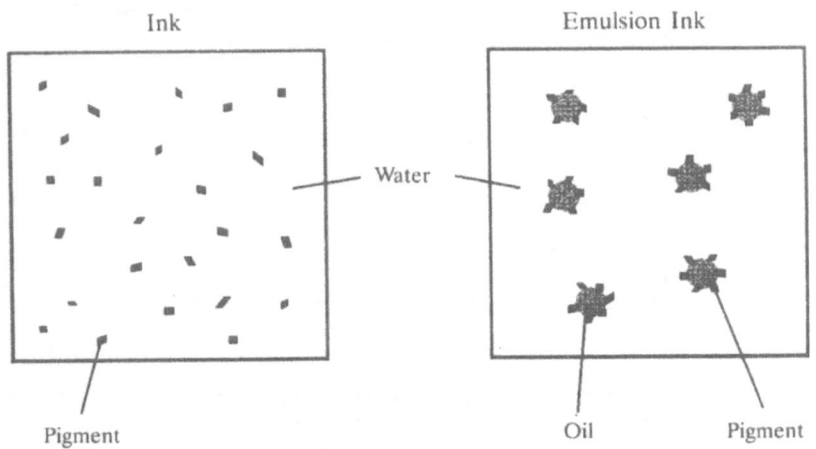

Fig. 19.   A Schematic Diagram for an Aqueous Ink With and Without Oil

Table 12. Several Additives and Their Functions in Coating/Printing Processes

Functions	Additives
Driers	Lead, Manganese, Cobalt, Calcium, Zirconium, Cerium, Zinc, Earth Metals, Tallates, Rosinates, Octoates, Linoleates, Naphthenates, Neodecanoates, Synthetic Acids, etc.
Defoamers	Waxes, Hydrophobic Solvents, Oils, Alcohols, Hydrophobic Silica, Surfactants
Anti-Skinning Agents	Butyraldoxime, Methyl Ethyl Ketoxime, Gelohexanone Oxime, Guaiacol
Viscosity Modifiers	Cellulosic derivatives, clays, alkali soluble polymers, natural gums, etc.
Biocides	Amical 48, Biobans, Cosans, Dowicide A, Tektamer, Nuosepts, Fungitrols, Vancides, Dowicil, etc.
Wetting and Dispersing Agents	Various classes of selected surfactants.

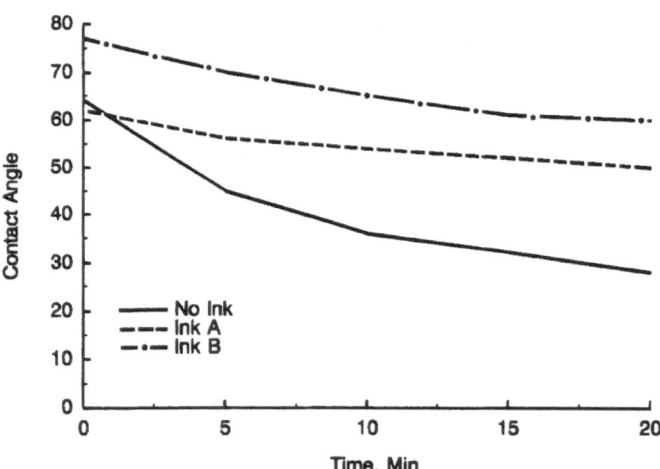

Fig. 20. Contact Angle of Water on Coated Paper Versus Time Before and After Printing with Different Aqueous Inks

hydrophobic oil is referred to as ink A, while ink formulated with
hydrophobic oil is referred to as ink B in further discussions. These
inks contain the same ingredients except one contains oil and the other
is without oil. The performance of these inks was evaluated on coated
paper and aluminum foil.

The contact angle of water on coated paper before and after printing
with ink A and ink B as a function of time is presented in Figure 20.
The contact angle on coated paper decreases with increasing elapse time,
whereas the contact angle remains almost the same after printing coated
paper with both inks. However, the contact angle of water on coated
paper after printing with ink B (e.g. emulsion ink) is higher than that
of ink A. These data suggest that ink B provides relatively higher water
resistance as compared to ink A. Therefore, the ink performance can be
modified by incorporating the desired additives in the system.

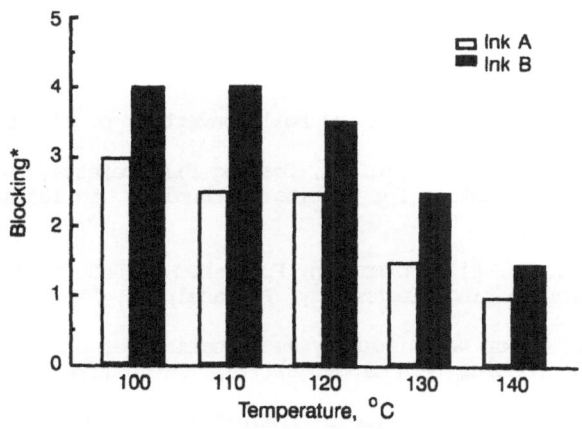

*1 = Poor; 2 = Fair; 3 = Good; 4 = Excellent

Fig. 21. Blocking Temperature for Different Aqueous Inks

The blocking temperature of these two inks is shown in Figure 21.
The experimental data show that ink B exhibits higher blocking
temperature than ink A. The water resistance of these two inks was also
examined by water spot test, and results obtained are plotted in Figure
22. The water resistance by incorporating hydrophobic oil is
significantly improved as compared to ink A (e.g. without hydrophobic
oil).

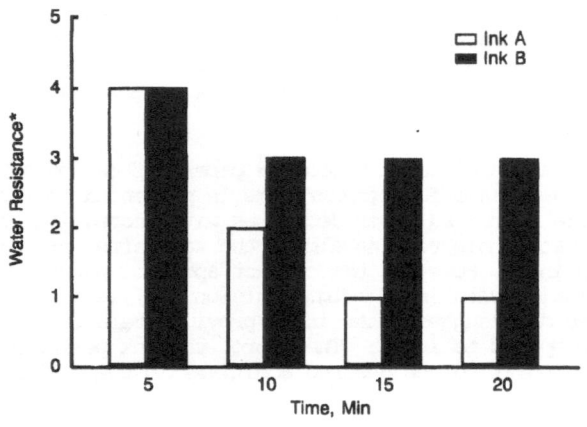

*1 = Poor; 2 = Fair; 3 = Good; 4 = Excellent

Fig. 22. Water-Resistance for Different Aqueous Inks

REFERENCES

1. Lodewyck, P.D., Paper, Film and Foil Converter, p.141, May (1982).

2. Pierce, P. E. and Schoff, C. K., Coating Film Defects, Published by Federation of Societies for Coatings Technology, Philadelphia, PA, Jan. 1988.

3. Wick, Jr., Z. W., Film Formation, Published by Federation of Societies for Coatings Technology, Philadelphia, PA, June, 1986.

4. Schnall, M., Color Nonuniformity as a Coating Defect: Prevention and Cure, J. Coating Technology, 61(773), 33-40 (1989).

5. Podhajny, R.M., TAPPI, Polymers, Laminations, and Coatings, Proceedings, Book 2, p.243 (1985).

6. Bassemier, R.W., American Ink Maker, p.17, Nov. (1980).

7. Nowak, M.T., High Solids Water-Based Inks and Coatings, High Solids Coatings, 7-10, June (1984).

8. Smith, N.C., Water-Based Coatings; Problems and Solutions, TAPPI Proceedings on Polymers, Laminations, and Coating Conference, Book 2, 583-585 (1987).

9. Seifried, Jr., C. G. and Mier, M. A., Surface Characterization of Corona Treated Polyethylene Films, ANTEC' 85, 269-272 (1985).

10. Blythe, A.R., Briggs, D., Kendall, C.R., Rance, D.G., and Zicky, V.J.I.; Surface Modifications of Polyethylene by Electrical Discharge Treatment and the Mechanism of Autoadhesion, Polymer 19, 1273-1278 (1978).

11. Landau, A., Properties of Water-Based Gravure Inks for Wall
    Coverings May Be Useful for Other Products as Well; American Ink
    Maker, 19-21, July (1989).

12. Margraf, D.A., Treatment Required for Printing with Water-Based
    Inks; TAPPI Proceedings on Polymers, Laminations, and Coatings,
    p.333-336 (1987).

13. Stoffer, J.O. and Gadodia, S.K., Development of Tests for Measuring
    Adhesion of Coatings; American Paint & Coatings J., 36-40, May
    (1989).

14. Watson, W.M., Adhering to Polyethylene with Water-Based Inks;
    American Ink Maker, p.38, October (1984).

15. Sonnlag, R.C., Hydrophilic-Lipophilic Balance System and the
    Emulsification of Coatings Components; J. Coatings Technology,
    60(757), 53-61 (1988).

16. Calbo, L.J., Editor, Handbook of Coatings Additives, Marcel Dekker,
    Inc. (1987).

17. Mercer, J.W., Water-Based Gravure Printing: Bureau of Engraving and
    Printing Experiences with Water-Based Gravure Postage Stamp
    Printing, TAPPI Journal, 30-34, December (1986).

18. Zisman, W.A.; Surface Energies of Wetting, Spreading and Adhesion,
    J. Paint Technology, 44(564), 41 (1972).

19. Owen, M.J., Surface Tension of Silicone Release Paper Coatings, J.
    Coatings Technology, 53(679), 65 (1981).

20. Hansen, C.M.; Surface Dewetting and Coatings Performance, J. Paint
    Technology, 44(570), 57 (1972).

21. Sharma, M.K., Shah, D.O. and Brigham, W.E.; A Correlation of Chain
    Length Compatibility and Surface Properties of Mixed Floating Agents
    with Fluid Displacement Efficiency and Effective Air Mobility in
    Porous Media; Industrial Eng. Chem. Fundamentals, 23, 213-220
    (1984).

22. Becker, P.; Emulsions: Theory and Practice, Reinhold Publishing
    Corp., New York, NY, p. 310-316 (1957).

23. Bateup, B.O.,; Surface Chemistry and Adhesion; Int. J. Adhes., 1(5),
    233-239 (1981).

24. ASTM Method D-2578-67 (reapproved in 1979) for measuring CST/Wetting
    Tension of Polyethylene and Polypropylene Films, p.469-472 (1988).

SURFACE PHENOMENA IN WATERBASED FLEXO INKS FOR PRINTING ON POLYETHYLENE FILMS

R. W. Bassemir and R. Krishnan

Sun Chemical Corporation
631 Central Avenue
Carlstadt, NJ   07072

In the Flexographic printing of polyethylene films with waterbased flexo inks, the partitioning of the surfactants between film/ink, pigment/water, ink/air interfaces plays a major role in determining the printability. In addition, in formulations containing nonionic surfactants the equilibrium surface properties are much different from the diffusion limited dynamic properties. Problems associated with the printability are examined from an analysis of the above surface chemical considerations.

INTRODUCTION

In the Flexographic printing process, the image area is raised above the surface of an elastomeric plate. Ink is applied to the image by an Anilox inking system and then is transferred directly to the printing substrate.

The conventional solvent based inks used in this process may contain volatiles such as alcohols, hydrocarbons, esters, and other solvents. Due to increased environmental and safety concerns, ink makers have developed water based inks for a number of substrates used in the packaging industry. [1-3] The dual requirement that the ink vehicle must be water dispersible and yet be completely water resistant when dry, seems contradictory. However, there are several approaches to accomplish this purpose. One is the use of polymers or resins with carboxyl groups, which in turn may be solubilized by an alkaline aqueous medium. (e.g. containing amines or ammonia) When this film dries it loses the volatile alkaline solubilizers which were holding the polymer in solution and it becomes water insensitive.

Another approach is the use of high polymer emulsions, which coalesce on drying into a relatively high molecular weight film, which is water resistant. [5-9]

Surface Phenomena and Fine Particles in Water-Based Coatings and Printing Technology
Edited by M.K. Sharma and F.J. Micale, Plenum Press, New York, 1991

Waterbased inks have been successfully used for printing on absorbent stocks such as paperboard and carton stock for many years. However, flexographic printing on impervious nonabsorbent substrates, such as plastic films or foils is more recent. It involves the surface chemistry of both inks and the substrates in order to obtain good printability, film properties and adhesion. [4] In addition, due to the speed of the printing operation, dynamic surface phenomena play a major role in obtaining good printability. The aim of this paper is to examine the types of surface chemical phenomena involved in several practical situations and to explain how they provided the clues necessary to solve ink or printing problems.

Experimental

Contact angles were obtained by the sessile drop technique on a Rame' Hart contact angle goniometer. HPLC grade water and analytical grade methylene iodide were used for contact angle measurements.

Advancing and receding contact angles were measured using a Cahn Dynamic Contact Angle Analyzer. The advancing contact angle is measured when a liquid is advanced over the solid during immersion and the receding contact angle during withdrawal. The mass changes during immersion and withdrawal are used to calculate the contact angles.

Dynamic Surface tensions were measured using a Sensadyne maximum bubble pressure tensiometer. The accuracy of the instrument is about 0.1 dyne/cm. Calibration of the instrument was done at the various bubble frequencies in the liquid whose dynamic surface tension is desired.

Results

A. Surface Energetics of Film and Flexographic Inks

The polyethylene film used for printing shopping bags is an opaque high density substrate. The film used for bread wrappers is transparent and has a different additive package. Inks are required to have good printability, gloss, adhesion, scratch and crinkle resistance, block and heat resistance, and must be resistant to moisture. In addition, they must be low foaming and be able to rewet the partially dried ink resident on the plate surfaces and on the Anilox roller.

The surface of these films is customarily treated by Corona discharge to improve the adhesion of the ink film to the substrate. Other types of treatment such as flame and oxidation are possible, but infrequently used in commercial practice. Both of these latter techniques are hazardous and/or environmentally unsafe. The Corona discharge technique produces an ion plasma which effects the chemical changes occurring on the plastic surface. Many functional groups have been identified by the use of sophisticated surface analytical techniques. These can include peroxy, hydroperoxy, carbonyl, carboxylic acids, hydroxyl, ether and others.

Both types of polyethylene film were characterized for surface energy and polarity by the use of the Rame' Hart Goniometer to measure contact angles of water and methylene iodide and calculating the surface properties from the raw data using a computer program (based on the geometric mean approximation for the interfacial tension)[10]. The inks were characterized by their contact angles on two model surfaces polyethylene and Teflon pelletized from pure powders, using a hydraulic press and pellet dies. The results were as follows:

Sample	Surface Energy (Dynes/cm)	% Polarity
Opaque Polyethylene	36	5
Bread Bag Polyethylene	35	8
Solvent Based Std. Ink	23	10
Water Based Ink A	36	22
Water Based Ink B	42	30

It is apparent that the waterbased inks in general have higher values of surface tension and relative polarity than the substrates. The former contributes to poor printability and the latter to poor adhesion to polyethylene films due to polarity mismatch.

B. Role of Dynamic Surface Tension of Inks in Printability

The chief components of a waterbased ink are a good, stable pigment, a suitable vehicle for dispersing the pigment, a letdown vehicle to achieve desired film properties, water as a solvent and additives to modify rheology and surface chemistry. Film formation is very important in flexible packaging printing and is largely dependent on the minimum film forming temperature of the vehicle in conjunction with the plasticizer and/or coalescent used in the formulation. The printability of the inks is governed by rheological and dynamic surface chemical properties of the ink.

In flexographic printing, surface renewal speeds of as low as 100-200 milliseconds are typically encountered. For example at a printing speed of 1500 feet per minute, with a plate circumference of 5 feet the surface renewal speed is less than 200 milliseconds (circumference/speed). It has been found that many nonionic surfactants, which are used to control surface chemistry, or may have been used in the vehicle synthesis, can exhibit strongly diffusion limited behavior. For example, several typical commercial flexo waterbased vehicles were measured using the Sensadyne Tensionometer at bubble rates (surface ages) of 200ms and 1000ms. The data obtained is shown in the following table:

| Vehicle | Surface Tension | |
	200ms	1000ms
A	24	22.5 (solvent based)
B	44	35.9
C	55	36.3

It is seen that whereas the surface tension of the solvent based inks are relatively insensitive to the surface age, the waterbased inks exhibit strongly diffusion dependent surface activities. In addition, the waterbased inks composed primarily of emulsion type polymers (ink C) are more dependent on the speed than those based on alkali solubilized dispersion polymers (ink B). This could be due to the diffusibility of the surface active agents used in the preparation of the latex polymers. In addition, the very small size of the dispersion polymers may inhibit diffusion less than the latex.

C. Role of Additives on Dynamic Surface Tension

The primary additives for this purpose are high diffusivity surfactants which will orient at the ink/air interface as rapidly as the surface is renewed during press operation. Diffusivity of material is governed by a number of factors such as, molar volume, temperature, viscosity, etc. In addition, the surface activity of these materials is dependent on solubility in the medium which in turn can be altered by coupling agents of both liquid and solid types. Some examples of this are given in the following table:

Substance	Dynamic Surface Tension (Dynes/cm) @ 200ms
Vehicle A	44
Vehicle A + Additive 1	32
Vehicle A + Additive 2	29
Vehicle + 2% Additive 3	30
Ink + 2% Additive	44

The ink vehicle which had the lowest dynamic surface tension was chosen for this study. Additive 1 is a coupling agent which by itself exhibits minimal or no surface activity. But when added to a vehicle system containing a surface active agent of limited solubility, it decreased the dynamic surface tension drastically (44 to 32). Apparently the effect of this additive is to increase the diffusivity of the primary surface active agent in the system by increasing its solubility.

In the case of additive 2, an anionic surface active agent, the dynamic surface tension of the ink was decreased to very low values. But the printability on polyethylene was poorer than the standard ink without the additive. The explanation for this is presumably that the surface active agent preferentially orients on the substrate rather than the ink interface. Thus the dynamic surface tension measurement of the ink does not reveal the true interaction between ink and the substrate. In this case, it appears that the surface energy of the polyethylene may have been decreased, causing the poor printability.

With Additive 3 the dynamic surface tension of the ink vehicle was also reduced. But when added to a finished ink containing pigment, the surface tension of the ink was not lowered as compared to ink without the additive. The reason for this behavior may be because the additive preferentially orients at the pigment/vehicle interface rather than the ink/air or ink/substrate interfaces. Therefore it has no effect on improving printability. Thus it can be seen that reduction of dynamic surface tension of an ink is a necessary, but not sufficient, condition for improved printability.

D. Water Resistance of Dried Films

One of the unfortunate side effects of surfactant addition can be a lowering of the water resistance of the dried ink film. This effect is undesirable and can cause an unsatisfactory print. The chemical nature of the surfactants and coupling agents chosen will affect this important property and both must be chosen with care, and tested for their effect on water resistance. Variations in alkyl chain length of the coupling agents, once a satisfactory surfactant has been identified, can be a quick route to obtaining a balance of printability and water resistance.

E. Specific Chemical Interactions Affecting Printability

If an ink which has reasonably good printability has been formulated using the techniques described in the above sections, it has been found that the addition of small quantities of certain specific chemicals can cause a marked increase in the printability and smoothness of laydown of waterbased inks on difficult substrates such as high slip polyethylene. This occurs even though the dynamic surface tension of the ink and its water resistance have been optimized. It appears to be a specific synergistic chemical effect with the vehicle of the ink and its interaction with the substrate. No rheological differences were observed when such an additive was added to a well formulated water based ink. No difference in dynamic surface tension or polarity were found. However, a test for receding and advancing dynamic contact angle using the Cahn Dynamic Tensionometer on certain model polymeric surfaces did reveal a significant differences between the two inks. This indicates that dynamic wetting angles can be a valuable technique if model substrates can be obtained upon which the studies can be performed. Some typical results that were found are given:

31

Substrate Polymer	Receding Contact Angles	
	Ink A	Ink A ± Additive
Polyethylene	65	60
Teflon	77	70
Polyamide	37	5

The polyamide substrate was used as a model for the amide additives usually present on the surface of high slip polyethylene.

It is apparent that the additive which led to marked improvements in printability did not cause any rheological changes in the ink (Fig. 1). In addition, the contact angles on polyethylene and Teflon were only affected marginally. So to try to elucidate the specific chemical interactions, a pure polyamide rod was used as the probe in a Dynamic Contact Angle Analyzer. A typical plot of dynamic contact angle is shown in Fig. 2. The ink with the additive exhibited a large decrease in dynamic contact angle. This is apparently due to some specific interaction between the additive and the polyamide. Indeed the slip additives used in the high slip polyethylene are usually waxes of the amide type. The differences in printability were absent in polypropylene, further confirming that specific interaction between the ink and the substrate was the cause of the remarkable improvement in printability and adhesion, using this additive.

It can be seen from this data that dynamic receding contact angles show a significantly lower value when good printability on the particular substrate (high slip polyethylene) in question is achieved.

It is also interesting to note that pure polyethylene without the slip additives did not show the difference in wettability in printing tests.

Conclusions

1. The printability of water based inks on polyethylene films can be improved not only by lowering the dynamic surface tension but also by lowering the surface polarity as much as possible. While these conditions are necessary, they may not be sufficient to assure good printability and transference.

2. Inks exhibiting strongly diffusion dependent surface tension can lead to poorer printability.

3. The surface tension should be measured at surface ages corresponding to printing speed.

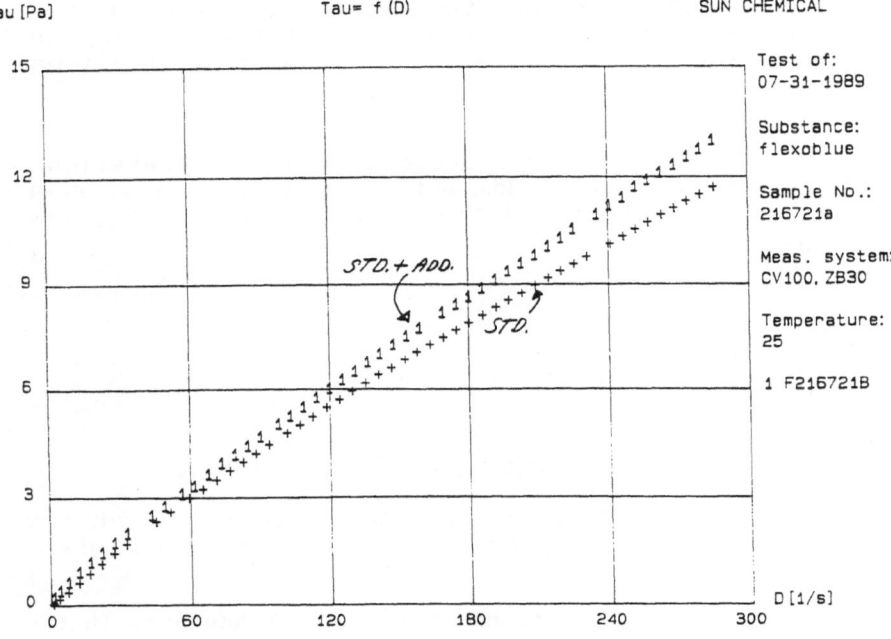

Figure 1

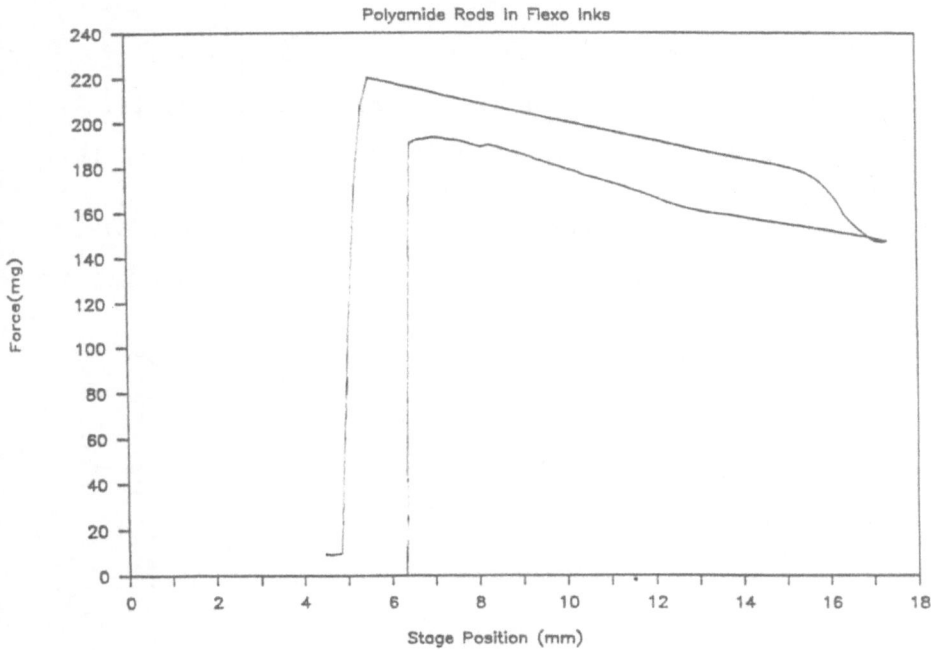

Figure 2. Dynamic Contact Angle Analysis.

4. The additives used to lower the dynamic surface tension of the inks should neither lower the free energy of the substrate nor decrease the water resistance of the dried film.

5. In addition to the above general consideration, specific surface chemical interactions between components of the ink and substrate can affect both printability and adhesion.

References

1. Hutchinson, G. H., JOCCA, 1985 (12) p.306-315.

2. Carlick, D. J., International Symposium on Flexography European Flexographic Technical Association, Birmingham, England, 1984, 6:1 to 6:8.

3. Vanderhoff, J. W., National Printing Ink Research Institute, Lehigh University Publication.

4. Podhajny, R., Flexographic Technical Association Annual Meeting, 1985.

5. Dunn, H., 8th TAPPI Graphic Arts Conference, Miami, Florida (1971).

6. Knodeor, S., American Ink Maker, 1980, 58 (3) 16.

7. Sauntson, B. J., British-Ink Maker, 1975, 18 (1) 26.

8. Knightley, J. A., Polymers Paint Col. J., 1977, 167, (326).

9. Firmin, D. M., Polymers Paint Col. J., 1975, 165, 577.

10. Owens, D. K., and Wendt R. C., J. Appl. Polym Sci 13 (1969).

WATER-DISPERSIBLE POLYESTERS FOR GRAVURE PRINTING INKS:

A NEW TECHNOLOGY

George J. O'Neill

Eastman Chemical Company
Kingsport, Tennessee 37662

Eastman has developed water-based (W-B) ink products for gravure printing that use novel water-dispersible polyesters for binders and film-formers. One of these linear, high molecular polyesters, which is known as Eastek 1100 polyester, imparts to such gravure products as inks, extenders and overprints an array of desireable properties-of-merit that does not exist in today's W-B materials: low odor, excellent rewettability, rapid drying and the absence of ammonia or amines to achieve water dispersibility. Since no amines are present the printer can run at a mild pH range of 5-7 without concern for maintaining the pH level because the amine has volatilized. This paper will describe selected chemical and physical properties of Eastek 1100 and certain performance character-istics of gravure printing inks made with this polymer.

INTRODUCTION

This year marks the 20th anniversary of the Clean Air Act of 1970[1]. As a result of this law and the 1977 compliance guidelines of the Environmental Protection Agency (EPA), the gravure industry made profound changes in its printing operations to substantially reduce the emission of volatile organic compounds (VOCs) from solvent-based inks[2].

In the 1970's, the best available technologies to reduce VOC emissions were solvent incineration and recovery. At the time, it was thought that water-based (W-B) ink technology would be an effective way to achieve EPA compliance but it could not meet the high performance and quality standards of the gravure industry.

In the 1980's the situation brightened considerably as ink chemists and formulators worked with printers to develop

Surface Phenomena and Fine Particles in Water-Based Coatings and Printing Technology
Edited by M.K. Sharma and F.J. Micale, Plenum Press, New York, 1991

35

W-B inks with improved press performance and print quality.
Now gravure printers have potentially lower cost alternatives
to buying or upgrading solvent incinerators or recovery units
to achieve EPA compliance guidelines.

One of the technical advances that enabled the
commercialization of W-B inks was the introduction of polymeric
binders based on styrene acrylic acid (SAA) copolymers, which
can be solubilized in water by reacting the acrylic acid
carboxyl groups with ammonia or a volatile amine to form
carboxylate salts (Fig. 1). Inks made with these versatile
resins have greatly improved press performance and print
quality especially in such markets as corrugated paperboard[3].
However, despite these improvements their acceptance as
replacements for solvent based inks in the gravure industry has
been slow because they suffer from the deficiencies shown
below[1]:

Press performance - Slow drying speeds
                                 Strong odors
                                 Difficult to wash up
                                 pH adjustment

Print quality - Poor lay
                                 Excessive dot skip
                                 Drying-in on the cylinder

FORMATION OF AMINE-CARBOXYLATE SALTS IN
STYRENE/ACRYLIC ACID COPOLYMERS

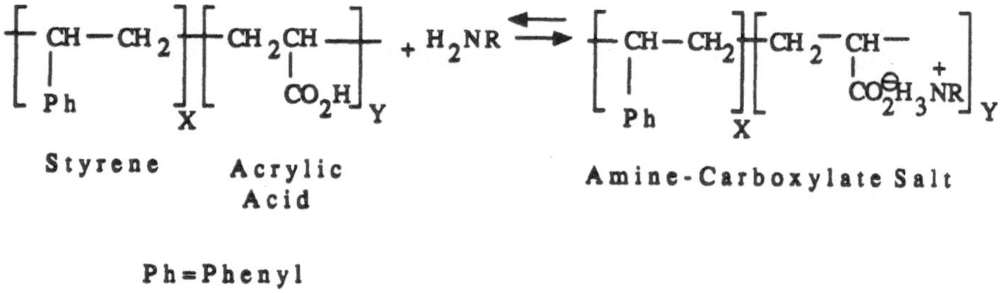

Styrene   Acrylic   Amine-Carboxylate Salt
            Acid

Ph=Phenyl

Figure 1

According to Podhajny[1] one of the major causes for the
deficiencies of W-B inks made with SAA resins is their
properties and performance are dependent on the pH of the
aqueous system. Figure 2 summarizes what happens to these inks
as the pH varies.

EFFECTS OF pH CHANGES ON PERFORMANCE OF
W-B STYRENE-ACRYLIC INKS

pH	Effects
6-7	Polymer precipitates
7-8	High viscosity, ink unstable
8-9	Good press performance and print quality
10-11	Poor drying, solvent retention, strong odor

Figure 2

In an effort to overcome the disadvantages of controlling the pH with ammonia or amines, Eastman introduced the Eastek polyester system in which the sodiosulfonate group ($SO_3^- Na^+$) is incorporated into the polymer backbone rather than the carboxylate salt group[1]. Fig. 3 illustrates the general structural formula for Eastek 1100 polyester. It can be seen that the chemical and physical properties of the water-dispersible polymer can be widely varied by the appropriate selection and concentration of the acids and glycols that are used to make them[5].

STRUCTURAL FORMULA AND PROPERTIES
OF EASTEK 1100

$$HO-G-O_2C-A-CO_2-G-O_2C-A-CO_2-G-O_2C-A-CO_2-G-O_2C-A-CO_2H$$

$$SO_3^- Na^+ \qquad\qquad SO_3^- Na^+$$

Melting point	None
Molecular weight (Mn)	14,000
Acid No.	< 2
Hydoxyl No.	< 10
Glass Transition Temp. (Tg, °C)	55
Min. Film Forming Temp. (MMFT, °C)	< 5

A = Aromatic
G = Aliphatic or alicyclic

Figure 3

By using the thermally stable sodiosulfonate group, there is no need to adjust the pH of W-B inks made with Eastek 1100 because no volatile amine is lost. Furthermore, none is released during drying nor is any retained in the printed product.

Selected Properties of Eastek 1100

This resin is a linear, amorphous thermoplastic and typical chemical and physical properties are shown in Fig. 3. The high molecular weight and glass transition temperature (Tg) are thought to impart sufficient cohesive strength and hardness for a coherent, abrasion resistant film. Furthermore, the low concentration of carboxyl end groups: Acid No.< 2 compares to 30-100 for SAA copolymers.

One of the unusual film-forming characteristics of Eastek 1100 is its minimum film forming temperature (MFFT) of < 5°C which is substantially lower than the polymer Tg of 55°C. This low MFFT is postulated to be the result of water serving as a volatile, coalescing agent for the polar, hydrophyllic polyester which enables the film to dry rapidly at unusually high press speeds for W-B inks.

EXPERIMENTAL

A 30% solids, aqueous dispersion of Eastek 1100 was prepared from polymer pellets by adding them to water at 70°C, with stirring. A clear, transparent dispersion with particle size in the range of 10-15 Å was obtained on cooling at room temperature. In order to improve the printing, 5-10% n-propanol or isopropanol was added to the aqueous polymer dispersion before formulation of ink for gravure printing.

RESULTS AND DISCUSSION

Selected Properties of Aqueous Dispersions of Eastek 1100

In general, aqueous dispersions of Eastek 1100 can be made without the use of ammonia, amines, surfactants or cosolvents. Typical properties of a 30% solids, aqueous dispersion of this polymer are shown in Fig. 4. The pH range is 5 to 7 which compares to 8 to 10 for SAA based inks. The dispersion viscosity is 42 cP and rises rapidly as the solids level increased. (Fig. 5).

Comparison of Printing Performance of Eastek 1100 and SAA W-B Inks

In production trials using commercial carton stock, a major gravure printer evaluated several W-B red inks including Eastman's to compare their press performance and print quality. The trial parameters are shown:

SELECTED PROPERTIES OF AN AQUEOUS DISPERSION
OF EASTEK 1100

% Solids	30
pH	5-7
Viscosity (cP, 100 rpm)	42
Shelf life	6 months
Freeze-thaw	pass, 3 cycles

Figure 4

PRINTING TRIAL PARAMETERS

Variables	Constants
Press speed: 400 & 800 feet per min (fpm).	Press and pressmen
Ink coverage - light to heavy	Day of trial
Inks	Dryer settings
	Carton stock

The responses that were measured to determine the press performance and print quality were:

> Drying (tracking or smearing)
> Lay of the ink
> Amount of dot skip

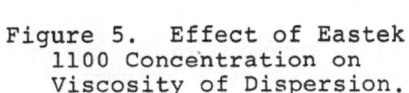

Figure 5. Effect of Eastek 1100 Concentration on Viscosity of Dispersion.

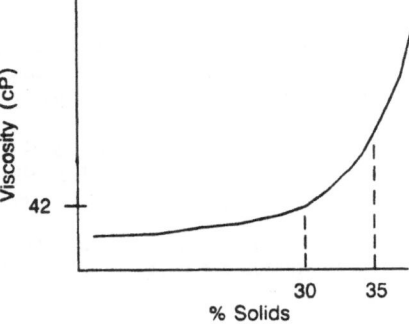

The trial results as judged by Eastman are summarized in Fig. 6. It can be seen that the production rates (drying), quality of lay and amount of dot skip were unacceptable for inks A and B when the ink coverage varied from heavy to medium regardless of the press speed. When the coverage was light, all inks dried well but A and B inks had fair quality of lay and medium to high levels of dot skip. The Eastman ink made with Eastek polyester gave

Printing Variables			Print Quality		
Supplier	Ink Coverage	Press Speed[1]	Dry	Lay	Dot Skip
A	Heavy	400	No	Fair	Low
B	↓	400	No	Fair	
Eastman[2]		800	Yes	Good	↓
A	Medium	400	Yes	Fair	Medium
	↓	800	No	Poor	Medium
B		400	No	Fair	High
Eastman[2]		800	Yes	Excellent	Low
A	Light	800	Yes	Fair	Medium
B	↓	↓	↓	Fair	V. High
Eastman[2]				Good	Low

1 = Feet per Minute
2 = Made with Eastek 1100

Figure 6. Comparison of Print Quality on Folding
Carton Stock of Selected Gravure W-B Inks.

excellent drying rates, good to excellent quality of lay
and low level of dot skip at 800 fpm.

In other commercial applications it has been found
that despite the rapid drying of Eastek polyester inks,
they do not present major problems in press clean up or
drying-in on the cylinder.

CONCLUSIONS

1.   Eastman W-B inks are based on novel binder resins:
     Eastek 1100 polyester which contains the $SO_3^- Na^+$
     group.

2.   These polyesters do not require ammonia or amines to
     be rendered water-dispersible and thus do not require
     pH adjustment nor do they have strong odors at the
     press or in the printed product.

3.   Eastek W-B inks give good quality gravure printing on
     carton stock at high production rates on a gravure
     press.

4.   These inks are easily cleaned up and do not have
     drying-in problems on gravure cylinders.

REFERENCES

1.   R. M. Podhajny, TAPPI Notes: 1988 Film Estrusion Short
     Course Seminar, p. 107-115, TAPPI Press, Atlanta, GA,
     USA.

2.   Anon., Paper, Film & Foil Converter, July, 1986, p. 46,
     R. H. Leach, Editor.

3.   The Printing Ink Manual, 1988, 4th Ed., R. H. Leach,
     Editor, Van Nostrand Reinhold Co. Ltd., London, p. 458.

4.   Ibid., p. 265

5.   N. I. Kotera, et al, U.S. Pat. 4,340,519, July 20, 1982.

SURFACE TENSION EFFECTS ON THE ADHESION AND DRYING OF WATER-BASED INKS AND COATINGS

Richard M. Podhajny

Graphic Arts Industries
P.O. Box 137
Sparta, New Jersey 07871

Water-based ink and coating use is reviewed with the emphasis on wetting and printing of film, metal, and metallized substrates. This paper addresses the effects of surface tension of water-based flexographic and rotogravure inks and coatings. The mode of drying in water-based technology is explored as well as static and dynamic surface tension of inks and coatings. Ink formulation and manufacturing considerations are reviewed to optimize surface tension effects for high speed presses. The role of substrate and ink transfer mediums are discussed relative to their impact on ink and coating drying rates. Corona treatment of film substrates is analyzed from the perspective of its effect on drying speed and ink adhesion. Suggestions are made to improve the quality performance of water-based ink and coatings through use of on-line surface tension equipment. Most importantly, the question "Why should I measure surface tension of my water-based ink?" is answered from the perspective of improving press'productivity and ink performance.

INTRODUCTION

The last twenty years have seen the greatest growth in water-based technology. This growth was stimulated by the Clean Air Act of 1970 which established The Environmental Protection Agency with the responsibility to issue regulations, clean up and protect the environment.

The 1970's can be considered the trial years for water-based technology. It was during the seventies that some printers took their first steps to running water-based inks and coatings. Ink companies found that very few resins were available with suitable properties. Pigments often were incompatible with water-based systems. Poor printability, inadequate adhesion, and poor drying rates were major problems that were encountered. Printers were ill equipped with conventional press equipment to deal with these new inks. They couldn't substitute the water-based ink for their solvent-based ink. If it didn't work, they had little patience to pursue the effort further.

However, things changed in 1977. It was the year the Clean Air Amendment was passed, giving the EPA the authority to enforce its regulations. Water-based inks were not just an interesting possibility for meeting long term compliance goals, but they became a short term compliance necessity for many printers. The 1970's ended with water-based inks being used on porous substrates. The full commercialization of water-based inks on non-porous substrates had to wait till the eighties[1].

Surface Phenomena and Fine Particles in Water-Based Coatings and Printing Technology
Edited by M.K. Sharma and F.J. Micale, Plenum Press, New York, 1991

WATER-BASED INK AND COATING APPLICATION PROCESSES
1970'S - THE TRIAL YEARS

* FLEXOGRAPHY

* ROTOGRAVURE

* OFFSET GRAVURE

FIGURE 1

The conventional flexographic process is composed of a fluid ink enclosed in an ink fountain and transferred from the fountain to the substrate by means of an "ink train" shown in Figure #2.

The ink is transferred to the substrate by means of a three roll ink train. This involves a rubber fountain roll, a metal anilox roll, and the plate cylinder. This process has been found to well suited for using water-based inks.

THE CONVENTIONAL FLEXOGRAPHIC PRINTING PROCESS

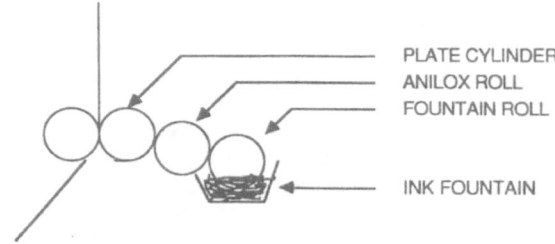

PLATE CYLINDER
ANILOX ROLL
FOUNTAIN ROLL

INK FOUNTAIN

FIGURE 2

The rotogravure process involves the transfer of fluid inks from engraved cells within a metal cylinder directly to the substrate as shown in Figure #3.

The gravure cylinder has an outer protective layer of chrome. The excess ink is removed by means of a steel doctor blade. When solvent rotogravure inks are used, the ink left on the land areas is dried before the print area contacts the substrate. The dry solvent ink layer readily redissolves as it passes into the fountain, and the wiping doctor blade process is repeated before the next print rotation. This is not the case with water-based inks. Water-based inks, either do not dry and lead to ink transfer in the non-image areas, or dry and are basically insoluble in the ink fountain at the next rotation. Obviously, this is a major and difficult problem for water-based inks. As a result, this process is better suited for soluble water-based resin systems.

Water-based coatings were best applied by offset gravure. The application process is shown in Figure #4.

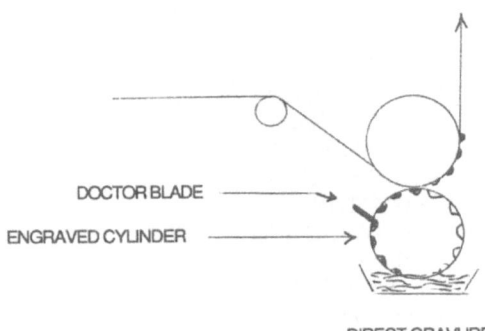

DOCTOR BLADE

ENGRAVED CYLINDER

DIRECT GRAVURE

FIGURE 3

Direct gravure had difficulty handling higher solids water-based coatings. In addition, the higher surface tension associated with the water-based coatings minimized the flow out characteristics of these coatings and often left the cell pattern on the substrate. Although roll coating application methods were shown to work, maintaining an appropriate thickness with low viscosity coatings was found to be difficult.

The 1970's ended a decade that was devoted to learning how to run water-based inks and coatings.

So, what was learned? (Figure #5). Water-based inks dried slower, had poor wettability and adhesion characteristics. Presses required significant modifications to run water-based inks. Inks were not stable and pH drift became a major problems for printers. Use of solvents were necessary to reduce ink surface tension.

It was clear if water-based inks were to be fully commercialized, the inks and equipment needed considerable improvement.

APPLICATION OF WATER-BASED COATINGS

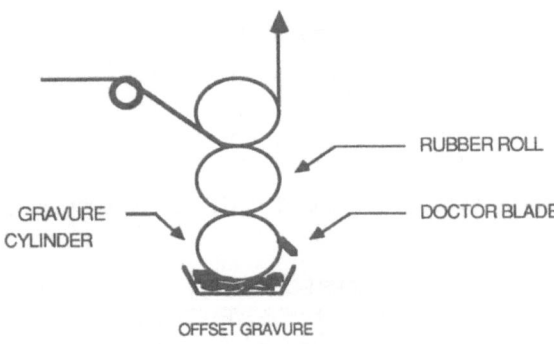

RUBBER ROLL

GRAVURE
CYLINDER

DOCTOR BLADE

OFFSET GRAVURE

FIGURE 4

WHAT WAS LEARNED?

* DRYING WAS TOO SLOW.
* WETTABILITY, ADHESION, AND RHEOLOGICAL CHARACTERISTICS WERE OFTEN POOR.
* THE DRYING MECHANISM WAS NOT THE SAME AS THAT OF SOLVENT-BASED SYSTEMS.
* PRESS EQUIPMENT HAD TO BE MODIFIED
* pH HAD TO BE CONTROLLED
* INKS ' /ERE NOT STABLE
* ODOR WAS DIFFERENT
* INK TRANSFER WAS UNPREDICTABLE
* PHOTOPOLYMER PLATES WERE BETTER
* INK ADHESION WAS POOR
* SURFACE TENSION WAS MORE IMPORTANT

FIGURE 5

II. WATER-BASED INKS TODAY

Without question, the last nine years have seen dramatic changes in the technology of water-based inks used by printers and converters. These improvements have come from different sectors of the printing industry.

Among the major technological advances responsible for the success of water-based inks today are: Press equipment, dryer improvements, resin and raw materials, the use of in-line corona treatment, and the introduction of many new films that are primed to accept water-based inks. (Figure #6).

Perhaps the most traumatic experience of the pressman in changing to water-based inks has been to understand the basic differences of these inks as shown in Figure #7.

Solvent-based inks are often based on nitrocellulose and a modifying resin such as polyamide. Water-based inks are formulated from a variety of soluble resins and emulsions. Some of the resin binders used in water-based inks and coatings today are illustrated in Figure #8.

It must be remembered that solvent-based inks utilize soluble resins.

TECHNOLOGICAL ADVANCES IN WATER-BASED INK USAGE

* PRESS EQUIPMENT
* DRYER IMPROVEMENTS
* WATER-BASED INK FORMULATIONS
* IN-LINE CORONA TREATMENT
* USE OF PRIMERS

FIGURE 6

COMPARISON OF WATER & SOLVENT-BASED INKS

SOLVENT	WATER
LOWER SURFACE TENSION	HIGHER SURFACE TENSION
ALCOHOL & ESTERS	WATER, AMINES, ALCOHOLS
pH NOT APPLICABLE	pH IS MAJOR FACTOR
RESINS ARE SOLUBILIZED	RESINS SOLUBILIZED/EMULSIONS
NITROCELLULOSE, POLYAMIDES	ACRYLICS/POLYURETHANES
LARGE PIGMENT SELECTION	LIMITED PIGMENT SELECTION
VISCOSITY STABLE	STABILITY IS VARIABLE
DRYING RATE - NO PROBLEM	DRYING RATE IS SLOWER
APPLICATION THICKNESS	1/2 THE THICKNESS OF SOLVENT

FIGURE 7

WATER-BASED BINDERS

* ACRYLICS

* SHELLACS

* POLYURETHANES

* WATER-REDUCIBLE POLYAMIDES

* POLYESTERS

FIGURE 8

Water-based inks attempt to mimic solvent-based inks using soluble resins as well as emulsions. Water-soluble resins often do not have the product resistance required. As a result, emulsions of various resins are used to obtain higher temperature resistance, scuff resistance, block resistance, water and/or alkali resistance.

One of the most significant differences between solvent and water-based inks is that of surface tension. Water-based inks have inherently higher surface tensions and as a result are more difficult to formulate for good wettability and printability. Solvent-based inks have a single fluid phase consisting of dissolved polymeric binders, lower viscosity, and lower surface tension. These inks wet corona-treated films mettalized surfaces and dissolve the migrating additives on these surfaces.

Water-based inks will wet a surface providing the ink surface tension is lower than the surface tension of the substrate. A comparison of ink surface tensions clearly identifies the problem with water-based ink wettability as shown in Figure #9.

INK SURFACE TENSIONS

INK	∂_c (dynes/cm)
FLEXO SOLVENT-BASED INK	∞ 23
FLEXO WATER-BASED INK	∞ 28-45

FIGURE   9

The surface tension of solvents generally follow the polarity of the solvent. The greater is the hydrogen bonding and association between molecules, the higher is the surface tension. Typical surface tension of some common solvents used in inks and coatings are shown in Figure #10.

TYPICAL SURFACE TENSION OF SOLVENTS

SOLVENT	∂_c (dynes/cm)
$CH_3CH_2CH_2CH_2CH_2CH_3$	18.4
CH_3CH_2OH	22.3
$(CH_3)_2CH\text{-}OH$	21.7
$CH_3CH_2CH_2\text{-}OH$	23.9
$CH_3\underset{\underset{O}{\|\|}}{C}\text{-}O\text{-}CH_2CH_3$	26.2 (20°C)
ETHYLCELLOSOLVE	27.7 (20°C)
ETHYLENE GLYCOL	47.5
PROPYLENE GLYCOL	29.2
CARBITOL	35.5 (20°C)
H-O-H	72.3

FIGURE   10

Water-based inks will not wet nor dissolve fatty amide slip agents or fatty acid salts commonly used in films and papers to improve machinability. The higher surface tension of water-based inks necessitates the use of co-solvents, slower amines, and surfactants to improve ink wettability and printability. Low boiling alcohols have low surface tensions, but do not adequately balance the water-based ink surface tension requirements under dynamic printing conditions. The inks must dry fast (loss of volatile alcohol and increase of ink surface tension) and have good film wettability (some higher boiling alcohol, amine, or surfactant is usually introduced). This area still remains the primary focus of today's R&D and future water-based ink improvements.

Water-based ink formulations require the use of organic co-solvents to achieve low enough surface tensions needed to wet treated films and metallized surfaces.

At a 20% concentration of solvent, water-based inks have surface tensions in the range of 28-45 dynes/cm. However, as the inks dry on the surface, the loss of solvent readily increases the ink surface tension before it completes its' wetting out phase. To improve the wettability, a slower solvent is incorporated into the ink formulation. Both the choice and quantity is critical to other key properties of the ink, and must be done with great care.

Today's water-based inks are based on alkali soluble resins and emulsions. Both soluble and emulsion resins are necessary to attain the product resistance requirements while at the same time balancing the press performance demands.

Figure #11 shows the general composition of a water-based laminating ink formulation.

WATER-BASED INK FORMULATION

* PIGMENT
* RESIN(S)
* CO-SOLVENT
* AMINES
* ANTI-FOAM AGENT
* ADDITIVES
* WATER

FIGURE 11

The basic formulation of an ink involves grinding the pigment in a resin to form the pigment dispersion, then letting down the pigment dispersion with suitable resins to meet rheological and functional properties. This includes additives to improve to improve end use properties.

We can compare the inherent differences between solvent-based and water-based inks by a specific example. For instance, a solvent-based laminating ink involves the use of several resins to obtain the adhesion and heat resistance. In addition, a crosslinking agent is added to increase the internal bond values of the laminated structure. A typical solvent-based laminating ink formulation is shown in Figure #12.

SOLVENT-BASED LAMINATING INK FORMULATION

```
35.5  PIGMENT
 2.8  RESIN A
 8.8  RESIN B
45.0  SOLVENT
 4.7  PLASTICIZER
 1.0  SURFACTANT
 2.1  CROSSLINKER
 0.1  ANTIOXIDANT
_____
100.0
```

FIGURE 12

WATER-BASED LAMINATING INK FORMULATION

```
30.0  PIGMENT
18.0  SHELLAC
 8.4  ISOPROPANOL
36.9  WATER
 2.6  AMMONIUM HYDROXIDE
 3.0  TRIETHANOLAMINE TITANATE
 1.0  DIMETHYLAMINO ETHANOL
 0.1  ANTIOXIDANT
_____
100.0
```

FIGURE 13

Water-based laminating ink formulations are similar. Figure #13 shows a specific water-based ink formulation of a laminating ink.

Water-based ink formulation are similar to that of solvent-based inks with the noticeable exception that they use alkaline components and have higher surface tension characteristics.

PRESS MODIFICATIONS

The progress of water-based inks has been the result of major improvements in the printing processes and drying systems. Flexography has had the most significant impact on the growth and use of water-based inks and coatings.

Figure #14 illustrates the flexographic process used today to print film on new high speed presses.

Flexography is not only the dominant water-based printing process, today, but is also the fastest growing printing process with nearly a 20% share of the total printing market.

DRYING OF WATER-BASED INKS

Drying of water-based inks at commercial speeds required modifications in ink application & formulation. Water-based inks are typically applied at half the wet ink thickness normally applied by solvent-based inks. This requires several ink and press changes. Figure #15 lists the key application and formulation changes necessary to dry water-based inks on these presses.

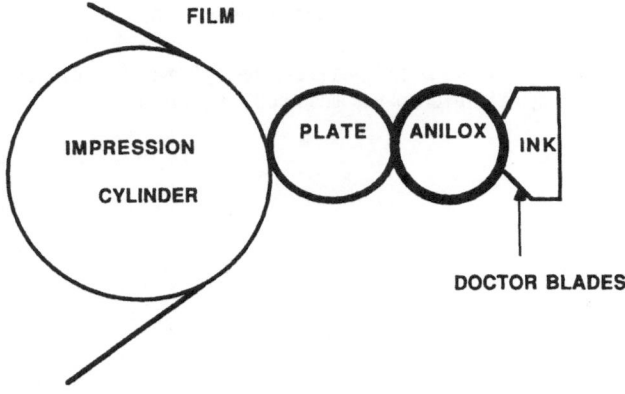

FIGURE 14

APPLICATION & FORMULATION MODIFICATIONS TO DRY WATER-BASED INKS

* INKS MUST BE MORE CONCENTRATED

* ANILOX ROLLS OR GRAVURE CYLINDER MUST DELIVER ABOUT HALF
  THE VOLUME OF INK COMPARED TO SOLVENT-BASED INKS.

* THE INK TRANSFER FROM THE PLATE TO CYLINDER MUST BE
  UNIFORM.

* SINCE THE AMOUNT OF INK DELIVERED IS CONSIDERABLY LESS, ALL
  PRESS SETTINGS AND NIP POINTS HAVE A LARGER EFFECT ON THE
  INK TRANSFERRED.

FIGURE 15

CHANGE OF pH DURING DRYING CYCLE

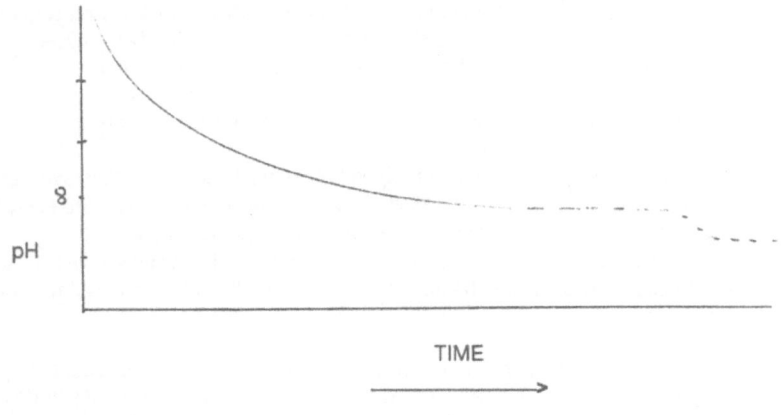

FIGURE 16

Unlike solvent-based inks which dry reversibly by solvent evaporation, water-based inks and coatings dry irreversibly as a result of pH change and evaporation. This is best illustrated by analyzing the pH during the drying cycle as illustrated in Figure #16.

In addition to pH, the drying of water-based inks is essentially irreversible, whereas, solvent-based ink drying cycle is reversible as illustrated in Figure #17.

DRYING OF SOLVENT & WATER-BASED INKS

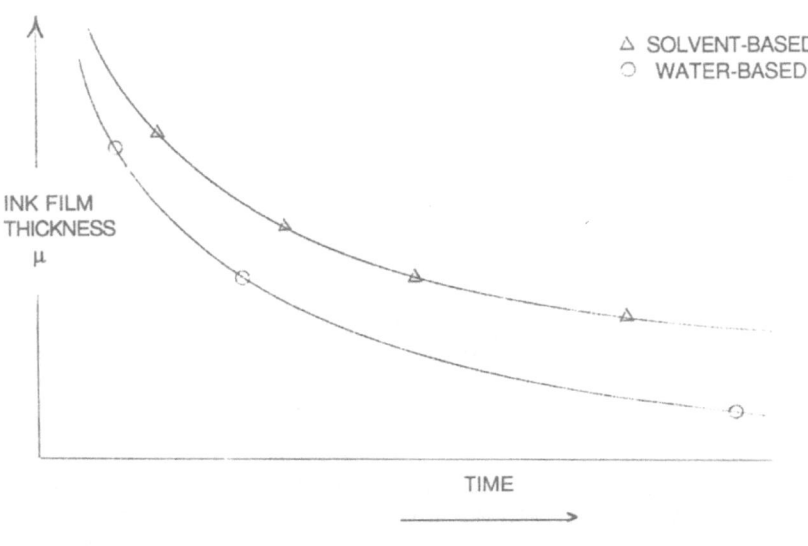

FIGURE   17

As the water-based inks dry, they become more and more insoluble as their drying sequence continues. As a result, water-based inks are more likely to skin during the drying cycle compared to solvent-based systems. As a consequence, clean up is more difficult.

Figure #18 illustrates what happens to printed web when the press speed is varied.

If the ink dries too rapidly the solvents can be trapped between this dried ink "skin layer" and the substrate. At low press speeds, this condition can occur if the oven temperature is too high. This "skinning" effect can lead to high solvent retention. This is particularly true of water-based inks which typically dry irreversibly. An adhesion tape test would be difficult to evaluate. The technician would assume he has dried the ink and the adhesion loss must be due to adhesion failure.

As the press speed is increased, an optimum condition is found which has the lowest solvent retention. A negative tape test at this point would suggest an adhesion failure. If you do have an adhesion problem here, it will usually be related to the choice of ink or lack of an adequate substrate treatment.

As the press speed is increased still further, solvent retention begins to rise and ink adhesion failure due to ink splitting can be expected. At press side, you may be able to identify the sound of wet ink transferring to the rollers. This is caused by ink cohesion failure as the ink mountain tops offset on the rollers, producing an "eye holing affect" in the printed area. I have seen this many times in printing water-based inks on various substrates. Improved flow out and ink leveling would have the greatest effect on improving your press speed. As the ink flow out is improved, the ink dries faster, and the solvent retained in the substrate drops.

SOLVENT RETENTION VS PRESS SPEED

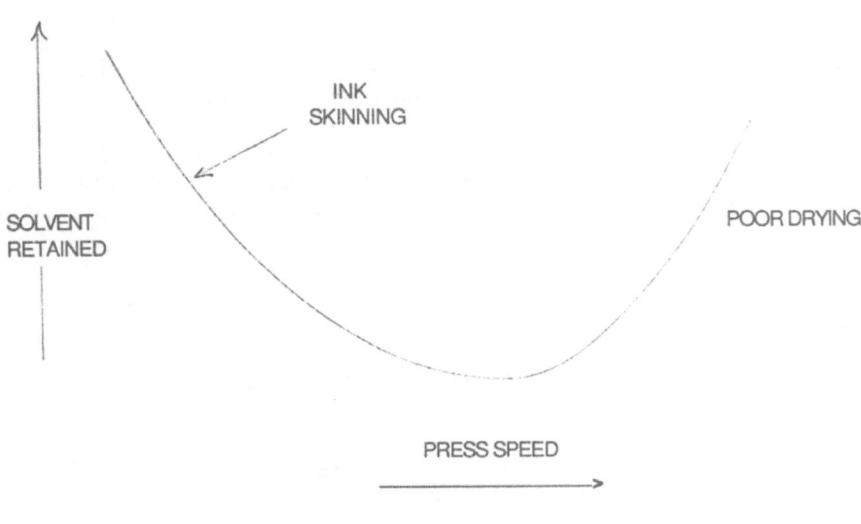

Figure   18

Skinning can contribute to higher solvent retention. At the same time, If the inks are thoroughly dried, water-based inks will resist solvent penetration in applications where the inks will be trapped or overprinted. This property is in sharp contrast to solvent-based inks and offers advantages in process printing applications.

Water-based inks combined with appropriate press modifications have allowed press speeds near 1,000 fpm to be realized. Web temperatures of 45-50°C or higher are usually required for full ink properties to be developed.

Higher substrate surface tension allow faster press speeds. This effect is due to apparent improvements in ink wetting which reduces both ink flow time and reduces both ink flow time and reduces the ink film thickness to be dried. In-line corona treatment configuration would be most desirable for optimum press speeds.

Commercial press speeds are attainable using water-based inks on non-absorbative substrates. To accomplish these drying speeds, several modifications are necessary.

The inks must be modified with enough alcohol to produce azeotropes that will evaporate rapidly from the ink surface. In addition, some slower alcohol and amines are necessary to assure good printability and gloss.

The gravure cylinder, or anilox roll, need to deliver a smaller amount of ink (compared to solvent-based inks), but cover the same area. To do this, the line screen of the cylinders needs to be increased, and the cells need to be shallow. A line screen minimum of 200 is common, with cell depths in the range of 25 microns.

The amount of ink transfer is controlled by the viscosity of the ink, its surface tension, the cell volume and shape, the line screen, and the surface tension of the substrate.

SUBSTRATE PARAMETERS

Printing on porous substrates with water-based inks has met with considerable success. Press speeds, printability, and adhesion have been quite good. Today, the focus is on improving ink mileage through the alteration of paper surface chemistry and formulation of water-based inks. Significant technological improvements have been realized in water-based inks for paper through resin binder modifications and the growth of the flexo newspaper industry. Their press speed objectives are in excess of 2,000 fpm. Water-based ink usage in newspaper flexo promises to be the fastest growing segment of the water-based ink market.

The printing of films and non-absorbative substrates is considerably more difficult with water-based inks and coatings. As a result, the substrate is primed or corona treated to achieve acceptable surface wettability characteristics. The effect of corona treatment on some common films is shown in Figure #19.

TYPICAL FILM SURFACE ENERGY PROFILES

FILM	∂_s	$\partial_s{}^d$	$\partial_s{}^p$
		(dynes/cm, 25°C)	
POLYETHYLENE	32	31	1
POLYETHYLENE (CORONA)	43	33	10
OPP	30	29	1
OPP (CORONA)	38	27	11

where $\partial_s{}^d$ is the dispersive component
and $\partial_s{}^p$ is the polar component of the
substrate surface energy, ∂_s.

FIGURE 19

Printing of non-porous substrates involves "gremlins" in the form of slip agents and lubricants used to improve the machinability of these materials. From the printing perspective, these chemicals introduce low surface energy boundary layers which must be penetrated by water-based inks to attain adequate wettability and adhesion. Typical slip additives used in films are shown in Figure #20.

SLIP ADDITIVES

$$R\text{-}C\text{-}NH_2$$
$$\parallel$$
$$O$$

* STEARAMIDE (WHERE R = 18 CARBONS & SATURATED)

* BEHENAMIDE (C = 22, SATURATED)

* OLEAMIDE (C = 18, UNSATURATED)

* ERUCAMIDE (C=22, UNSATURATED)

FIGURE   20

These slip agents are dipolar molecules and will orient themselves on surfaces that are at high energy. As a result, you can anticipate surface tension of corona treated films to dissipate with time as a function of the concentration of these additives and their mobility within the film. This effect is particularly important when metallizing, since the fresh aluminum provides a very high surface energy which will orient dipolar additives as if by a strong magnet, forming very tight and highly populated boundary layers.

To pressmen, these slip agents are the gremlins in the press operation. They can cause numerous printing problems. The effect of slip agents is to provide a low surface energy boundary layer which lowers the film's printability and adhesion characteristics. The degree of this effect depends on the film, concentration of slip agents and the type of slip agent used.

Water-based inks normally cannot wet or dissolve these boundary fatty acid amides. This is in sharp contrast to solvent-based inks, which dissolve the boundary layer of amides and allow their binders to adhere to the film substrate.

SURFACE TREATMENT

Printers realize that corona treatment of non-porous substrates offers improved ink adhesion characteristics. It is common to pretreat films and metallized substrates to improve ink wettability and adhesion.

Corona treatment increases the polar characteristics of the film surface. In most cases, the treatment is reversible and the rate of reversal is temperature and humidity dependent.

With water-based inks, corona-treatment is important for the necessary ink wetting and effective drying speed. Polyethylene and polypropylene must be corona-treated to at least 40 dynes/cm to assure proper ink wettability, adhesion, and drying rate. The polar resins in the water-based ink will form ionic and hydrogen bonds with the corona-treated surface.

The effect of corona on drying speed is less obvious. When the corona-treatment of a film is high, the rate of wetting (or flow out) is more rapid. The net result is that the ink can dry faster since it needs to dry a thinner layer of ink film. The effect of corona on drying speed can be quite substantial. Speed increases of 100-200 fpm have been realized with water-based inks as well as solvent-based inks. This is one way to improve press productivity.

The mechanism of corona treatment of films is complex [2], however, the most popular interpretation is based on consumption of oxygen generating surface polar functional groups. Figure #21 illustrates the likely mechanism of corona treatment.

PROBABLE MECHANISM OF CORONA TREATMENT

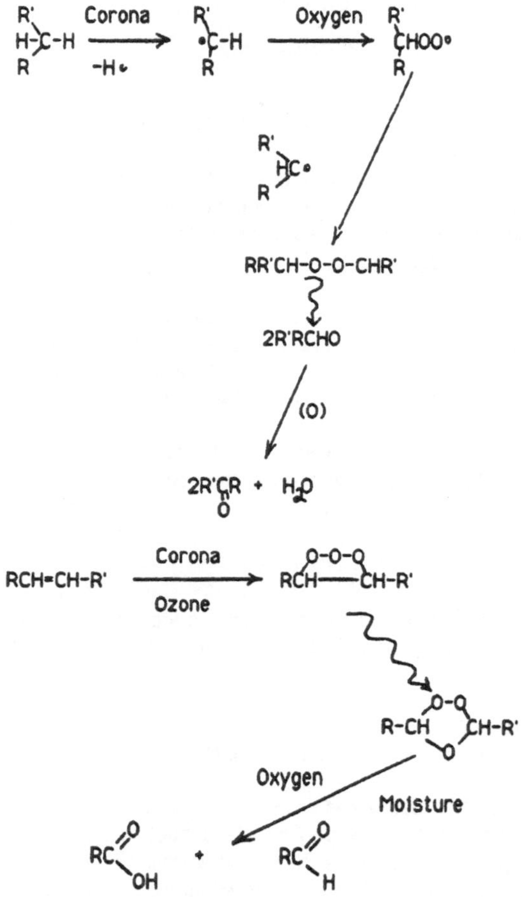

FIGURE 21

INK ADHESION

Ink adhesion is affected by many variables. Some of these key variables affecting ink adhesion are shown in Figure #22.

INK ADHESION VARIABLES

* INK COHESION
* SUBSTRATE SURFACE TENSION
* INK COMPOSITION
* INK SURFACE TENSION
* DRYING RATE

FIGURE 22

Of these variables the most common source of ink adhesion failure is ink film cohesion failure due to incomplete drying. Cohesion failure is most easily noticeable by ink splitting when subjected to the usual tape adhesion test. Complete or area lifting to the tape is usually indicative of low surface tension and is common ink adhesion failure.

These two forms of ink failure are usually lumped together under the umbrella of "poor ink adhesion". However, it is very important to distinguish which problem you are up against so that the right remedy can be applied.

Polyethylene, polypropylene, polyester, and metallized films and paper are typically corona treated to increase their critical surface tensions to assure good ink adhesion and substrate wettability. Typical values are shown in Figure #23.

TYPICAL VALUES OF CORONA TREATMENT OF SUBSTRATES

SUBSTRATE	CRITICAL SURFACE TENSION (∂_c in dynes/cm @ 25°C)	
	UNTREATED	CORONA TREATED
POLYETHYLENE	30	40
POLYPROPYLENE	29	40
POLYESTER	46-48	54
METALLIZED PAPER	32-50+	50+

FIGURE 23

The increase in substrate surface tension has several significant consequences. These are:

* INKS PRINTED ON THESE TREATED SUBSTRATES HAVE BETTER ADHESION AND SUBSEQUENT BONDS IN LAMINATIONS.

* CORONA TREATMENT OF SOME FILMS CAN CAUSE THE FILM TO CROSSLINK, MAKING IT MORE CRYSTALLINE AND INCREASING ITS SOFTENING POINT.

* IMPROVED INK PRINTABILITY AND INCREASED INK FLOW OUT

If we take a closer look at what happens at the plate/substrate nip as the ink is split under dynamic press conditions, we can see how surface tension effects influence the ink film forming process on the substrate. Figure #24 illustrates what happens as the ink splits under dynamic conditions.

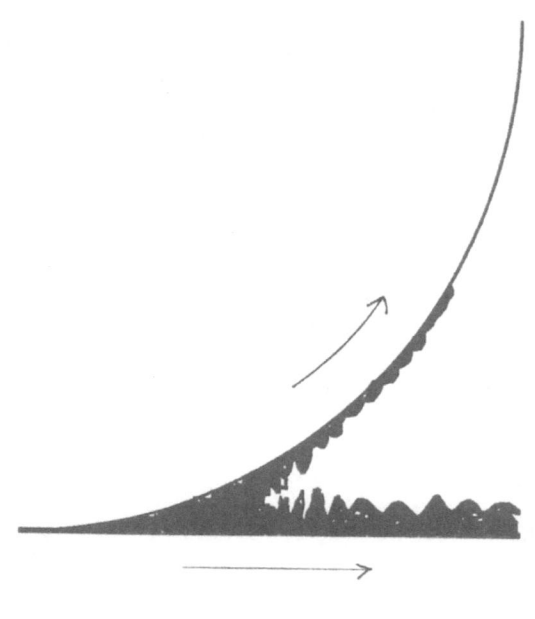

SUBSTRATE

FIGURE 24   INK SPLITTING

The ink is split at high press speed. Some of the ink will transfer to the film and the rest of the ink will stay on the transferring medium. In flexo, this is the printing plate. The splitting causes the ink to come apart. This separation is resisted by the cohesive forces in the ink trying to keep the ink together. Obviously, the high sheer forces of the press will dominate. However, this resistance causes the ink film surface transferred to be very irregular. The net result, is that, the surface is made up of high and low points.

The drying speed of an ink will be limited by the height of these "mountain" peeks.

In a dynamic printing operation, the ink is transferred rapidly to the substrate. As a result, the ink does not have much time to flow out and become uniform before entering the drying chamber. The rate of ink flow out and uniformity before entering the drying chamber has a significant effect on the press drying rate.

This phenomenon is particularly important today for the successful commercialization of water-based inks. These inks, unlike their solvent-based counterpart, have higher surface tension characteristics and have poorer wetting properties on non-porous substrates. For commercial press speeds to be attained, the surface treatment should be high to allow for faster flow out and better drying. As a result, in-line corona treating is preferred.

The limiting factors controlling press speeds are (1) the ink adhesion to the substrate, and (2) the ink cohesion failure due to retained solvent.

The concept of improving press speed by increasing the surface tension of the film is not new. Some of us may remember when, back in the 1970's, we had an energy shortage. In those days we were attempting to produce energy consumption and dry our inks. We found that solvent-based inks could dry without heat if adequate air turbulence was provided and the substrates were corona treated. This concept was never really pursued because, in those days, it meant you could lose your gas allotment.

Use of in-line corona treatment to increase ink flow out time can increase your optimum press speeds. Actual results have varied from printer to printer, but increased press speeds of 50-200 fpm can be realized.

Today, however, we need to improve our press productivity just to be competitive. The struggle with water-based inks has been to maintain production at speeds similar to what was expected of solvent-based inks.

III. CONCLUDING REMARKS

Our understanding of water-based inks and their commercial use has come a long way. We are able to print water-based inks on paper, films, as well as on metallized substrates with good printability and adhesion. This has required new inks, significant modification of press and dryer equipment as well as in-line corona treatment.

Water-based inks are commercial in various segments of the printing industry, but as one printer put it[3], "they are statistically out of control". It lacks the predictability that solvent-based inks have achieved over the last forty years.

One element of water-based technology that would improve the commercial consistency of performance is to control surface tension of inks, plates, substrates, and anilox rolls. Until we can effectively control these inks and printing surfaces, the processes will continue to lack consistency.

Control of surface tension parameters will provide one link in the chain required for consistency in ink wetting, drying rate, and ink transfer. We need better process control equipment that can measure these surface properties in real time during the printing operation.

Today, we need to investigate the static and dynamic surface tension effects in the printing of water of water-based inks. These effects need to be quantified and verified under actual press conditions.

The role of surface tension in water-based ink transfer needs to be clarified. To date, there have been conflicting results[4].

As we look toward the nineties, the role and understanding of surface phenomena in the dynamic printing process promises to make a major contribution to statistically improving the printing of water-based inks.

REFERENCES

1. Anon, Paper, Film & Foil Converter, Feb: 41 (1981)

2. Podhajny, R.M., J. of Plastic Film & Sheeting, 4, 177-188 (1988)

3. "State of Current Technology in Using Water-Based Inks on Films", FTA International Ink Conference Proceedings, Dec, 1988, Flexographic Technical Association.

4. Surface Tension Session, 1989 FTA National Forum Proceedings, April, 1989, Flexographic Technical Association.

HIDING POWER OF ALUMINUM PIGMENTS IN PRINTED INK FILMS

George M. Kern

Silberline Manufacturing Co., Inc.
R.D. #2, Hometown
Tamaqua, PA 18252

Fortunato J. Micale, Diego P. Valenzuela, and
Jean S. Lavelle

Zettlemoyer Center for Surface Studies
Sinclair Laboratory, Bldg. #7
Lehigh University
Bethlehem, PA 18015

Several leafing and non-leafing pigments varying in
average particle size were formulated into toluene
based inks containing 5, 10, and 15 percent aluminum
pigment. The inks were drawn down on coated paper
and transparent plastic film using #3 and #6 wirewound rods and a 140 line gravure handproofer. The
plastic prints were weighed to determine ink film
thickness and aluminum concentration per unit area on
the print. Using the same applicator and inks with
the same concentration of aluminum, the amount of ink
transferred to plastic film was independent of pigment type and particle size. Ink opacity or hiding
power was evaluated using conventional optical density measurements on the paper prints and a novel
photographic technique to quantify light transmission
through the ink films on plastic. Results from the
paper and plastic prints indicated that the opacity
of both leafing and non-leafing pigments decreased as
the average particle size of the aluminum pigment
increased. An economically important saturation
value could be calculated for each pigment which
signified the minimum concentration of aluminum per
square centimeter of ink film on clear plastic required for complete opacity. Saturation value de-
pended upon whether the pigment was leafing or non-
leafing and upon average particle size.

Surface Phenomena and Fine Particles in Water-Based Coatings and Printing Technology
Edited by M.K. Sharma and F.J. Micale, Plenum Press, New York, 1991

INTRODUCTION

Economically, aluminum pigments are priced on the basis of opacity or hiding power which is, theoretically, inversely proportional to pigment particle size. Smaller particles are more expensive because they require longer milling times elevating production and labor costs. Differences in the hiding power or opacity of aluminum pigments are usually determined by visual comparisons of drawdowns in a clear lacquer, letdowns in a vehicle containing titanium dioxide, or coverage determinations in paint systems(1).

Described in this article is a method for quantifying opacity by measuring light transmission through ink films drawn down on clear plastic. The procedure involves dispersing in the same vehicle a series of leafing and non-leafing aluminum pigments varying in average particle size, preparing photographic contact prints from the drawdowns, and quantifying light transmission by making optical density measurements on the contact prints. Opacity can be related to pigment type (leafing or non-leafing), particle size, and concentration of aluminum in the ink and per square centimeter of ink film.

MANUFACTURE OF ALUMINUM PIGMENTS

Aluminum pigments are produced by a wet ball milling process. Three raw materials are charged into the ball mill containing steel balls: aluminum metal, mineral spirits, and a suitable fatty acid. The mill is operated at a speed that will permit cascading of the balls onto the metal, thus flattening it and eventually breaking it into tiny flakes. The lubricant prevents cold welding of the metal, and a sufficient amount of mineral spirits is charged into the formulation to maintain a wet slurry. The length of time for ball mill operation depends on the grade being manufactured and the particle size distribution desired. When the grinding operation has been completed, the slurry is discharged from the ball mill and is passed over the appropriate mesh screen separating out the undesirable flakes. After screening, the remaining aluminum flakes pass into a filter press which removes excess solvent. The filter cake is then homogenized in a mixer along with the addition of solvent to bring the material to a pastelike consistency.

Aluminum pigments have a lamellar geometry and it is this geometry, along with the metallic nature of the pigment, which provides a combination of aesthetics and properties unlike other pigments. When incorporated into a coating, the flakes orient themselves in a multilayered position parallel to the substrate and film surface. Leafing grades may vary from flat, silvery white to a highly polished, chromelike appearance, while the non-leafing grades provide metallescent or polychromatic effects with varying degrees of sparkle and color.

Leafing aluminum pigments have the ability to cover the surface with layers of flakes. This leafing phenomenon is a function of the fatty acid lubricant used in the milling process. Generally a saturated fatty acid is employed, which is absorbed onto the metal surface providing both hydrophobic and oleophobic characteristics. It is these nonwetting characteristics that provide the leafing properties of the pigment. Any vehicle system that tends to "wet" the flakes will decrease the leafing characteristics. Evaporation of solvents from an applied leafing pigmented coating sets up convection currents which permit movement of the flakes to the surface of the film. Once at the surface of the film, the flakes are held there by a combination of factors including the interfacial tension between flakes and vehicle and increasing system viscosity.

Non-leafing aluminum pigments are distinguished from leafing grades by their orientation in a coating film. Rather than orienting at or near the film surface, they are distributed evenly throughout the entire film. This distribution is due to the unsaturated fatty acid employed during the milling process. The properties of tint strength, color, sparkle, gloss, etc. are closely controlled and are dependent to a large degree on particle size and particle size distribution.

EXPERIMENTAL AND RESULTS

INK AND PRINT PREPARATION

Inks based on the same vehicle and the aluminum pastes described in Table I were formulated to contain 5, 10, or 15 percent aluminum. A 50/50 mixture of resin and solvent was blended on rollers turning at 150 rpm and on a high speed impeller mixer. DIDP plasticizer was added with stirring. The aluminum pastes were diluted slightly with ink solvent to reduce viscosity, and the resin mixture was slowly stirred into the paste with a spatula. The final inks, described in Table II, were mixed overnight on the rollers.

To evaluate opacity, the inks were drawn down with #3 and #6 wire wound rods on 5x7 inch Leneta white coated paper with a 1 1/2 inch wide black bar printed across the bottom.

Standard weighed prints were prepared by drawing down the inks on 4 mil polyester photocopy transparency stock using #3

Table I. Identification of Aluminum Pigments

Grade	Particle Size							
	Avg. (um)	Cumulative Percent Less Than						
		1	10	25	50	65	100	125um
Leafing								
Eterna Brite[R] 651-1	5.6	5	89	100				
Eterna Brite[R] 601-1	9.0	3	69	98	100			
Eterna Brite[R] 301-1	9.5	3	62	94	100			
Silvar A	20.0	1.4	29	59	86	91	97	100
Non-Leafing								
Sparkle Silver[R] 7500	10.4	2	59	92	100			
Sparkle Silver[R] 5500	15.6	1	35	77	100			
Sparkle Silver[R] 3500	24.8	0.4	16	48	94	100		

Table II.   Ink Ingredients

Ingredient	Leafing			Non-Leafing		
			Wt. %			
Al pigment	5	10	15	5	10	15
Mineral spirits	2	4	6	3	5	8
Resin[a]	24	24	24	24	24	24
DIDP[b]	3	3	3	3	3	3
Solvent[c]	66	59	52	65	58	50

[a]Univrez polyamide from Union Camp Inc.
[b]Di-isodecyl phthalate plasticizer (DIDP).
[c]65% toluene, 35% isopropanol.

and #6 wire wound rods and a 140 line gravure handproofer. The polyester transparency was first wiped with isopropanol to remove any surface film. A 6x7 inch piece was cut from the center of the sheet, weighed on a five-place analytical balance, and reinserted in the original sheet. After drawing down the inks, the center piece was removed and weighed to determine the amount of dried ink deposited.

OPACITY MEASUREMENTS

Optical Density

Optical density of the ink films on Leneta paper was measured over the white and black backgrounds using a Macbeth Densitometer. An optical density ratio (DOR) was calculated by dividing the value obtained over white by the value over black. As ink opacity increases, the DOR value increases up to a value of 1.0 which represents complete opacity.

Photographic Technique

To quantify opacity and hiding power of aluminum based ink films, a novel technique based on light transmission was developed. In this technique, ink draw downs on clear plastic film were used as photographic negatives. A standard photographic enlarger was used to impress the projected image on photographic paper as shown schematically in Figure 1. The optical density of the resulting photographic print was then measured using a Macbeth Densitometer.

All the results presented in this paper were obtained by using the same enlarger magnification, exposure time, f-stop, type of photographic paper, developing parameters, etc. The photographic conditions were chosen which maximized the resolution of the technique.

Figure 2 illustrates the photographic prints prepared from draw downs made with #3 and #6 wirewound rods using inks with different aluminum concentrations. The darkest print on the bottom left represents the drawdown of resin solution which transmitted the maximum amount of light. The photographic prints made from #3 wire wound drawdowns became progressively lighter as the aluminum concentration in the ink increased and prevented light transmission through the film. Results from the thicker #6 wire wound films indicated that complete opacity was reached with inks containing only 10 percent aluminum.

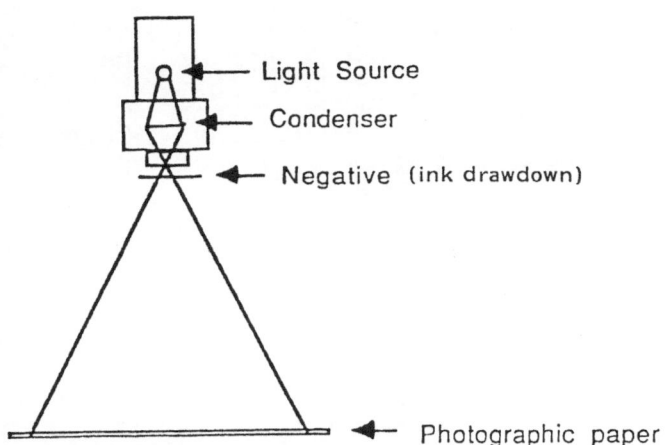

Figure 1. Schematic of Photographic Apparatus for Hiding Power Measurement.

INK RHEOLOGY

Ink viscosity on the Brookfield Viscometer at 25°C using the UL Adapter ranged from 100 to 300 cp. At shear rates of approximately 1 sec^{-1} on the Brookfield and on the Bohlin Rheometer, inks with leafing pigment were slightly shear thinning compared to the Newtonian non-leafing inks. All the inks exhibited Newtonian behavior above 10 sec^{-1}, and ink viscosity appeared to be dependent on aluminum concentration but independent of particle size.

PRINTABILITY AND OPACITY

Despite the complex reflectance pattern from metallic films, results from the Leneta paper prints shown in Figure 3 indicate an essentially linear decrease in DOR and, therefore, in opacity as pigment particle size increases. The same relationship between film opacity and aluminum particle size was reported by Ferguson and Rolles (1,2). The decrease in opacity appeared to be independent of pigment type. For all inks con-

#3 WW   #6 WW

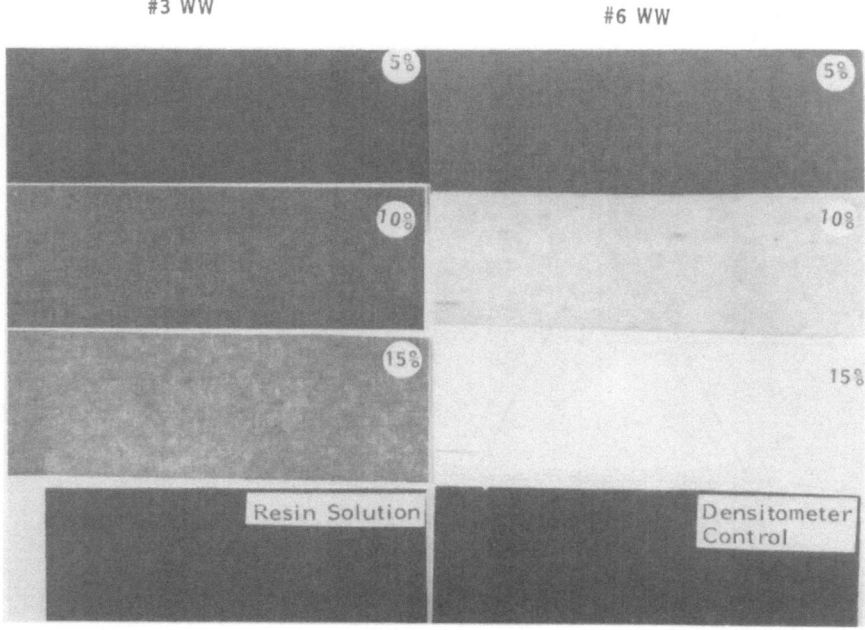

Figure 2.   Typical Photographic Contact Prints

taining 5 percent pigment and for those containing 10 percent of larger particles, the thicker films from the #6 rod were more opaque. The opacity of inks containing 10 percent of smaller particles with an average diameter of 10 μm remained relatively constant irrespective of film thickness or pigment type. DOR values larger than 1.0 were attributed either to complex reflectance or to film non-uniformity.

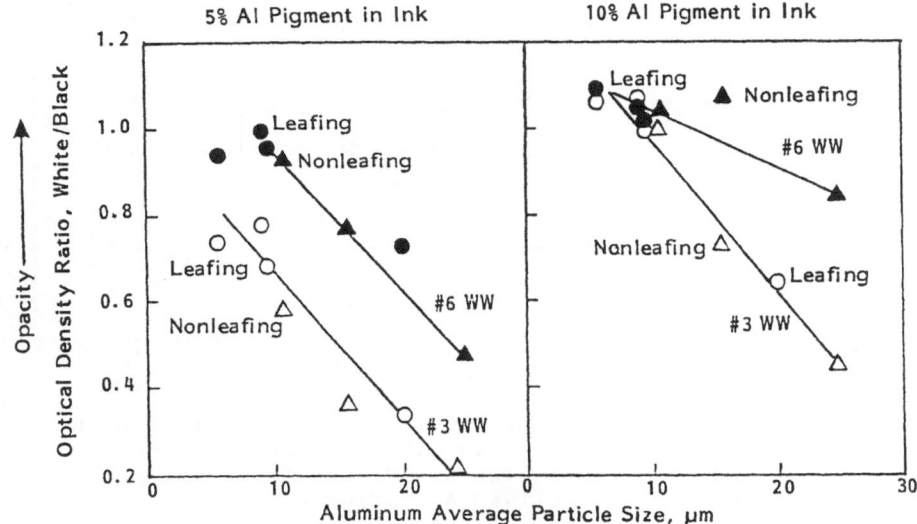

Figure 3. Optical Density Measurements on Ink Drawdowns
Over White and Black Backgrounds on Leneta Paper

Paper is difficult to weigh because it is hygroscopic.
The polyester film could easily be weighed to determine the a-
mount of dried ink deposited as a function of aluminum concentration in the ink, pigment type, and mode of ink application.
Duplicate prints were always made, and ink content agreed within 10 percent. As shown in Figures 4 and 5, more ink was deposited with each incremental increase in pigment concentration.
The same amount of vehicle or ink was deposited by the #6 wire
wound and handproofer, but significantly less was deposited by
the #3 wire wound. Visual observations confirmed these results
and also indicated that the #3 wire wound films were least uniform and the #6 wire wound films most uniform. Aesthetically
based on sparkle, gloss, and uniformity, the best paper and
polyester prints were made with 651 or 601 leafing pigments
and the 7500 non-leafing pigment.

Theoretically, a more accurate analysis of opacity should
be obtained from light transmission measurements through ink
films on transparent polyester. Quantifying light transmission
by measuring optical density on the corresponding photographic
contact print provided information which agreed with results
from the Leneta paper prints. As shown in Figure 6, opacity
was enhanced by increasing pigment concentration in the ink
from 5 to 10 percent, by depositing a thicker film with the #6
wire wound, and by using smaller size leafing and non-leafing
pigments.

The availability of weighed prints permitted the evaluation of opacity as a function of aluminum concentration per
square centimeter of print. Fortuitously, for inks with the

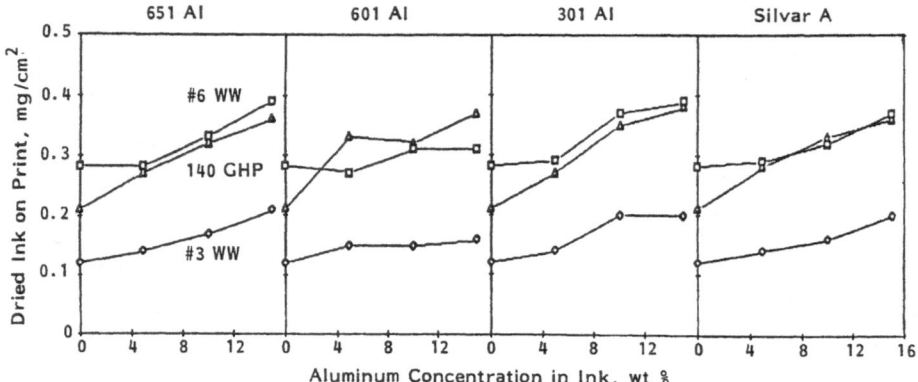

Figure 4. Ink Transfer to Polyester Film Using #3 and #6 Wire Wound Rods, 140 Line Gravure Handproofer, and Inks containing 5, 10, or 15 Percent of 651, 601, 301, or Silvar A Leafing Pigments.

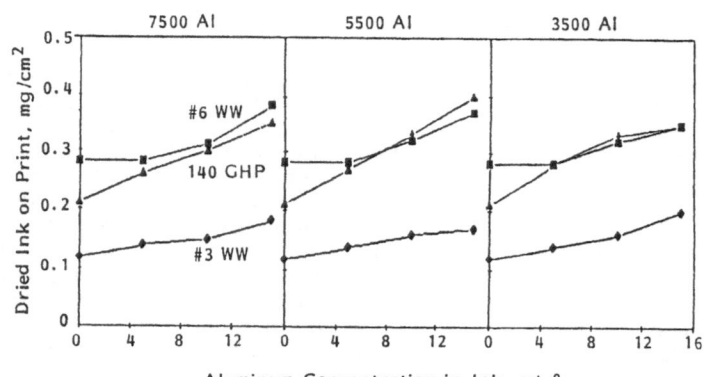

Figure 5. Ink Transfer to Polyester Film Using #3 and #6 Wire Wound Rods, 140 Gravure Handproofer, and Inks containing 5, 10, or 15 Percent of 7500, 5500, or 3500 Non-Leafing Pigments.

same concentration of aluminum, the amount of ink deposited on the polyester was independent of pigment type and particle size. Therefore, any difference in the opacity of ink films with constant aluminum concentration per square centimeter can be attributed to variations in pigment particle type and size. As shown in Figure 7 representing the experimental points and a dotted line obtained by regression analysis, opacity increased as the concentration of leafing aluminum pigment in the ink was increased from zero to 15 percent. The slope of the dotted line was much less for Silvar A than for 651 or 601 indicating the inefficiency of larger size particles to promote film opacity.

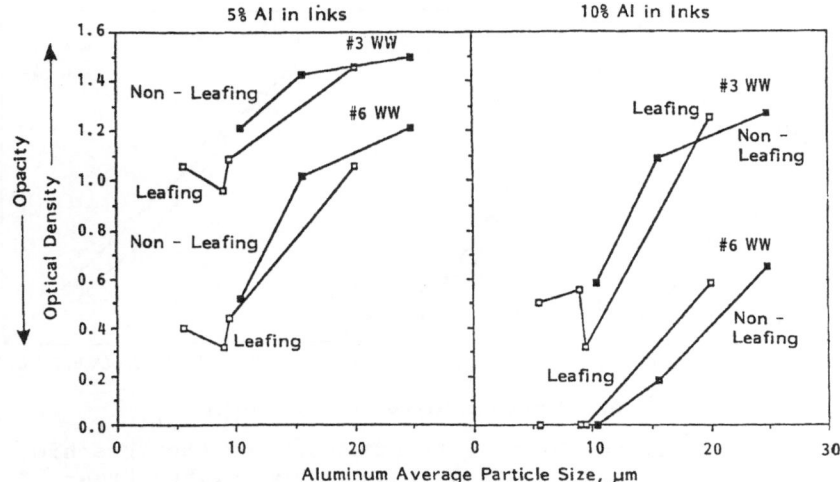

Figure 6.   Optical Density Measurements on Photographic Contact
Prints Prepared from Weighed Drawdowns Using Inks
Containing 5 or 10 Percent Leafing or Non-Leafing
Pigments Varying in Average Particle Size.

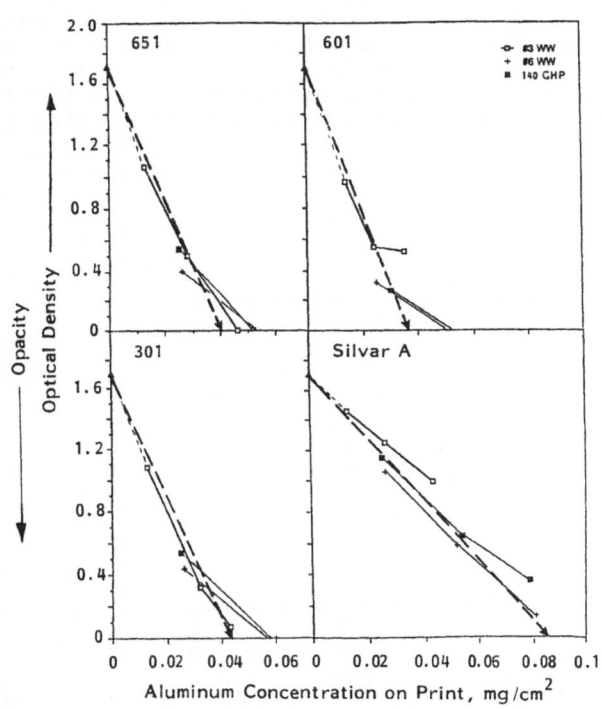

Figure 7.   Optical Density Measurements on Photographic Contact
Prints Prepared from Weighed Drawdowns Using Inks
Containing 5, 10, or 15 Percent Leafing Pigments.

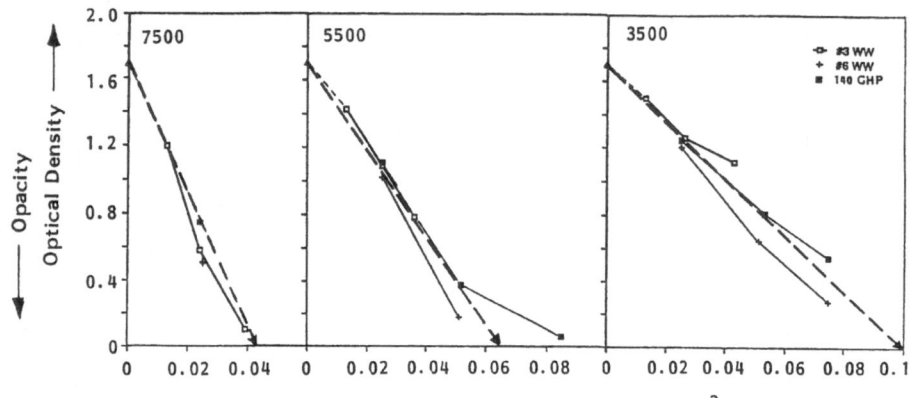

Figure 8. Optical Density Measurements on Photographic
Contact Prints Prepared from Weighed Draw-
downs Using Inks Containing 5, 10, or 15
Percent Non-Leafing Pigments.

As illustrated in Figure 8, the non-leafing pigments performed
in the same manner.

The point at which the dotted line intersects the x-axis
is called the saturation value and represents the minimum concentration of aluminum in the ink film required to produce complete opacity under these testing conditions. The availability
of these values permit the ink manufacturer and printer to use
aluminum pigments cost effectively. As illustrated in Figure 9,
saturation value increased linearly with increasing pigment particle size at the same rate for leafing and non-leafing pigments.

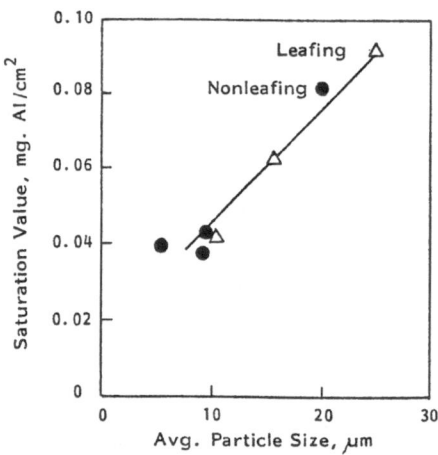

Figure 9. Saturation Value of Aluminum Pigments as a Function
of Average Particle Size (Minimum Concentration of
Aluminum per Square Centimeter of Ink Film on
Polyester Required for Complete Opacity).

CONCLUSIONS

The observation that smaller particles of leafing and non-leafing aluminum pigments enhance opacity has been confirmed and quantified by measuring light transmission through ink films on clear plastic which contain a known amount of aluminum per unit area of print. An economically important saturation value can be derived which indicates, for a given average particle size, the minimum concentration of leafing or non-leafing aluminum pigment per unit area of ink film required to provide complete opacity under specified test conditions.

REFERENCES

1. Ferguson, Russel L., *Pigment Handbook*, Volume I, Properties and Economics, Second Edition, Edited by Peter A. Lewis, John Wiley & Sons, Inc., New York, 1988.

2. Rolles, Rolf, *Pigment Handbook*, Volume I, Properties and Economics, First Edition, Edited by Temple C. Patton, John Wiley & Sons, Inc., New York, 1973.

AQUEOUS POLYMERIC DISPERSIONS FOR FILM COATING

OF PHARMACEUTICAL SOLID-DOSAGE FORMS

Stuart C. Porter

Colorcon
415 Moyer Boulevard
West Point, PA 19486

Film coating as a method of coating pharmaceutical solid-dosage forms (for example, tablets, granules, capsules) has been in common use for three decades. Early success with this technology resulted from the use of highly volatile organic solvents, which permitted the employment of relatively simple processing methodologies. Regulatory concerns over the use of such solvents, however, coupled with the development of sophisticated processing equipment, have resulted in water becoming the solvent/vehicle of choice.

A growing dependence on film coating as a means of modifying drug-release characteristics from the dosage form requires that essentially water-insoluble polymers must be used. Consequently, in order to make use of aqueous processing technology, the development of aqueous polymeric dispersions has been necessary.

Such polymeric systems, presented as either dry, water-dispersible polymers or aqueous dispersions of the same, require special consideration during processing if their functionalities (as coatings) are to be reproducible. The coalescence process for film formation, and those factors which directly influence this process, must be well understood.

This paper will discuss those polymer systems currently available as aqueous dispersions that are suitable for use in the pharmaceutical film-coating process. In addition, it will describe many of those factors relating to the formulation of the substrate, the coating iteself, and the coating process that influence the performance of the final dosage form.

Surface Phenomena and Fine Particles in Water-Based Coatings and Printing Technology
Edited by M.K. Sharma and F.J. Micale, Plenum Press, New York, 1991

INTRODUCTION

A study of the marketplace for pharmaceutical oral solid-dosage forms will show that a significant proportion are coated. Examination of the coating technologies practiced in the pharmaceutical industry will confirm that the wheel has gone full circle, as shown in Fig. 1.

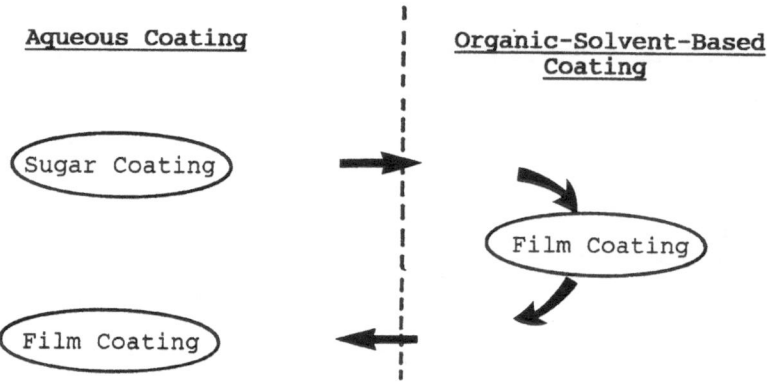

Fig. 1. Evolution of Pharmaceutical Coating Processes

Although sugar coating has long since lost its dominant position as the coating process of choice, it is still practiced, and occasionally we see it being used for new products (e.g. Advil). There is no doubt, however, that aqueous film coating is the preferred process and will be the focus of this discussion.

Over the course of time, coating processes have developed from the art of earlier years to those that are more technologically advanced and controlled such that compliance with GMP's is facilitated. The design of new equipment, the development of new coating materials, and recognition of the impact of applied coatings on subsequent release of drug from the dosage form, have all contributed to improved products.

Changes that have occurred in coating processes reflect a desire to:

- Consistently obtain a finished product of high and reproducible quality.

- Develop processes in which the economics are maximized, particularly with respect to process times and equipment utilization.

What is Film Coating?

Film coating is quite a complex process that draws on technologies associated with polymer chemistry, industrial adhesives and paints, and chemical engineering. The process of film coating can be simplified to represent one that involves the application of thin (in the range of 20 to 200 µm), polymer-based coatings to an appropriate substrate (tablets, beads, granules, capsules, drug powders and crystals) under conditions that permit:

- Balance between, and control of, the coating liquid addition rate and drying process.

- Uniformity of distribution of the coating liquid across the surface of product being coated.

- Optimization of the quality (both visual and functional) of the final coated product.

While film coatings can be applied by manual ladling techniques, they now almost always utilize a spray-atomization technique.

A schematic outline of the pharmaceutical film-coating process is show in Fig. 2.

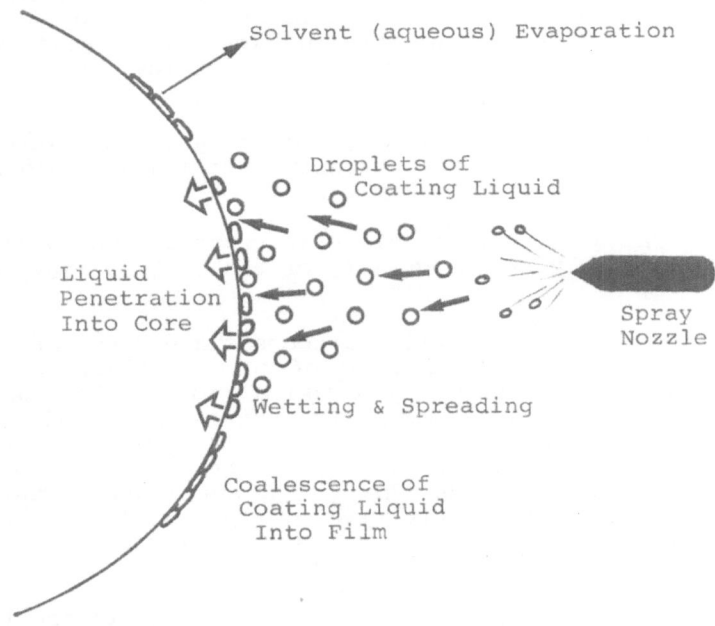

Fig. 2. Schematic Outline of Film Coating

In the spray-application process, bulk coating liquids are finely atomized and delivered in such a state that droplets (of coating liquid) retain sufficient fluidity to wet the surface of the product being coated, spread out and coalesce, to form a film. Because of the highly adhesive (or "tacky") nature of partially dried droplets, it is imperative that these droplets dry almost instantaneously the moment they contact the surface of the substrate, otherwise sticking and picking will occur. Hence a need to strike an appropriate balance between liquid application rate and the drying process.

Because of the rapid drying that typically takes place during the application of film coatings, uniformity of distribution of the coating is controlled both by uniformity of application of the coating liquid (i.e. the number of spray guns used, types of spray patterns used, and fineness of atomization of coating liquid) and tne uniformity of mixing (controlled by pan speed, baffle design, tablet size, and shape) of the product being coated.

From the description of film coating thus far, it is apparent that the pharmaceutical process is predicated on technologies "borrowed" from other industries. Indeed it is, but some major constraints exist to make the pharmaceutical process different, namely:

- The substrate being coated is small (100 μm to 20 mm), and is coated as a batch consisting of tens of thousands to several million units that are in contact with one another and yet must be discretely coated.

- Coating materials are substantially restricted to a limited list of ingredients that are approved for ingestion.

- The quality of health-care treatment may well hinge entirely on the quality of the coating applied.

Most current film-coating processes are considered to be continuous; that is, application of coating liquid continues uninterrupted until all the coating material has been applied. However, it is perhaps more appropriate to consider film coating as a discontinuous process since each tablet (or granule, etc.) receives only a small fraction of its total coating each time it passes through the spray zone. Thus, film coatings are generally built up as a series of layers one upon another so that the final coating may structurally be far from homogenous.

Over the years, as film coating has grown in popularity, use of organic solvents (with their associated flammability, toxicity and environmental pollution hazards) has proven somewhat limiting. Fortunately, significant improvements in processing equipment have facilitated the introduction of aqueous-based coating formulations. The result is that aqueous film-coating processes are used preferentially by a majority of manufacturers (of film-coated products). A useful review of aqueous coating systems has been given by McGinity[1]

Major Types of Pharmaceutical Film Coatings

The applications for use of film coatings are quite diverse, and while most require some functionality of the coating, it is not uncommon to see film coatings described as either functional or non-functional. Functionality in this context relates specifically to an ability to modify drug-release characteristics. Non-functional (or conventional) film coatings are typically reserved for situations where it is necessary to improve product appearance, ease of swallowing, and drug stability; they are also used for taste masking. Functional film coatings are used when drug-release characteristics need to be modified, and are represented by enteric coatings and sustained- (or controlled-) release coatings.

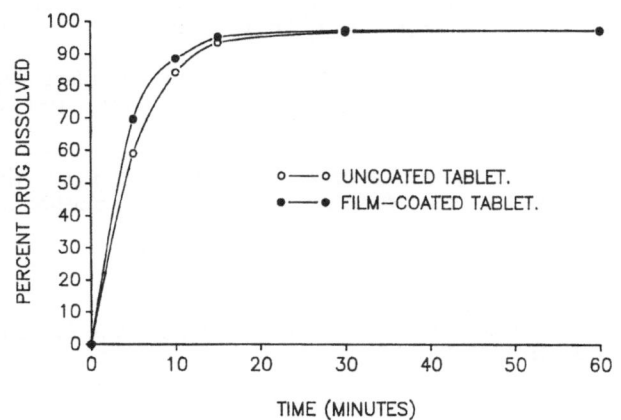

Fig. 3.   Dissolution characteristics of uncoated aspirin tablets and those that have been aqueous film coated with a non-functional coating.

Non-functional coatings are usually applied as aqueous solutions of polymers such as hydroxypropyl methylcellulose (H.P.M.C.), hydroxypropyl cellulose (H.P.C.), and methyl-cellulose (M.C.). Since it is desirable that these coatings do not retard drug release from the dosage form, use of readily water-soluble polymers is desirable. A typical example of the performance of a non-functional coating is shown in Fig. 3.

On the other hand, the performance characteristics expected of functional coatings dictates that water-insoluble polymers be used. The growing popularity of functional coatings, and the strong desire to utilize aqueous processing technology, have formed the basis for the introduction (into the pharmaceutical industry) of AQUEOUS POLYMERIC DISPERSIONS.

While such polymer systems have long been used outside the pharmaceutical industry, pharmaceutical formulators and process engineers have only recently started to come to grips with the complexities of film formation with dispersed polymer systems, as described by Bindschaedler et al[2] and Wicks.[3]

Specific requirements of the pharmaceuticl coating process have conspired to make life difficult, since:

- Product being coated should remain discrete (i.e. no agglomeration) during the coating process.

- Uniformity of distribution of the coating is of paramount importance.

- Complete coalescence of the coating during the time-frame of the application process is critical.

- Time-dependent changes in coating performance (up to the time of ingestion by the patient) are unacceptable.

Thus the quality of a coating applied from an aqueous polymeric dispersion is very sensitive to process conditions used.

APPLICATIONS FOR AQUEOUS POLYMERIC DISPERSIONS IN THE PHARMACEUTICAL INDUSTRY

As stated earlier, because they utililze water-insoluble polymers, aqueous polymeric dispersions are of primary importance in the production of oral solid-dosage forms where there is a strong desire to modify the release characteristics of the active ingredient.

While modern pharmaceutical technology makes possible the design of dosage forms that exhibit modified time of release or rate of release (or both) of the active ingredient, a plethora of terminology (relating to these kinds of dosage forms) exists that confuses formulators, prescribers, and consumers alike.

The U.S.P./N.F. has simplified this terminology somewhat by defining a modified-release dosage form as one in which "the drug-release characteristics of time course and/or location are chosen to accomplish therapeutic or convenience objectives not offered by conventional dosage forms ....."

Under this umbrella definition, the U.S.P./N.F. recognizes two types of modified-release dosage form, namely:

- EXTENDED RELEASE - one that permits at least a two-fold reduction in dosing frequency as compared to the situation where the drug is presented as a conventional dosage form (extended-release dosage forms are often called sustained release or controlled release).

- DELAYED RELEASE - one that releases the active ingredient at some time other than promptly after administration (an enteric-coated product is an example of this type of dosage form).

Enteric Film Coatings

By definition, enteric coatings are those which remain intact in the stomach (and exhibit low permeability to gastric fluids), but break down readily once the dosage form reaches the small intestine. The prime uses of such coatings are:

- To maintain the activity of drugs that are unstable when exposed to the gastric milieu (examples of such drugs include erythromycin and pancreatin).

- To minimize either nausea or bleeding that occurs with those drugs that irritate the gastric mucosa (examples of these types of drug are aspirin and certain steroids).

Currently-used enteric coatings are usually formulated with synthetic polymers that contain ionizable functional groups that render the polymer water soluble at a specific pH value. Such polymers are often referred to as "poly-acids".

Examples of commonly used enteric-coating polymers (including those introduced more recently) are listed in Table 1.

Table 1. Examples of Enteric Coating Polymers

Type	Polymer Example
1. Cellulose esters	Cellulose acetate phthalate (C.A.P.) Cellulose acetate trimellitate (C.A.T.)
2. Cellulose ether/esters	Hydroxypropyl methylcellulose phthalate (H.P.) Hydroxypropyl methylcellulose acetate succinate (H.P.M.C.A.S.)
3. Vinyl esters	Polyvinyl acetate phthalate (P.V.A.P.)
4. Acrylics	Poly (MA-EA)* 1:1 Poly (MA-MMA)* 1:1 Poly (MA-MMA)* 1:2

* MA = methacrylic acid   EA = ethylacrylate
  MMA = methyl methacrylate

Since many of these polymers are esters, they may be subject to degradation (as a result of hydrolysis) when exposed to conditions of elevated temperature and humidity. Such hydrolysis can result in a substantial change in enteric properties. Delporte has described the effects of aging on the physical properties of the enteric polymer, cellulose acetate phthalate.

While many of the polymers shown in Table 1 have been used for many years in enteric-coating formulations, the special aqueous-solubility requirements for an enteric polymer have delayed the routine employment of aqueous enteric-coating technology. More recently, various approaches to aqueous enteric coating have been introduced, and examples are shown in Table 2.

Table 2. Examples of Aqueous Enteric Coating Systems

Polymer	Form	Trade Name
Poly (MA-EA)* 1:1	Latex dispersion	Eudragit L 30 D
	Spray-dried latex	Eudragit L-100-55
C.A.P.	Spray-dried pseudo latex	Aquateric
	Dry powder	Eastman C.A.P.
C.A.T.	Dry powder	Eastman C.A.T.
P.V.A.P.	Dry powder	Coateric
H.P.	Dry powder	HP-F
H.P.M.C.A.S.	Dry powder	H.P.M.C.A.S.

* MA = methacrylic acid
  EA = ethacrylic acid

As these examples suggest, many of the coating systems exist as dry powders, with the coating liquid being prepared shortly before use by dispersing (or dissolving) the polymer in water. The reason for supplying many enteric coating systems as dry powders is to avoid problems of poor stability (due to hydrolysis) when these polymers are exposed to water for extended periods.

The performance of enteric-coated dosage forms has often been open to question. Much of the uncertainty can be related to the earlier common use of "natural" polymers (such as shellac) and simplistic coating procedures. The use of synthetic, predictable polymers and adoption of modern processing technology should have done much to dispel these concerns. However, problems still exist today. Unfortunately, many of the factors that can dramatically affect the performance of enteric coatings have long gone unrecognized. Ozturk, et al[5] recently presented information on some of the important factors that can influence the behavior of enteric coatings. These factors include:

- The nature of the drug in the dosage form (the presence of aspirin, for example, can greatly influence dissolution of the coating).

- The quantity of coating applied (application of excessive quantities of coating can substantially delay release of drug from the dosage form) as shown in Fig. 4.

- The presence of imperfections in the coating (fissures or "pick" marks will destroy the integrity of the coating).

- The dissolution pH of the polymer used in the coating (see Fig. 5).

- The effect of "in vitro" test conditions (dissolution of the coating, and ultimate drug release, can be affected dramatically by the pH and ionic strength of the test solutions and the agitation rate), an example of which is shown in Fig. 6.

Finally, while most enteric products are in tablet form, it has been demonstrated that enteric-coated tablets are influenced significantly by G.I. transit. Focus has thus begun to shift towards using enteric-coated pellets or granules, which can give greater reproducibility[6] (with respect to release and absorption of drug).

Sustained or Controlled-Release Film Coatings

Film-coating techniques to produce sustained-release dosage forms have been utilized since the late 1940's, when SmithKline used a pan-coating process to apply various mixtures of fats and waxes (dissolved in organic solvents) to drug-loaded beads. Since that time a variety of materials and coating processes have been used for the same purpose. Drug release from such sustained release products is moderated by the film coating which acts as a membrane that allows infusion of G.I. fluids and the outward diffusion of dissolved drug. In some instances, the release process may be augmented by a coating that slowly dissolves (for example, shellac) or is subject to digestion by enzymes (for example, fats and waxes).

As with enteric coatings, pharmaceutical formulators are showing a growing interest in using aqueous polymeric dispersions, a list of which is shown in Table 3.

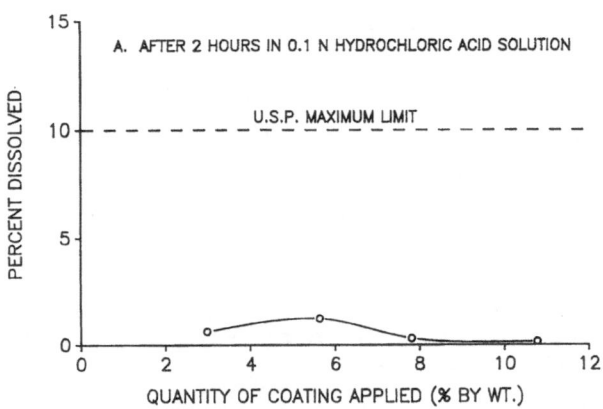

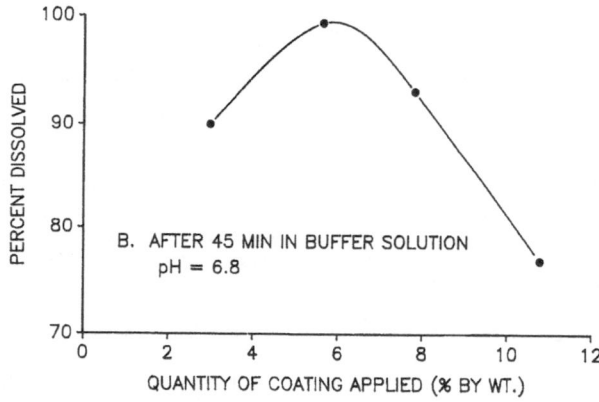

Fig. 4.   Effect of quantity of an enteric coating applied on
release of aspirin in artificial gastric juice (0.1N
HCl solution) and artificial intestinal fluid (buffer
solution pH = 6.8)

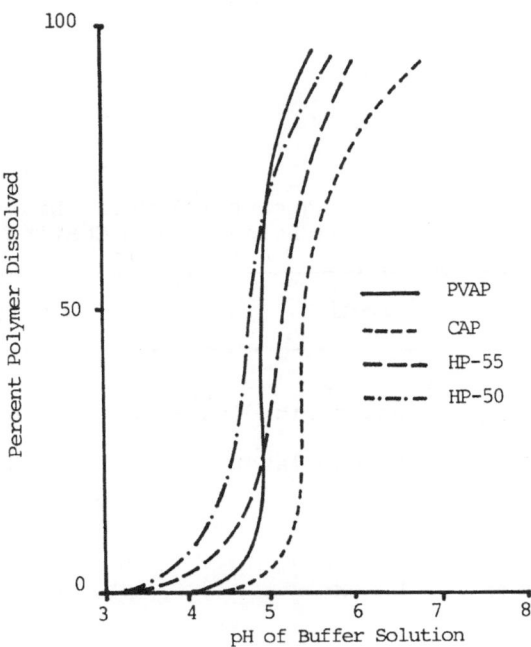

Fig. 5.   Effect of pH on the dissolution of various
          enteric-coating polymers

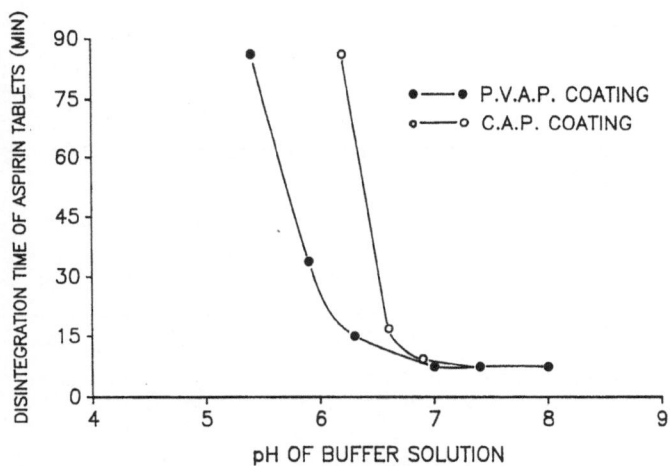

Fig. 6.   Effect of pH of test fluid on the disintegra-
          tion time of tablets enteric coated with either
          polyvinyl acetate phthalate (P.V.A.P.) or
          cellulose acetate phthalate (C.A.P.)

Reprinted from Ref. (1), p. 341 by courtesy of Marcel Dekker,
Inc.

Table 3. Examples of Aqueous Polymeric Dispersions
for Sustained-Release Film Coating

Polymer	Form	Trade Name
Ethylcellulose	Polymeric dispersion	Surelease
	Pseudo latex	Aquacoat
Silicone elastomer	Latex	---
Poly (EA-MMA)* 2:1	Latex	Eudragit NE30D
Poly (EA-MMA)* Triethyl ammonio- ethyl methacrylate chloride 1:2:0.2	Polymeric dispersion	Eudragit RL30D
Poly (EA-MMA)* Triethyl ammonio- ethyl methacrylate chloride 1:2:0.1	Polymeric dispersion	Eudragit RS30D

* EA = ethyl acrylate
  MMA = methyl methacrylate

Various pharmaceutical forms may be used as substrates for sustained-release film coatings. These may generally be classified as:

- Tablets.

- Multiparticulates (for example, drug-loaded beads, granules, crystals, powders, drug/ion-exchange resin complexes).

While both general types of substrates are in current use, the preference now is trending toward multiparticulate systems which are perceived to have advantages such as minimization of risk of dose dumping (should membrane rupture occur) and optimization of gastro-intestinal transit.

Although multiparticulates (especially drug-loaded beads) were once commonly film-coated in pans, the wide variety of multiparticulate systems coated today often requires specialized processing techniques that involve the use of fluid-bed coating equipment.

Dosage forms where an applied film coating provides the main means for rate control for drug release are commonly termed "membrane-moderated, controlled-release systems" or "reservoir systems". With such a system, the drug diffuses from the core material, through the rate-controlling membrane into the surrounding environment. Typically, such factors as membrane porosity, tortuosity, geometry, and thickness play an important part in determining the rate at which the drug can pass through the coating. A discussion of these factors (and appropriate mathematical treatment) has been given by Tojo, et al.[7]

A simplified model which is commonly used to describe drug release from a reservoir system and utilizes a derivation of Fick's Laws of diffusion is shown in Fig. 7.

While the expected behavior of reservoir systems can be depicted by such relatively simple mathematical treatment, this treatment often ignores the impact of structural irregularities in the membrane that can be related to the substrate on which the coating is applied, formulation of the coating itself, and the impact of critical processing factors.

Structural modifications to the membrane (some deliberate, but others totally unexpected) may influence the system such that drug release occurs as a result of:

- Transport of drug through a network of capillaries (filled with dissolution media) created by the leaching out from the film of a water-soluble component.

- Transport of drug through a hydrated, swollen film as the result of the presence in the coating of a water-soluble component (usually high-molecular weight) that cannot be readily leached out.

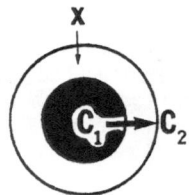

Drug release is given by:

$$\frac{dM}{dt} = A\ D_m\ P\left[\frac{C_1 - C_2}{x}\right]$$

where: $\frac{dM}{dt}$ is drug release rate

        A is surface area

        D_m is drug diffusion coefficient

        P is polymer/dissolution fluid partition coefficient

        x is film (membrane) thickness

        C_1 is drug concentration in core

        C_2 is drug concentration in fluid immediately adjacent to surface of membrane

Fig. 7. Model predicting drug release from a reservoir.

- Transport of drug through flaws in the membrane (often present as stress-induced cracks in a coating of low mechanical strength).

- Transport of drug through pores that result from incomplete coalescence of a coating prepared from an aqueous polymeric dispersion.

The membrane formulation for a reservoir system will have a major impact on membrane structure and thus will greatly affect the performance of the final product.

It is critical that the formulation chosen yields a coating whose properties remain little changed throughout the life of the product. Consequently, we have to be concerned with mechanical properties of the polymer and how these are impacted by additives such as plasticizers and fillers.

The presence of leachable ingredients can be counted on to change the permeability characteristics of the coating.

There has been some discussion[8] of the importance of molecular weight of the polymer on the behavior of controlled-release film coatings. In these studies, it was determined that low molecular weight ethylcellulose will produce coatings with mechanically inferior properties that are likely to influence drug release.

Recently, studies have been conducted[9] into examining what happens when an aqueous polymeric dispersion is applied as a controlled-release coating to a pharmaceutical substrate.

These studies suggest that:

- In the early stages of film deposition, flaws exist in the coating so that drug release occurs through pores in the membrane structure.

- As film coating progresses, sufficient coating is applied such that any flaws are no longer contiguous between the substrate and the surface of the coating; drug release is now barrier controlled rather than pore controlled.

- A mathematical model can be generated to predict at what level of applied coating the drug-release mechanism changes from pore control to membrane control.

- The level of applied coating at which this change in mechanism occurs varies depending on the core formulation and (and its method of, preparation) and the coating formulation.

A schematic representation of the formation of a membrane under these conditions is shown in Fig. 8.

It is very difficult, when considering aqueous polymeric dispersions, to say which product gives the best results (with respect to amount of coating that needs to be applied to produce a given target release profile). The chemical characteristics of the membrane, for example, may well influence interaction with a particular drug entity and thus determine how quickly that drug diffuses through the coating.

The performance of such coatings may well be determined by a variety of factors, including those relating to:

- The substrate.

- The coating formulation.

- The coating process.

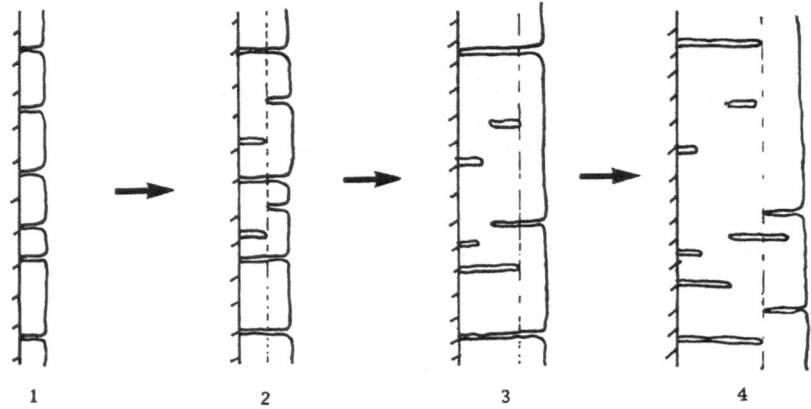

Fig. 8.   Schematic representation of development of a film-coating membrane.

Substrate Factors

The material to which a coating is applied is a key component in determining the behavior of the final coated product.   Substrate variables and their impact on product performance are likely to be extensive; however, some of the variables commonly experienced are:

- Surface area (which is affected by size, shape, size distribution, and surface roughness).   Variation in surface area may be exemplified by change in particle size of a multi-particulate system. Some typical results obtained when varying particle size are shown in Fig. 9.

- Substrate friability. Coating processes are usually sufficiently stressful to cause abrasion of the substrate. Drug powder so generated may well be distributed throughout the coating, thus affecting the behavior in which the drug is ultimately released from the system.

- Substrate porosity. A highly porous substrate may permit water from the latex coating to penetrate into the substrate, causing drug to be leached out. Additionally, removal of excessive water from the latex (by penetration into the substrate) may influence ultimate coalescence of latex particles into a coherent film.

- Chemical factors. The chemistry of the substrate is determined by the nature of the drug and other excipients used. Drug release will be determined by drug solubility in the release fluid, drug solubility in the membrane, drug diffusivity in the membrane, and molecular size of the drug. Release may also be impacted by osmotic effects.[10] A typical example of how the nature of the drug may influence drug release is shown in Fig. 10.

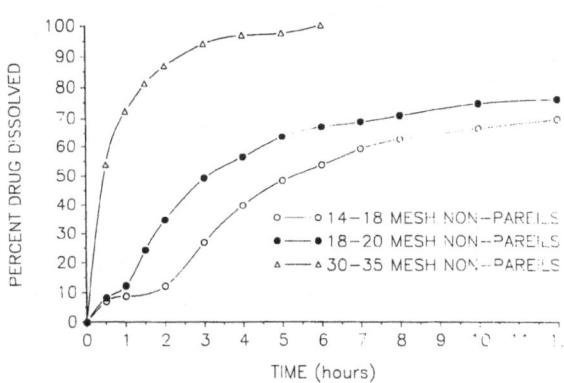

Fig. 9. Effect of mesh size (of drug-laden beads) on the release of chlorpheniramine maleate from beads coated with Surelease (10% theoretical coating level).

Reprinted from Ref. (14), p. 1512, by courtesy of Marcel Dekker, Inc.

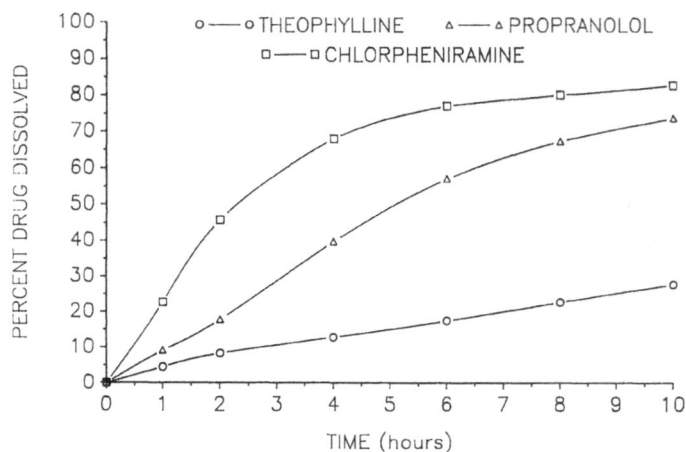

Fig. 10. Effect of drug type on drug release from
non-pareils coated with Surelease (10%
theoretical coating applied).

Reprinted from Ref. (14), p. 1511, by courtesy of Marcel Dekker,
Inc.

An approach that may be used to minimize the effects of
substrate variability (on drug release) involves the
application of a seal coat to the core prior to applying the
release-modifying coating. Use of such a seal coat tends to
smooth the substrate surface (thereby minimizing surface area
variations), seal off substrate porosity, minimize abrasion,
and reduce drug leaching. A typical result obtained when a
seal coat is applied is shown in Fig. 11.

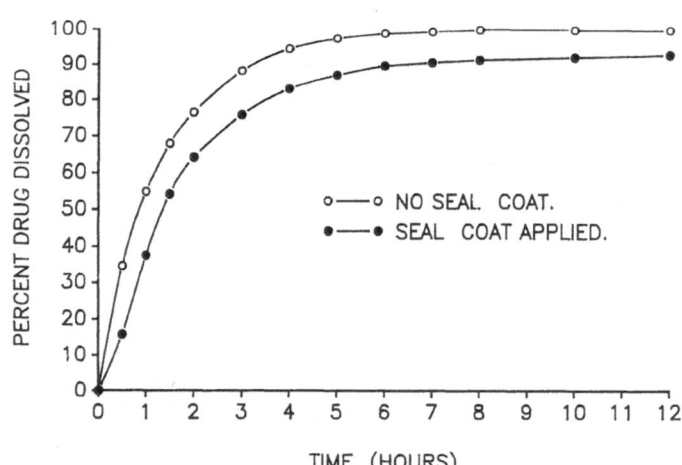

Fig. 11. Effect of precoating drug-loaded beads
with a seal coat on the release of chlorpheniramine maleate from beads ultimately
coated with Surelease (6% theoretical
coating level applied).

Coating Formulation Factors

The applied coating has a major influence on the rate at which the drug will be released from the dosage form. The major factors to be considered are:

- The nature of the polymer. In this regard, we are concerned with polymer chemistry, molecular weight[8] and glass transition temperature. Figs. 12a and 12b illustrate some typical results that may be obtained with the various aqueous polymeric dispersions that can be utilized for the purpose of preparing controlled-release film coatings.

- The presence of additives. Additives which can affect the results obtained may well be used as formulation aids in the preparation of the latex dispersion.[11] Additional effects may be obtained when including processing aids (such as anti-tack agents) designed to facilitate application of the coating. Special effects can also be created by adding water-soluble polymers as shown in Fig. 13.

- Coating thickness. Various factors influence the final coating thickness, including surface area (of substrate) to be covered, uniformity of distribution of coating and coating process efficiency. Of course, a major factor is the quantity of coating[12] applied as shown in Fig. 14.

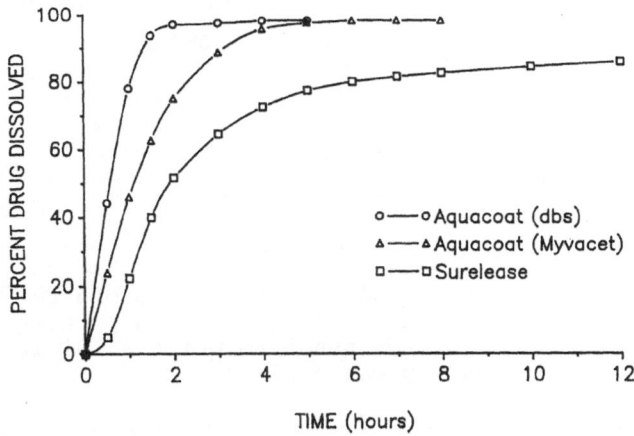

Fig. 12a. Release of chlorpheniramine maleate from beads coated with various ethylcellulose coating systems (10% theoretical coating level applied).

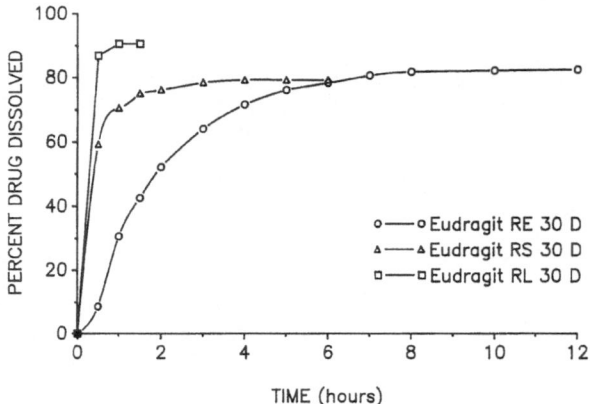

Fig. 12b. Release of chlorpheniramine maleate
from beads coated with various acrylic
coating systems (10% theoretical coating
level applied).

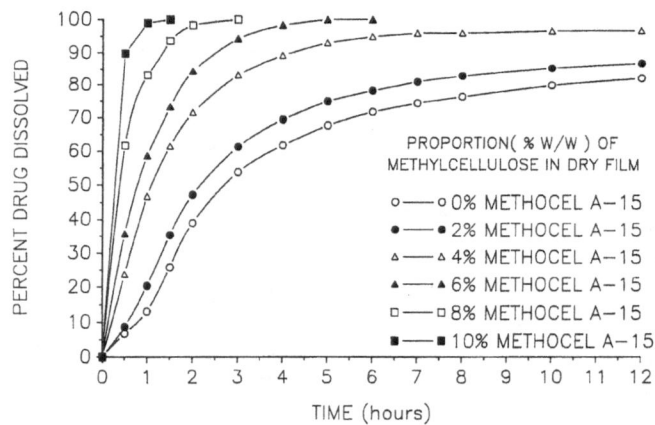

Fig. 13. Effect of adding methylcellulose to
Surelease on release of chlorpheniramine
maleate from beads to which a 10% level
of coating was applied.

Reprinted from Ref. (14), p. 1508, by courtesy of Marcel Dekker,
Inc.

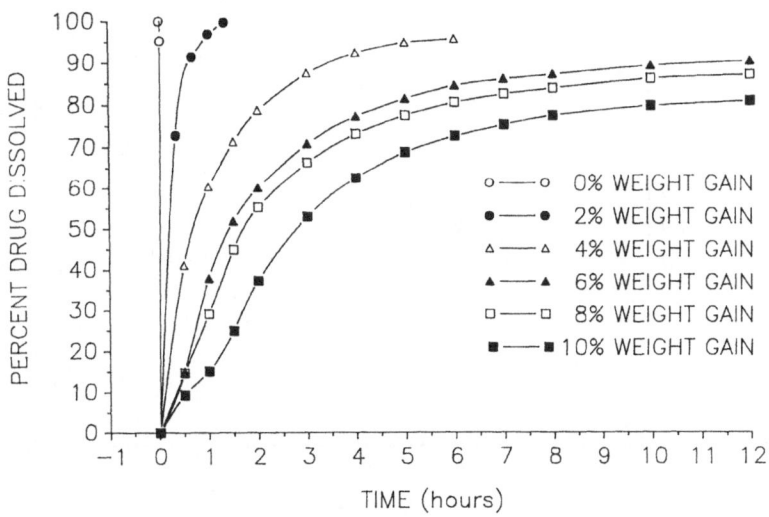

Fig. 14. Effect of quantity of coating applied (% by weight of Surelease) on release of chlorpheniramine maleate from drug-loaded beads.

Coating Process Factors

The idiosyncracies of the pharmaceutical film-coating process are such that the method of application of the coating can have a major impact on final quality of that coating.

Issues that may have to be considered are:

- Type of process used. Common pharmaceutical processes utilize both coating pans and fluid-bed coating units. Since use of multi-particulate materials as substrates is common, fluid-bed coating processes are often used. Even so, various processing approaches (top spray, bottom spray or tangential spray) may be considered, and it has been demonstrated that each may produce different results.[13] Nonetheless, appropriate understanding of the coating processes used may help to minimize differences in results obtained when using different processing techniques as exemplified by the data illustrated in Fig. 15.

- Uniformity of distribution of the coating material. Processing factors that might be important here include coating liquid addition rate (spray rate - see Fig. 16) and concentration of solids in the coating liquid (see Fig. 17). Such factors as these are not only likely to affect distribution of the coating, but because they also influence the relationship between rate (and quantity) of liquid added and liquid removal rate (drying), they may have an impact on coating structure.

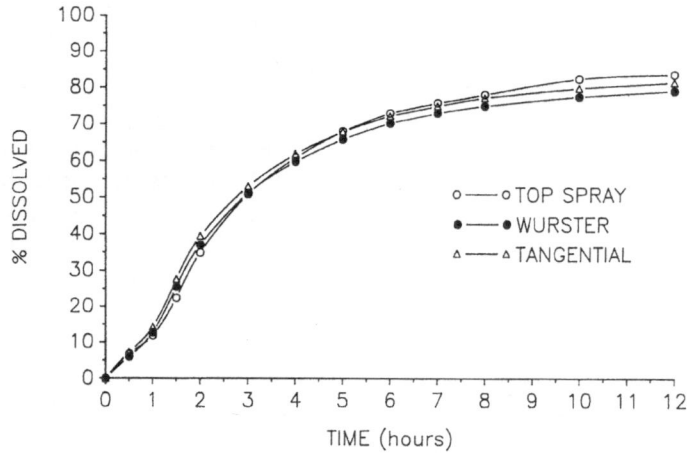

Fig. 15.   Release of chlorpheniramine maleate from
           beads coated (with 10% Surelease coating)
           in three types of fluid-bed process.

Reprinted from Ref. (15), p. 143, by courtesy of Marcel Dekker,
Inc.

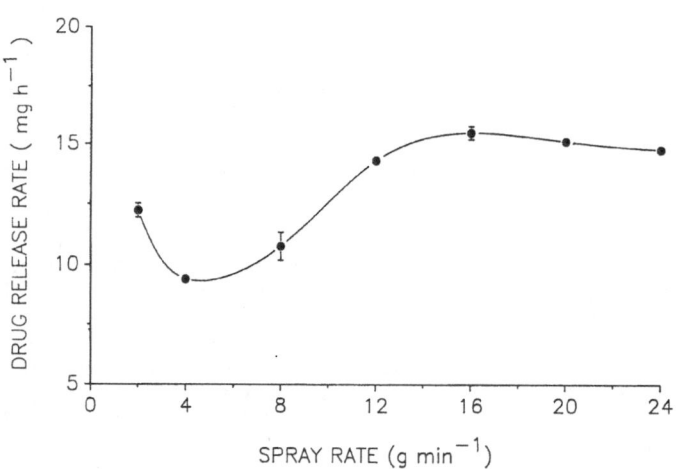

Fig. 16. Effect of spray application rate on release
         of chlorpheniramine maleate coated with
         Surelease (10% coating level applied).

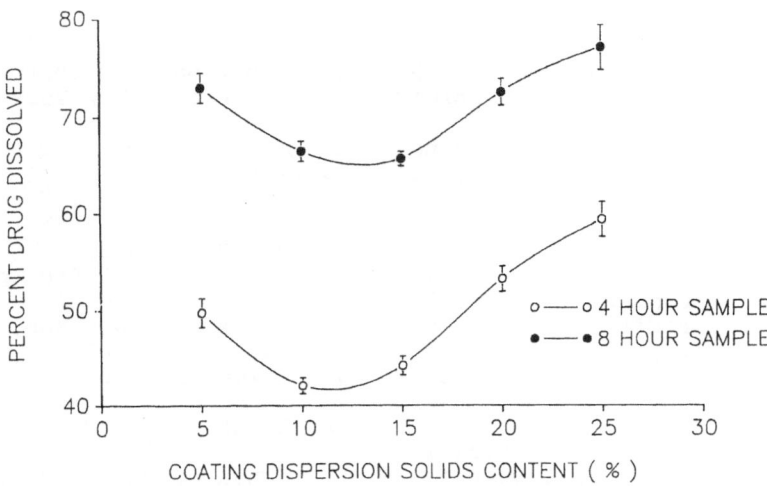

Fig. 17. Effect of coating dispersion solids on
release of chlorpheniramine maleate from
beads coated with Surelease (10% coating
level applied).

- Drying process (air flow, temperature, and humidity).
  This can influence both coalescence of a latex
  coating and tackiness during application. Other
  factors to be considered are the influence of the
  drying process on drug leaching from the substrate
  (which can occur if the drying rate is too low) and
  coating process efficiency (typically related to the
  actual amount of coating deposited, which can be
  detrimentally affected when the drying rate is so
  high that premature spray drying of the coating
  liquid occurs).

SUMMARY

Pharmaceutical formulators are gradually becoming familiar
with the use of aqueous polymeric dispersions. However, such
coating systems are used to produce highly functional coatings
that are designed to have a critical impact on the rate at
which the drug is released from the dosage form.

Consequently, the coating must be fully formed during the
time frame of the coating process if reproducible results (with
respect to drug release characteristics) are to be achieved,
and time-dependent changes are to be avoided. Achievement of
such an objective will be dependent on the characteristics of
the material being coated, the coating formulation, and the
coating process used.

BIBLIOGRAPHY

1. J. W. McGinity, "Aqueous Polymeric Coatings for Pharma-
    ceutical Dosage Forms", Marcel Dekker, (New York),
    1988.
2. C. Bindschaedler, R. Gurney & E. Doelker, Labo-Pharma -
    Probl. Tech. 31, (331), 389 (1983).
3. Z. W. Wicks, Jr., J. Coatings Technol., 58 (743), 23
    (1986).
4. J. P. Delporte, Pharm. Ind., 41 (10), 984 (1979).
5. S. S. Ozturk, B. O. Palsson, B. Donohoe & J. B. Dressman,
    Pharm. Res. 5 (9), 550 (1988).
6. C. Bogentoft, I. Carlsson, G. Ekenved & A. Magnusson,
    Europ. J. Clin.. Pharmacol., 14, 351 (1978).
7. K. Tojo, K. Miyanami & L. T. Fan, Powder Technol., 35, 89
    (1983).
8. R. C. Rowe, A. D. Kotaras & E. F. T. White, Int. J.
    Pharm. 22, 57 (1984).
9. G. Zhang, J. B. Schwartz & R. L. Schnaare, Pharm. Res. 5
    (10), S-53 (1988).
10. I. Ghebre-Sellassie, R. H. Gordon, R. U. Nesbitt & M. G.
    Fawzi, Int. J. Pharm., 37, 211 (1987).
11. R. K. Chang, C. H. Hsiao & J. R. Robinson, Pharm. Tech.,
    11 (3), 56 (1987).
12. F. W. Goodhart, M. R. Harris, K. S. Murthy & R. U.
    Nesbitt, Pharm. Tech. 8 (4), 64 (1984).
13. S. T. Yang, J. Weiss & I. Ghebre-Sellassie, Pharm. Tech.
    (suppl.), 6 (9), S-64 (1989).
14. S. C. Porter, Drug Dev. & Ind. Pharm., 15 (10), 1495,
    (1989).
15. S. C. Porter & C. H. Bruno, "Coating of Pharmaceutical
    Solid Dosage Forms" in Pharmaceutical Dosage
    Forms:Tablets 3, Marcel Dekker, (New York), In
    Press.

PERFORMANCE CHARACTERISTICS OF AQUEOUS COATING FOR PHARMACEUTICALS

M. K. Masih, J.L. McHan and S. Z. Masih

Pharmaceutical Services Enterprise Inc.
6008 Fort Henry Drive
Kingsport, TN 37663 (U. S. A.)

Surface coating of medicinal tablets, capsules,
pellets and suppositories with aqueous polymeric
dispersion called pseudolatex, has much in common with
surface coatings by the paint and adhesive industry which
use a latex system. However, there are stringent regulatory,
safety, processing, and performance constraints which
put surface coating of pharmaceutical products in a
special category. Since much of the theoretical
considerations and the mechanism of film formation or
physical chemistry of various polymers is being
addressed by other contributors, this communication is
restricted to illustration of the application of aqueous
coating within the pharmaceutical practice, with special
emphasis on performance evaluation using caffeine tablet
as a model substrate. According to the regulatory
compliance guidelines, the facility, equipment, coating
procedure, coating conditions, and method used to evaluate
performance parameters such as disintegration time, drug
release profile as a function of pH, and stability
evaluation are presented.

INTRODUCTION

The primary aim of surface coating of pharmaceutical dosage forms is
to impart a protective membrane around the entire surface of a tablet,
capsule, drug crystal, and granules or pellets of drug and excipient blend
so as to encase the drug unit within a membrane. The pharmaceutical
formulator, the clinical pharmacologist, and the product manager for
selling a product line has specific reasons for applying the coating.
Some of the reasons for coating of dosage form including the primary
desire to target the delivery of medicaments to a specific part of the
body are listed in Table I.

The coating practice in the Drug Industry dates back to ancient times.
Until recently, the coating polymer or resin solution in an organic
solvent was the primary means to apply coating on the surface of the
dosage forms. However, the strict enforcement of regulations by
Environmental Protection Administration (EPA), and the Occupational
Safety Hazard Agency (OSHA) as well as the desire of the management to
eliminate explosive destruction of plants and human lives as well as

Surface Phenomena and Fine Particles in Water-Based Coatings and Printing Technology
Edited by M.K. Sharma and F.J. Micale, Plenum Press, New York, 1991

TABLE I. REASONS FOR COATING PHARMACEUTICAL PRODUCTS.

TO PROVIDE A RATE LIMITING MEMBRANE FOR DIFFUSION OF MEDICATION

TO PROVIDE TASTE MASKING FOR BAD TESTING MEDICATIONS

TO IMPART STRENGTH TO SOFT AND FRIABLE TABLETS FOR WITHSTANDING
THE STRAINS DURING PACKAGING AND TRANSPORTATION

TO PROTECT MEDICATION FROM PHOTOLYTIC DEGRADATION

TO PROTECT MEDICATIONS FROM HYDROLYTIC DEGRADATION FROM MOISTURE.

TO PROVIDE ESTHETIC APPEAL AND TO ENCOURAGE PATIENT COMPLIANCE

TO PROVIDE A MEANS FOR PRODUCT AND BRAND IDENTIFICATION

TO AID IN SWALLOWING

TO SEPARATE INCOMPATIBLE DRUGS

TO PROVIDE SITE SPECIFIC DELIVERY OF DRUGS IN THE GASTROINTESTINAL TRACT

---

management's effort to provide protection to workers from toxic exposure
to organic solvent vapors, have forced the industry to explore the use of
water-based coating systems. During the past years, several polymeric
dispersion systems have been developed and employed for water-based film
coating of solid dosage forms[1-20]. Generally, the water-based polymeric
dispersions for coating pharmaceutical products are referred to as
pseudolatex to distinguish it from true latex. Whereas, the true latex is
prepared by emulsion polymerization process involving monomers or monomer
blends, surfactants, initiator and water to produce water-insoluble latex
particles of 0 - 1.0 micrometer in size, the pseudolatex is prepared by
dissolving the thermoplastic water-insoluble, FDA approved polymer in
organic solvent and emulsifying it with the aqueous solution of surfactant
and other additives, without the help of initiator. The organic solvent is
then removed by vacuum distillation leaving the aqueous dispersion of
polymer for coating application. If the polymer in latex dispersion is
prone to hydrolysis, the aqueous dispersion is spray dried. The spray
dried pseudolatex powder is redispersed in water containing surfactant,
color or other additives just before coating. The particles, in spray
dried powder form, are 0.5 to 30 micron in size, with the average particle
size being below 10 micron. Although these particles can not be considered
truly fine particles, they provide an excellent means to cover the surface
of the substrate with a thin film without the use of organic solvents
containing plasticizers.

From the human health and safety point of view, the true polymeric latex is not suitable because the latex can not be demonstrated to be free from the monomers and traces of initiator used during emulsion polymerization[5]. The monomers, are often implicated to be carcinogens and therefore, must be avoided in pharmaceutical products for internal use.

Despite a large number of polymers available for paint and adhesive applications, only a few selected polymers, generally recognized as safe (GRAS), or which have been proven to be safe through extensive toxicological studies are allowed in the pharmaceutical products. Table II lists the commonly used polymers and some commercial water-based coating systems for pharmaceutical dosage forms. These pseudolatex or polymeric dispersions in water are gaining popularity because they can carry more

TABLE II. COMMERCIAL WATER BASED COATING SYSTEMS

BRAND NAME	CHEMICAL NAME	TYPE	MANUFACTURER
AQUATERIC	CELLULOSE ACETATE PHTHALATE	SD	FMC
EUDRAGIT L 30 D	POLYMETHACRYLIC ACID COPOLYMER	SD	ROHM PHARMA
COATERIC	POLYVINYL ACETATE PHTHALATE	SD	COLORCON INC.
AQUACOAT	ETHYLCELLULOSE	LD	FMC
SURELEASE	ETHYLCELLULOSE	LD	COLORCON
EUDRAGIT RS 30 D	ACRYLIC & METHACRYLIC ACID COPOLYMER	LD	ROHM PHARMA
EUDRAGIT E 30 D	DIMETHYLAMINOETHYL METHACRYLATE & NEUTRAL METHACRYLIC ACID ESTER	LD	ROHM PHARMA

SD = SPRAY DRIED WATER DISPERSIBLE

LD = LIQUID DISPERSION IN WATER.

than 30 - 40 % polymer in coating dope[2-3] without increasing the viscosity beyond the handling capabilities of the spray pumps and spray guns employed. It is also reported that the water vapor transmission rates (WVRT's) of the pseudolatex film is one-half to one-third of the film generated from the solution of polymers in organic solvents. This property is valuable for coating moisture sensitive drugs. The mechanism of film formation from the water-based pseudolatex coating dope is very unique as illustrated in Figure 1. Instead of the usual layer of film produced from the evaporation of solvent from the solution of the polymer, the pseudolatex droplets from the coating dispersion are deposited on the surface of the tablet as segregated spheres. As water loss occurs at a zero order rate through evaporation, the interfacial tension between the water and latex spheres decreases, the spheres pack together closely, the inter-particle contact increases and ultimately the particle coalesce or fuse together to form continuous film. Generally, the rate of solvent loss from polymeric coating solution is fast in the initial stages, but as the evaporation progresses the concentration of polymer increases, the viscosity increases and the vapor pressure drops, which decreases the rate of solvent loss. The overall effect on the processing time from solvent system versus the aqueous dispersion may not be significantly different since more solvent-based coating has to be applied to get the comparable thickness of polymer film resulting due to the high solid content water-based coating dope.

Addition of a suitable plasticizer to the latex dispersion is, generally, necessary to assist in reducing the cohesive force along the polymer chain, and to decrease the tensile strength, T_g, and minimum film forming temperature to form a good flexible non-brittle film. The plasticizer is also necessary to promote swelling and softening of the polymeric latex sphere, to facilitate coalescence and ultimate film formation. Information on the commonly used plasticizers is shown in Table III.

TABLE III. COMMONLY USED PLASTICIZERS

PLASTICIZER	BP (^{O}C)	SOL. PAR.	INTRINSIC VISCOSITY (dl g^{-1})
DIETHYL PHTHALATE	282	20.5	1.38
DIBUTY SEBACATE	345	18.8	-
TRIETHYL CITRATE	127	-	-
TRIACETIN	258	-	-
ACETYLATED MONO-GLYCERIDES	150	-	-

The aim of this communication is to report performance characteristics of selected FDA approved water-based pseudolatex coating systems for pharmaceutical products. The pharmaceutical product selected for study is in the form of tablets containing caffeine as a active material. The caffeine release profiles were developed for coated tablets with various water-based pseudolatex coating systems.

EXPERIMENTAL METHODS

All chemicals, excipients, and reagents including the model drug caffeine were received from approved suppliers. Each material was, where available, of USP grade. The facility and equipment were operated in compliance with the Current Good Manufacturing Practice Regulation[21] . Selected quality control tests, with respect to identity, purity, and potency of each material, were performed according to the United States Pharmacopeia. XXI[22] .

Preparation of Coating Sustrate : The tablets of Caffeine, to be used as the substrate for aqueous coating evaluation, was prepared by the conventional wet granulation process. Tablets were compressed in a Stokes D3 Rotary Tablet Press using 3/8 inch biconvex punch. The composition of the granules and tablets is shown in Table IV.

TABLE IV. COMPOSITION OF GRANULES AND TABLETS

INGREDIENTS	GRANULES (Wt %)	TABLETS (Wt %)
CAFFEINE ANHYDROUS	77	75
LACTOSE	14	13.6
MICROCRYSTALLINE CELLULOSE	7	6.8
SODIUM CARBOXY METHYL CELLULOSE	2	1.9
TALC	-	1.0
MAGNESIUM STEARATE	-	0.5
MODIFIED CELLULOSE GUM	-	0.5
TOTAL WEIGHT OF ONE TABLET (mg)	476	
AMOUNT OF CAFFEINE PER TABLET (mg)	357	

Coating of Tablets: Although, a number of different coating equipment such as Accela Coata with perforated pan to permit drying air to pass through the perforations into the tablet bed, or a fluidized bed Wurster coater to permit fluidization of the tablets in the hot air stream while the coating dope is sprayed onto the tablets, is now used more frequently to coat pharmaceutical products. A pear shaped 36" conventional coating pan equipped with hot air intake and exhaust ports, temperature monitoring gauge, and Bink Spray Gun was used for coating experiment. Not less than a 10 Kg of tablets were loaded into the coating pan and warmed with hot air at 70°C for 5 minutes. The pan rotated at 15 RPM to tumble the tablets, and the coating dope was pumped continuously into the spray gun by Master Flex peristaltic pump. Compressed air at 20 psi provided the automization of the coating dispersion or solution. The spray coating was continued until a 8% or 15% increase in weight of the tablet was obtained. Pertinent details of the coating condition is shown in Table V. The coating dope was pumped at a constant rate of about 17 ml/min. After applying desired coating, the coated tablets allowed to tumble for about 15-30 min to remove excess water, and then removed from the coating pan. These coated tablets were cured at 40-45°C for about 8 hours.

TABLE V. COATING APPLICATION EQUIPMENT AND CONDITIONS

PEAR SHAPED FITZPATRIC STAINLESS COATING PAN	36 " SIZE
PAN ROTATION	15 RPM
BAFFELS	3, 2 " WIDE
NOZZLE HEIGHT FROM BED	14 "
AUTOMIZING AIR	20 PSI
INLET TEMPERATURE	70 C
EXHAUST TEMPERATURE	40 C
TABLET BED LOAD	10 KG
PUMP (MASTERFLEX)	PERISTALTIC
SPRAY GUN	BINK
SPRAY RATE	17 ML / MIN
CURING TIME AND TEMPERATURE	8 HOURS 45 C

Preparation of the Coating Dope : Spray dried pseudolatex powder of cellulose acetate phthalate available as Aquateric, or poly vinyl acetate phthalate available as Coateric were dispersed in water containing a suitable surfactant, plasticizer, color, antifoaming agent and an agent to prevent tackiness of the tablets during the coating process. The pseudolatex available as liquid dispersions such as Aquacoat (dispersion of ethyl cellulose), or Eudragit (dispersion of methacrylic acid copolymers, or methacrylic ester copolymers) were mixed with a suspension of talc, FDC color, a plasticizer, and or a suitable surfactant so as to provide 15 - 20 % solids in the coating dispersions. The mixed dispersions were screened through a number 30 mesh sieve to remove larger particles which were found to clog the spray gun. The composition of dope employed in coating the pharmaceutical dosage forms is given in table VI.

TABLE VI. GENERAL COMPOSITION OF AQUEOUS COATING DOPE

COMPONENT CATEGORY	COMPONENT PARTICULAR	PERCENT RANGE
POLYMERIC DISPERSION	AQUACOAT, EUDRAGIT, SURELEASE AQUATERIC COATERIC (SOLIDS)	10 - 15
SURFACTANT	TWEEN 80, 60, 85, OTHERS	0.5 - 1.0
PLASTICIZER	CITROFLEX, TRIACETIN, DBS, DEP	2.0 - 4.0
TACKINESS BARRIER	TALC, KAOLIN	3.0 - 6.0
ANTIFOAM AGENT	DIMETHYLPOLYSILOXANE, SILICA GEL	0.5 - 1.0
COLOR	FDC DYES OR LAKES	QS
DISPERSION MEDIUM	DISTILLED WATER TO VOLUME	100

VISCOSITY RANGE : 30 - 100 cPS.

QS = SUFFICIENT QUANTITY TO GET DESIRED COLOR INTENSITY

Drug Release Study: A Pharma Test automated dissolution apparatus, equipped with computer controlled constant temperature water bath, a sampling valve station, and a UV spectrophotometer with flow through cell was used to evaluate the coating performance by monitoring the rate and extent of caffeine released from the coated tablets. The coated tablets were placed into 900 ml of simulated gastric fluid (pH 1.2) or simulated intestinal fluid (pH 6.8) according to USP method[22]. The rotation of the paddle was set at 50 RPM to provide agitation in the flask. The amount of caffeine release was monitored every fifteen minutes over 1-8 hour by measuring the absorbance of the dissolution medium at 210 nm. Efficiency of coating was judged from the percent of caffeine released as a function of time. Amount of caffeine release from the uncoated caffeine tablets served as control.

Stability Evaluation: The coated tablets were packaged in a high density poly ethylene child resistant screw capped bottles with safety seals. The packaged samples were stored at room temperature ($20^{O}C$), $37^{O}C$, and $37^{O}C$ at 75% humidity for 12 weeks. The dissolution profile of tablets stored at various conditions was determined initially and after 4, 8, and 12 weeks of exposure to various storage conditions. The tablets were also examined microscopically for any cracks, deformation or change in color or physical appearance of the tablets. If no change in physical appearance of the tablets or release profile were noticed, the coating was considered to have passed the test.

RESULTS AND DISCUSSION

Consistent with the contents of table I, the primary reason for coating caffeine tablets was to provide sustained release delivery system. Caffeine is a common constituent of coffee, tea, and many soft drinks. It is, therefore, not considered as a drug. It is estimated that a cup of coffee provides 60 mg of caffeine to human subject. The caffeine tablets, not being regulated by the FDA, is very popular among students and truck drivers who use it as stimulant to keep awake. Although, 60 mg caffeine combined with aspirin and codeine is sold as a prescription drug, as large a dose as 357 mg in one tablet is freely available in the market. Since this high dose, from the immediate release tablets, may not be reasonably safe, the present study attempted to design a sustained release tablet of caffeine through polymeric aqueous coating approach. Enteric coating was attempted to demonstrate that excessive dose dumping occurs once the drug passes from the stomach into the upper intestine.

In order to promote appreciation of the results of this investigation, table VII lists the ideal performance characteristics of an aqueous coating dispersion for pharmaceutical application. While Tables II and III list the common polymeric dispersions and plasticizers approved for coating of human medications, Figure 1 illustrates mechanism of film formation from the aqueous polymeric dispersions.

Since the nature and composition of coating substrate is of considerable interest to painters and polymer physical chemists involved in the study of the interaction between the fine particles and the surface of the coating substrate, the composition of the caffeine tablet is shown in Table IV. The surface of the tablets was fairly smooth and the tablets were hard enough to withstand the rough tumbling of tablets in the coating pan which is known to create dust and roughness on the coating surface in poorly formulated tablets. Optimized and appropriate coating conditions, which are often specific to the coating equipment used, are very important to guarantee reliability and reproducibility of the coating process for batch operation production. Table V summarizes the conditions

TABLE VII. IDEAL PERFORMANCE CHARACTERISTICS OF AQUEOUS COATING
DISPERSIONS

---

* THE COATING POLYMER MUST BE NON-TOXIC TO HUMAN.

* THE COATING POLYMER MUST BE COMPATIBLE WITH MEDICAMENT,
  FDC APPROVED DYES AND LAKES, TALC, CABOSIL, KAOLIN AND,
  FDA APPROVED PLASTICIZERS, AND PRESERVATIVES FOR INTERNAL
  USE IN HUMAN.

* THE COATING POLYMER MUST BE FDA APPROVED.

* THE POLYMERIC DISPERSION MUST BE FREE FROM MONOMERS AND
  TRACES OF INITIATOR.

* THE PARTICLE SIZE OF LATEX SPHERES IN THE AQUEOUS
  DISPERSION MUST BE WITHIN THE RANGE OF 0.5 -10 MICRONS
  AND NO PARTICLE MUST BE LARGER THAN 35 MICRON.

* THE VISCOSITY OF THE COATING DISPERSION CONTAINING
  20 - 40 % SOLIDS MUST NOT BE GREATER THAN 70-100 cP.

* THE SPRAY DRIED PSEUDOLATEX MUST BE EASILY DISPERSIBLE
  IN WATER AND THE AGGREGATES OF LATEX SPHERES MUST BREAK
  INTO DISPERSED PARTICLES OF 0.5 - 30 MICRON, PREFERABLY
  WITHIN 18 - 20 MICRON SIZE.

* THE COATING DISPERSION MUST REMAIN IN SUSPENSION WITH A
  MINIMUM AGITATION AND MUST NOT SETTLE OR CAKE ON STANDING.

* THE COATING DISPERSION MUST BE EASILY SPRAYABLE WITHOUT
  BLOCKING THE SPRAY NOZZEL, AND MUST BE EASY TO PUMP INTO
  THE SPRAY GUN.

* THE FILM PRODUCED MUST BE FLEXIBLE, AND MUST NOT DISSOLVE
  BETWEEN pH 1 - 4 FOR ENTERIC PROTECTION APPLICATION, BUT
  MUST DISSOLVE WITHIN 15-30 MIMUTES AT pH 5.5 - 6.8.

* FOR SUSTAINED RELEASE APPLICATION, THE FILM PRODUCED MUST
  BE FLEXIBLE AND ALLOW DIFFUSION OF DRUG AT A DESIRABLE
  AND CONTROLLED RATE INDEPENDENT OF THE pH OF THE
  DISSOLUTION MEDIUM.

* THE FILM MORPHOLOGY OR PHYSICAL PROPERTIES MUST NOT CHANGE
  UNDER ACCELERATED STABILITY TEST STORAGE CONDITION.

---

under which acceptable coating of the tablets was obtained in the conventional coating pan at the PSE facility. The conditions listed, in our experience, provided continuous trouble free coating for several hours of operation during an 8 hour shift. A typical pan coating system was employed in coating caffeine tablets. Although, the regular pan coating technology is very old, it is still popular because a number of manufacturers have established facilities with this system, and because it is considerably less expensive system than the fluid bed coating and the perforated pan coating. However, the precise electo-mechanical or computerized controls available in these modern advanced equipment has made the coating process a science rather than an art which is highly dependent on the skill and experience of the coater.

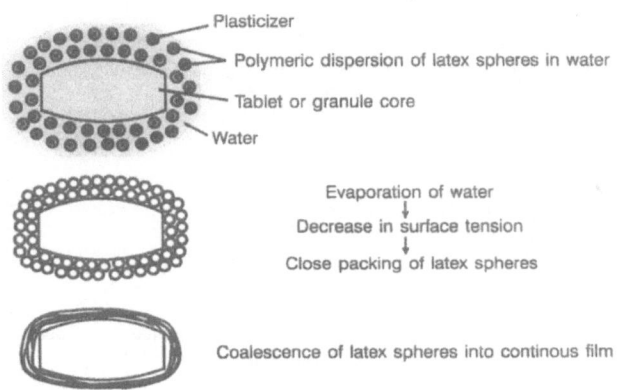

Figure 1. A schematic illustration of coating and film formation on solid dosage forms with pseudolatex systems.

Despite the fact that the disintegration test offers a quick means for evaluation of the performance of coating, it is not always a reliable indicator. A tablet may not disintegrate and imply a good coating performance, but the drug may not release out of the tablet to provide the desired medicinal effect through absorption. On the other hand, a tablet may fail disintegration test because of the excessive turbulence, and yet the coating may be satisfactory with respect to rate and extent of release. Dissolution tester was used in measuring the rate and extent of drug release as a means to evaluate the performance and efficiency of aqueous coating dispersions. Generally, the results of dissolution tests have been found to correlate with the data from the actual absorption of drug in the human system. The caffeine release as a function of time is shown in figure 2. It is observed that several commercially available tablets release caffeine at different rates.

Figure 3 shows the dissolution profile of caffeine tablets coated with the enteric polymeric dispersions, which must either protect the contents of the tablet from destruction in the acidic environment of pH 1-4 found in human stomach, or prevent its release in the stomach so as to avoid undesirable side effects, such as nausea and vomiting, associated with some drugs. Except for Aquateric, the other two enteric dispersions passed the performance criteria listed in Table VII. The failure of Aquateric dispersion is attributed to insufficient polymer in the dispersion after screening through 30 mesh sieve. The dissolution profile for caffeine tablets coated with sustained release polymeric dispersions

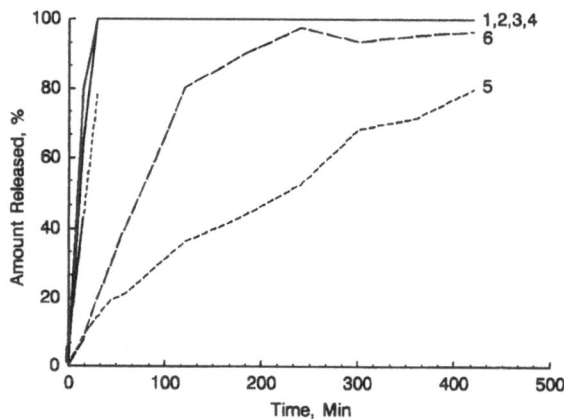

Figure 2. Dissolution profile of caffeine from various commercially available tablets (1) Nodoz tablet-100 mg, (2) Alert, 225 mg, (3) Vivarin, 200 mg, (4) Stimulent, 357mg, (5) SR200, 200 mg, and (6) SR357, 357 mg

is shown in Figure 4. Each of the three polymeric dispersions tested provided a controlled rate of release of caffeine over 8 hour period, thus indicating that frequent dosing with 60 or 120 mg tablet is not necessary and that a single dose administration of 357 mg dose of caffeine will not result into undesirably high blood level and lead to nervous hyperactivity. As expected, the increase in the coating load from 8 -15% coating, slows the release rate of caffeine.

The caffeine tablets coated with different water-based systems were studied for accelerated stability as described in the experimental section, and data obtained are recorded in table VIII. Results show that the caffeine tablets coated with Aquateric and Coateric did not meet accelerated stability test, while the tablets coated with Eudragit L30D passed the accelerated stability test.

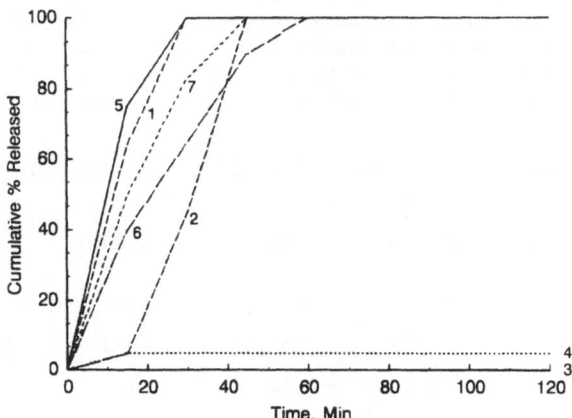

Figure 3. Release profile for cafeine from aqueous enteric coated
tablets. (1) Uncoated pH 1.2, (2) Aquateric pH 1.2,
(3) Eudragit L30D pH 1.2, (4) Coateric pH 1.2,
(5) Aquateric pH 6.8, (6) Eudragit L30D pH 6.8, and
(7) Coateric pH 6.8

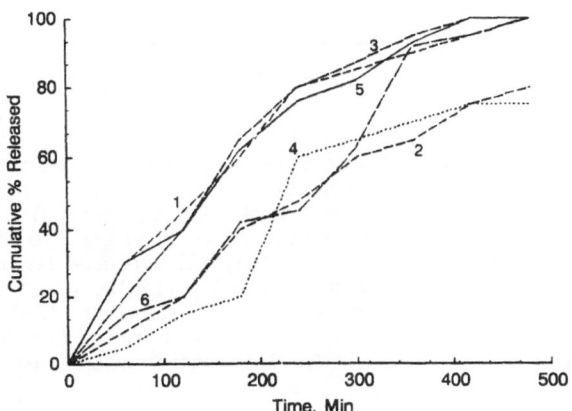

Figure 4. Release profile for caffeine from sustained release
coated tablets with water-based systems. (1) Aquateric
8% coat pH 1.2, 1hr, then pH 6.8, (2) Aquacoat 15% coat
pH 1.2, 1hr, then pH 6.8, (3) Eudragit RS30D 8% coat
pH 1.2, 1hr, then pH 6.8, (4) Eudragit RS30D 15% coat
pH 1.2, 1hr, then pH 6.8, (5) Surelease 8% coat pH 1.2,
1hr, then pH 6.8, and (7) Surelease 15% coat pH 1.2,
1hr, then pH 6.8

The study offered a meaningful way to evaluate the performance of aqueous polymeric dispersions for coating medicinal tablets and also demonstrated that a logical sustained release caffeine tablet dosage form can be produced. No conclusion about the actual sustained release performance of caffeine tablet can be made without a controlled human blood level study following a single and multiple dosing schedule.

TABLE VIII. SUMMARY OF STABILITY STUDY RESULTS

TEST PARTICULARS	WEEK 0	WEEK 4	WEEK 8	WEEK 12
ASSAY % ROOM TEMPT.	95-105	95-105	95-105	95-105
ASSAY % 37 C	"	"	"	"
ASSAY 75% RH	"	"	"	"
GASTRIC PROTECTION AQUATERIC	.... FAILED			
COATERIC	---------------- PASSED			
EUDRAGIT L30D	.... PASSED			
% DISSOLVED pH 6.8	GREATER THAN 70 % FOR ENTERIC IN 60 min NO CHANGE FOR SUSTAINED RELEASE			

CONCLUSIONS

The pharmaceutical solid dosage forms can be coated successfully with water-based systems under optimum conditions. In general, the aqueous coating systems tested performed well with limited manipulations. The failure of aquateric to pass the gastric protection test is attributed to inadequate amount of polymer in the coating film because the dispersion had large amount of agglomerated undispersed particles which were removed from the dispersion after passing through 30 mesh sieve. Without removal of larger particles, the spray guns were blocked during the coating process. It is possible that after further optimization this product may perform well.

The coating with water-based system for sustained release dosage form works well. The coating process requires various adjustments in order to prevent tackiness during coating the caffeine tablets.

REFERENCES

1. Banker, G. S., "Film coating theory and practice; J. Pharm. Sci., 55, 81-98 (1966).

2. Harris, M. R., Ghebre-Sellassie, I. and Nesbitt, R. U., A water-based coating process for sustained release; Pharm. Technology, 102-107, sept. (1986).

3. Banker, G. S. and Peck, G. E., The new, water-based colloidal dispersions; Pharm. Technology, 5(4), 55-61 (1981).

4. Goodhart, F. W., Harris, M. R., Murthy, K. S. and Nesbitt, R. U., An evaluation of aqueous film-forming dispersions for controlled release; Pharm. Technology, 8(4), 64-71 (1984).

5. Onions, A., Film from water-based colloidal dispersions; Manufacturing Chemist, 55-59, March (1986).

6. Osterwald, H. P., Properties of film formersand their use in aqueous systems; Pharm. Res., 2(1), 14-18 (1985).

7. Ghebre-Sellassie, I., Banker, G. S. and Peck, G., Water-based controlled release drug delivery systems; in Pharm. Tech. conference proceedings (Pharmceutical Technology Publications, Springfield, Oregon, pp. 234-241 (1982).

8. Vanderhoff, J. W., U. S. Patent 4,177,177 (1979).

9. Lehmann, K., Application of acrylic latices from redispersible powders for controlled release drug formulations; Proceedings International Symposium Controlled Release Bioactive Materials, 12, 361 (1985).

10. Testing of polymer and copolymer aqueous dispersions; Deutsche Industrie Normen No. 53787, February (1974).

11. Sipos, T., German Patent Application No. 2,626,109 (1976).

12. Chehade, J., et al., Aqueous coating with the laboratory apparatus Hi-Coater HCT-20 Mini.; Acta Pharm. Technology, 28(2), 141-148 (1982).

13. Fricke, H. and Krohn, E., Erste erfahrungen mit der Japanischen coating maschine des typs Hi-Coater HCF-130; Pharm. Ind., 39(10), 1016-1018 (1977).

14. Chopra, S. K.; Aqueous alcoholic dispersions of pH sensitive polymers and plasticizing agents. a process for preparing the same, and a solid medicinal oral dosage form containing an enteric coating prepared by using same; European Patent No. 0072,021 A2 (1983).

15. Porter, S. C., Film coating equipment; Int. J. Pharm. Technology and Prod. Mfr. 3(1), 27-32 (1982).

16. Gumowski, F., Doelker, E. and Gurny, R., The use of anew redispersible aqueous enteric coating material; Pharm. Technology, 11(2), 26-32 (1987).

17. Onions, A., Films from water-based colloidal dispersions-part-2; Manufacturing Chemist, 66-67 April (1986).

18. Kildsig, D. O., Nedich, R. L. and Banker, G. S., Theoretical justification of reciprocal rate plots in studies of water vapor transmission through films; J. Pharm. Sci., 59, 1634-1637 (1970).

19. Ortega, A. M., Latices of cellulosic polymers: manufacture, characterization and applications as pharmaceutical film coatings; Ph. D. Diss., Purdue University (1977).

20. Aquacoat Handbook (FMC Corporation, Philadelphia, Pennsylvania, 1982).

21. Blois, D. W., Regulatory requirements for clinical studies, Paper presented at the 10th International GMP Conference, Athens, Georgia, February 25-27 (1986).

22. United States Pharmacopeia XXI/National Formulary XVI (United States Pharmacopeial Convention, Rockville, Maryland) pp.1244 (1985).

It is with deep regret that we inform you of the passing of Dr. Shabir Z. Masih, one of the authors of this article. Dr. Masih died of heart attack at his residence in Kingsport, Tennessee, on January 25, 1990. He was well known for his contributions in the field of pharmaceutical sciences. He was member of several pharmaceutical societies: American Pharmaceutical Association; American Association of Pharmaceutical Scientists; American Men & Women of Science; American Society of Clinical Pharmacology & Therapeutics; International Society of Pharmaceutical Engineers; Controlled Release Society; International Pharmaceutical Federation, Licensing Executive of America.

Dr. Masih served College of Pharmacy, Boston, MA from 1974 - 1979 as a Professor of Industrial Pharmacy. In 1979, he joined Reid Provident Laboratories in Atlanta, Georgia as Vice President of Scientific and Technical Affairs. He later joined the Eastman Chemical Company as a Senior Research Chemist in 1985, and promoted to Research Associate in 1987. A great scientist, inventor, researcher, and friend has left all of us.

There was much more in Dr. Masih's life. He was a warm person, a constant friend, and a wonderful father and husband. His colleagues gladly pay him the highest tribute for his contributions to the pharmaceutical science. His survivors include Mrs. M. K. Masih, wife, and four sons.

DR. SHABIR Z. MASIH

SURFACE PROPERTY MODIFICATION VIA WAX EMULSIONS[*]

P. Marshall Wiseman

Michelman, Inc.
9089 Shell Road
Cincinnati, Ohio 45236

Wax emulsions are used extensively in aqueous polymeric coatings
for a multitude of reasons. In order to establish guidelines for
formulators, a quantitative study of the effectiveness of wax
emulsions was pursued. Three different wax emulsions were
incorporated into both a PVDC and a styrenated acrylic latex, and
surface properties of these coating mixes on paper were evaluated
as follows:

> Effect of Wax Concnetration on COF, Pick Temperature,
> Gloss and Wetting Tension
> Effect of Post-Cure on COF and Wetting Tension
> Effect of Cure Temperature on COF

It was found that the selection of the correct wax emulsion and use
at the proper level will greatly enhance the properties of water
based polymeric coatings. There are three mechanisms submitted for
explanation of the results: the ball bearing theory, the bloom
theory and a new concept entitled surface continuity.

INTRODUCTION

As the need for waterborne coatings increases, we find many
different conditions under which waterborne coatings are dried and
used. While conditions vary from application to application and these
variances are multiplied by the conditions under which these coatings
are used, several general performance parameters emerge as most
important. Coefficient of friction, blocking temperature, gloss, and
surface energy are four important surface phenomena which can be
modified by the addition of wax emulsions.

Previous work with solventborne PVDC coatings (1) and LDPE
extrusions (2) indicated significant migration of waxy materials
occurred and varied with temperature and wax species. The presence of
wax at the surface of any coating can produce large changes in the
performance parameters mentioned. Except for blocking temperature,
standardized tests exist for these parameters, so a project was
established to measure surface modification of two latex systems, a
styrenated acrylic and a polyvinylidene dichloride. The goals of the
project were to provide insight into the mechanism of the function of

---

[*]Originally Published in Tappi Journal, April 1989.

wax emulsion in these waterborne coatings and to provide practical formulary guidelines for their uses.

THEORETICAL

The ratio of the forces involved in the passage of two surfaces across each other is known as the coefficient of friction. Keeping the perpendicular force constant, the lubrication effect of additives in a coating can be measured directly by measuring the decrease in force needed to move the surfaces against each other. Conversely, measuring the coefficient of friction of two surfaces provides an indication of the amount and configuration of lubricant particles at the interface of the surfaces.

Blocking tendency is a measure of the cohesive forces between two coated surfaces. Reduction of the force needed to separate these surfaces or an increase in the temperature needed to cause such thermoplastic coatings to exhibit cohesion are indications of intimate polymer contact. The presence of interfering particles on the surface of the coatings give a corresponding decrease in the blocking force or an increase in the blocking temperature of such thermoplastic coatings.

The specular reflection of light is an indication of the smoothness of a coating's surface. Coating glassine paper with an unmodified polymer latex results in an increase in gloss. The addition of additives which might come to the surface of the coating might also change the gloss of the coating if the particles are numerous and/or large enough.

If the surface energies of a bulk polymer and a bulk wax are sufficient different, the presence of such wax on the surface of a polymer coating can be inferred by changes in surface energy. By measuring the wetting tension of various coating formulations, the printability and waterbead characteristics of the coatings can be predicted.

Measurement of the above properties can be compared with the actual surface wax concentration as determined by direct measurements such as attenuated total reflectance. Changes in the above properties can be compared with the changes in surface wax concentration, to see if a correlation exists.

EXPERIMENTAL

Three wax emulsions, containing the same emulsifier system used at the same level, were prepared at 25% total solids. The wax used in the first emulsion was #1 Carnauba Wax. The second wax emulsion contained a blend of #1 Carnauba Wax and a fully refined paraffin; melt point, 147°F. The third wax emulsion contains a blend of #1 Carnauba Wax and a microcrystalline wax; melt point, ca. 185°F. Formulations based on a styrenated acrylic and a PVDC latex were adjusted to 40% solids in all cases and fortified with 0 - 15% wax solids. The formulations were coated on 30-pound glassine, using a #3 Meyer Rod, dried in a convection oven, and conditioned before surface property determinations were made. A coating thickness of 11 micron was used for the ATR work. All other coatings were 3 microns thick.

RESULTS

The coefficient of friction (c.o.f.) of coatings fortified with increasing levels of waxes were then measured. (See Figures 1-4.)

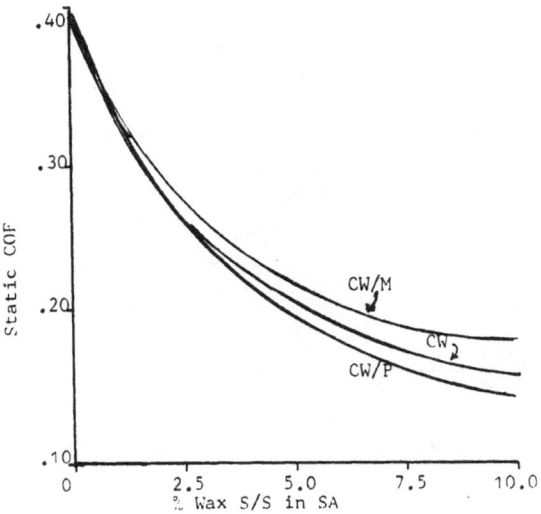

Fig. 1 Static COF vs % Wax in SA
(ctd substrate cured 20 sec,110°C)

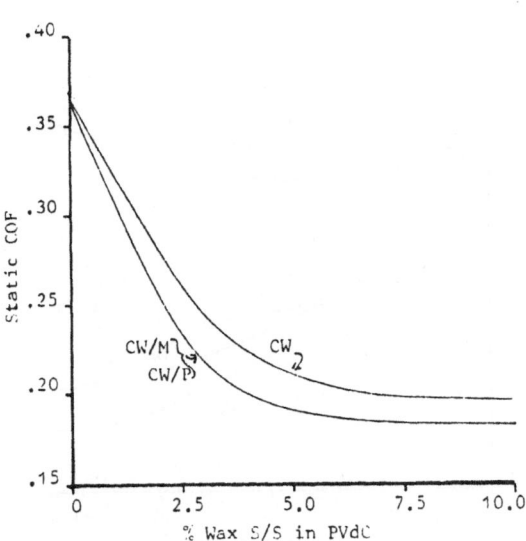

Fig. 2 Static COF vs % Wax in PVdC
(ctd. substrate cured 20 sec, 110°C)

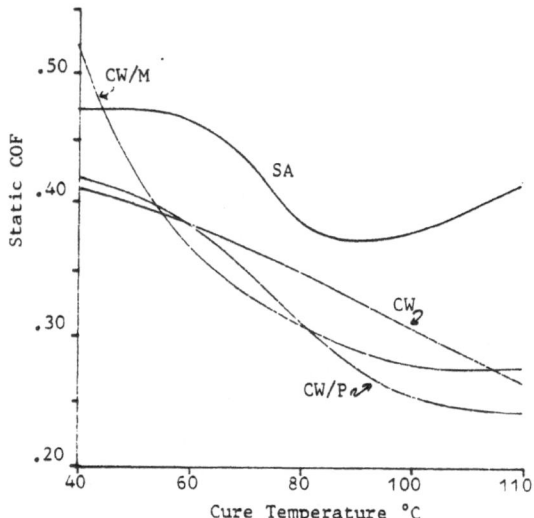

Fig.3 Static COF vs Cure Temp.,3% Wax/SA
(ctd substrate cured 20 sec,110°C)

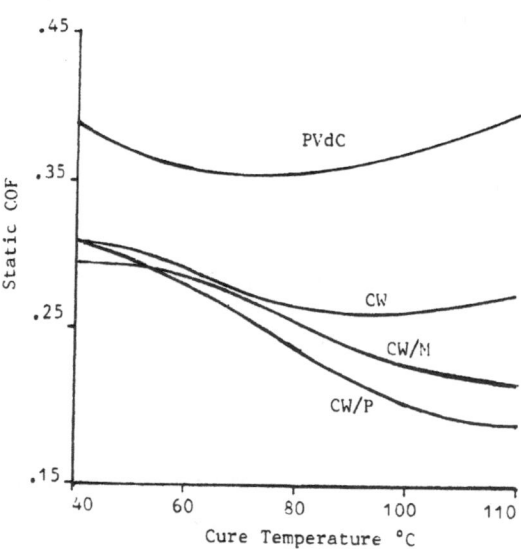

Fig.4 Static COF vs Cure Temp,3%Wax/PVdC
(ctd substrate cured 20 sec.,110°C)

Cure conditions for all coatings were the same. A decrease in c.o.f with increasing wax concentration up to about 4% was noted in the PVDC coatings. Above 4% no further decrease in static c.o.f. was noted. In the styrenated acrylic the same decrease was noted with a slightly higher level of wax needed before an asymptote was reached. We also considered the effects of post-curing and studied c.o.f. on an hourly basis for the first 48 hours and a daily basis for the next 90 days. Little change in the c.o.f. was noted, after the initial conditioning period.

The block resistance of the prepared coatings was measured by subjecting the coated surfaces to face-to-face pressure in the heated jaws of a Sentinel Seal Tester. While the temperatures for the block condition test were run in step-wise fashion, interpolation was made by noting the number of samples in a group of ten which showed fiber picking or blocking. (See Figures 5 & 6, referred to as Pick Temperature.) In the PVDC coated samples no asymptote was reached even at a level of 10% wax solids on system solids, with the pick temperature being increased as much as 50° above the unfortified polymer. In the styrenated acrylic system an increase in block resistance was achieved very quickly with the straight carnauba wax, but all waxes showed an asymptotical limit of increase - about 30° in pick temperature at the 10% wax level.

Change in gloss was measured to determine if the wax was present on the surface in sufficient quantities to affect the esthetics of the coatings. Poor correlation of wax concentration versus change in gloss was found for both systems.

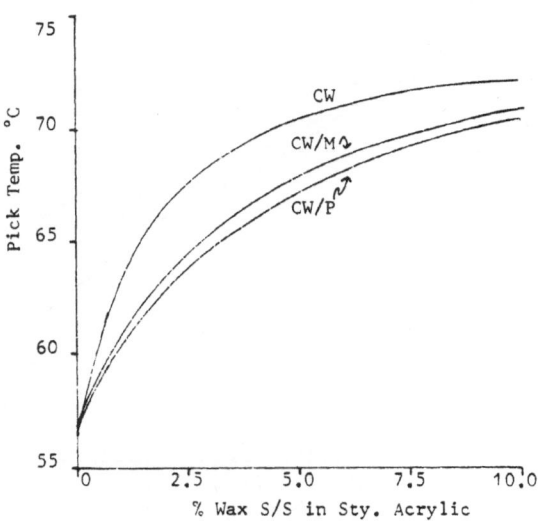

Fig.5 Pick Temp. vs. % Wax in SA
(cured 20",110°C; Tested @ 30 psi,2.5")

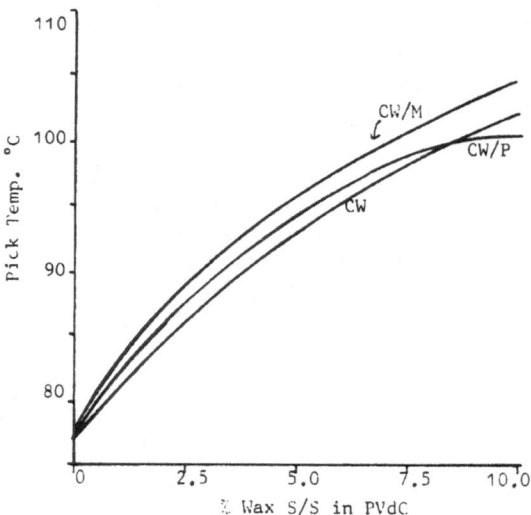

Fig.6   Pick Temp. vs. % Wax in PVdC
(cured 20",110°C; Tested @ 30psi,2.5")

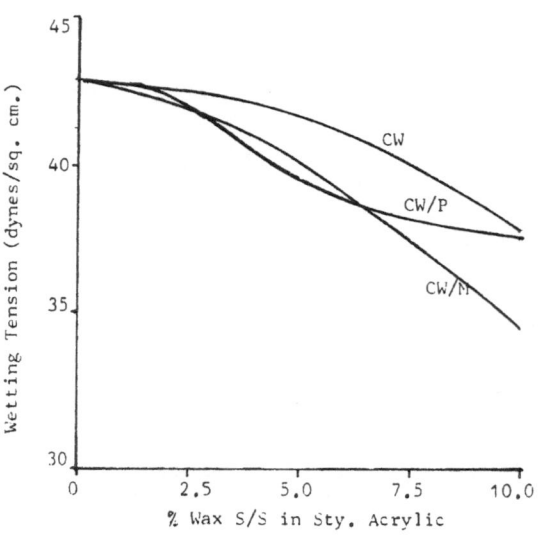

Fig.7   Wetting Tension vs % Wax in SA
(cured 20" @ 110°C,aged 2 weeks @ RT)

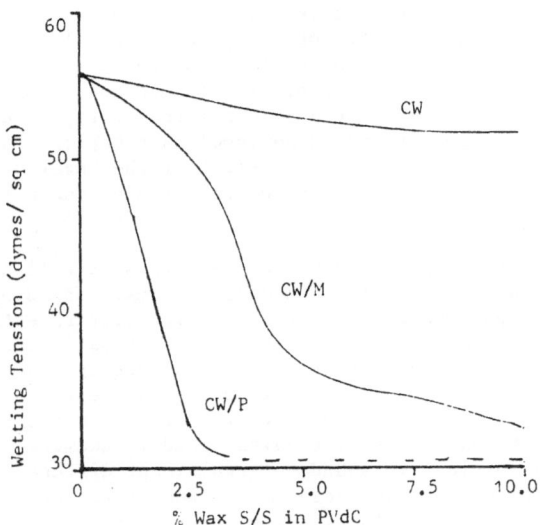

Fig. 8 Wetting Tension vs %Wax in PVdC
(cured 20" @ 110°C, aged 2 weeks @ RT)

The printability or recoatability of coated surfaces is a prime consideration in many industrial applications. We measured the wetting tension of aged samples over the entire range of wax incorporation. (See Figures 7 & 8.) Since the bulk surface energy of paraffin wax is about 27 dynes/cm, that was the lowest wetting tension we could expect. Using wetting tension solutions to indicate surface tension as low as 30 dynes/cm, we saw a marked difference in the response of the coatings to the various wax emulsions. This was anticipated since the surface tension of carnauba is reported between 40-50 dynes/cm and surface tension of the microcrystalline wax reported to be about 33 dynes/cm. If Roth's observations on the apparent migration of wax were applicable to latex systems, one should observe the surface population of wax particles change with the cure temperature. In investigating various cure temperatures at 3% wax loadings, we did observe a decrease in the coefficient of friction for both the PVDC and styrenated acrylic systems. At this point in our investigation, we had carried the work as far as our instrumentation allowed. We then formed our conclusions about the mechanism of wax emulsions in latex coatings.

DISCUSSIONS AND CONCLUSIONS

The coalescence of latex particles is the same in almost all latex systems. Heat brings about evaporation of water and plastic deformation of the latex particles to form a more or less continuous film. We know that there are imperfections in the film formation since the film properties of polymers made from latex do not match those of the film formed from bulk polymer. We believe that the coalescence of latex particles in the presence of additives takes place with little interruption of the overall particle-to-particle bonding, since properties such as tensile and modulus are not changed by the small amount of wax emulsion added. Our observations lent some support of various theories which have been proposed to explain how wax emulsions work in latex systems.

A simple ball-bearing theory advances the mechanism that protruding particles, either from an emulsion or from inorganic fillers such as talc, silica, or clay, separate the surfaces of the coatings sufficiently to reduce the coefficient of friction and bring about some block resistance. This is supported by our measurements of coefficient of friction versus wax concentration but could not explain the increase in block resistance well above the melt point of the waxes or of the continuing decrease in the coefficient of friction versus cure temperature.

If the bloom theory applies to aqueous coatings as well as solvent coatings, it explains several of these observations, including independence of the polymer's T_g and the wax migration rate (3). We extended the original experimental design when FTIR equipment became available. We then prepared similar coatings at higher coat weights to investigate the actual surface concentration of the coatings. In correlating the calculated bulk concentration of wax in the prepared PVDC coating with the experimental results found in examining the top one-quarter of the coating, we found no significant difference in concentration. Note that these measurements were taken on coating films prepared at room temperature. Even so, over the entire range of measurement, no difference was observed between the top one-quarter of the coating and the entire coating under these conditions.

We then measured the effect of heating these coatings to determine if there was an increase in surface concentration, but again observed no significant difference. If the differences in c.o.f. versus cure temperature were not the result of migration, we believe that it is purely a matter of morphology. If no wax is present on the surface, the unevenness of the surface with varying cure temperature is apparently insignificant as indicated by the lack of change of c.o.f. of the unmodified polymer coatings versus varying cure temperatures. If wax is present on the surface, then the evenness of the surface can be increased due to the plastic nature of the wax. If the surface characteristics change because the wax is more evenly spread, the observations made in this study can be explained as a combination of wax population and morphology.

REFERENCE

(1)   Roth, S.F.; King, S.S.;Herr, J.E.; "Distribution of Additives in Saran Coatings on Cellophane", Dow Chemical, USA, 1965.

(2)   Glover, J.H.; "1987 Polymer, Laminations and Coatings Conference Proceedings", TAPPI Press, Atlanta, p.231

(3)   Owens, D.K.; "J. of App. Polymer Sci."; Vol. 14, p.191

MOISTURE PERMEABILITY OF EMULSION PAINT FILM WITH MICROGEL

T. Yagi

Nippon Paint Co., LTD.
Department of Architectural Coatings
Neyagawa-city, Osaka, 572 Japan

Housing paint requires a high moisture permeability to remove excess amount of water out of concrete structure. It also requires an enough flexibility of paint for the purpose of maintaining an external appearance regardless of the formation of cracks on the concrete walls. Furthermore, it is beneficial for housing paint to have a property of gas barrier against CO_2 gas to prevent the neutralization of concrete. Microgel (MG) having hydrophilic functional groups on their surface can provide a high moisture permeability in the film formed from emulsion without sacrificing the properties of water resistance, flexibility and CO_2 gas permeability.
The moisture permeability coefficient of its film is about 5.5 (cc(stp)cm/cm² sec cmHg) and about ten times as larger as the coefficient of the film without MG. Its moisture permeability can be controlled by the amount of MG in the film and the particle size ratio of MG and emulsions.
The percolator structure such that MG spread like canals in the film, was observed with a transmission electron microscope.

1 Introduction

The technology to make elastic emulsion paints that are resistant to water but transmit moisture is one of the major challenges for coating researchers. The need for elasticity of films covering wall cracks and for good waterproof properties have forced the manufacture of low PVC wall paint. Generally, moisture does not transmit through the low PVC wall paint film, this is not preferable for lengthening the building's life or allowing the paint films to perform for a long time. We have successfully used a hydrophilic microgel as an additive for formulating moisture transmittable emulsion wall paints with low PVC. The effect of the microgel on the moisture transmittance was evaluated by comparing it with the effect of conventional surfactants. The general properties of the low PVC emulsion paint containing the microgel were examined. Moisture transmittance of the film containing the microgel was found to be ten times larger than that of the film without the microgel.
I am going to discuss the moisture vapor transmission mechanism of the film containing the microgel. Chain linkage of microgels was observed by an electron microscope picture of a vertical section of the paint film.

Surface Phenomena and Fine Particles in Water-Based Coatings and Printing Technology
Edited by M.K. Sharma and F.J. Micale, Plenum Press, New York, 1991

117

2 Experimental

2-1 Preparation of emulsion

A typical recipe of acrylic emulsion is indicated in Table-1. Emulsion polymerization was carried out at 80℃ under nitrogen atmosphere using a batch process. Theoretical solid content in all the formulation was 50 wt% and generally the conversions were always better than 98%. All acrylic emulsions in this study were produced with surfactants such as N-271A and N-504. The structure of these surfactants are shown in Table-1. N-271A is a 45% aqueous solution of ammonium salts of disulfonated dodecyl phenylphenylether, of which molecular weight is 531. N-504 is 20 moles ethylene oxide adduct of nonylphenol, of which molecular weight is 1100. The initiator was ammonium persulfate. To start emulsion production, the surfactant and initiator were added in a reactor. After ten minutes delay, the aqueous solution, in which monomers were emulsified with additional surfactants, was added. Thirty minutes later, the rest of the initiator aqueous solution was added. The feed rates of addition streams were set such that both the monomer mixture and the initiator solution lasted 3 hours. After completion of the addition, the reactor was maintained at 85℃ for 2 hours, and then allowed to cool. Below 40℃, 28 % ammonium hydroxide was added into the reactor to adjust the pH over 8. Particle diameters of these emulsions were measured as an

Table 1.   Polymerization Recipe of Acrylic Emulsion.

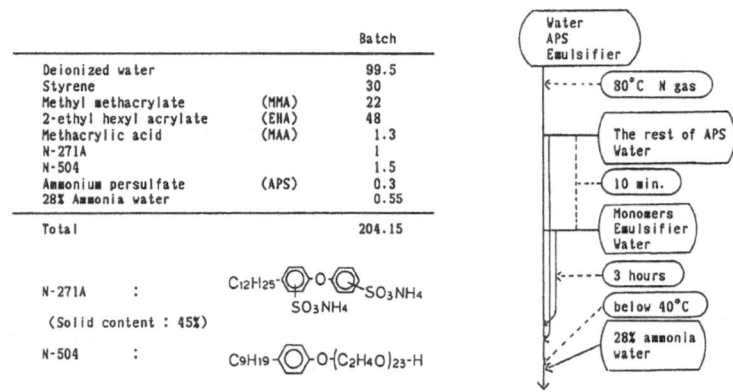

		Batch
Deionized water		99.5
Styrene		30
Methyl methacrylate	(MMA)	22
2-ethyl hexyl acrylate	(EHA)	48
Methacrylic acid	(MAA)	1.3
N-271A		1
N-504		1.5
Ammonium persulfate	(APS)	0.3
28% Ammonia water		0.55
Total		204.15

N-271A   : $C_{12}H_{25}$—⬡—O—⬡—SO_3NH_4
                                      |
                                   SO_3NH_4

(Solid content : 45%)

N-504   : C_9H_{19}—⬡—O—$(C_2H_4O)_{23}$—H

Table 2.   Preparative Method of Emulsion with Various Diameters.

Run No	Particle Diameter of Emulsion	% of intial surfactant addition
1	100 nm	50 %
2	180 nm	10 %
3	350 nm	0 %

average particle diameter by quasielastic light scattering method. The particle diameters of emulsions prepared were 100 nm, 180nm, and 350nm as listed in Table-2. The monomer component of emulsions was fixed in all samples. Particle size varied by changing the emulsion polymerization process. These monomers were commercial grade. The glass transition temperature of all emulsion polymers was adjusted to be -5℃.

2-2 Preparation of microgel

The emulsion polymerization process used for the preparation of the microgel was similar to the process for acrylic emulsions. Particle diameter of the microgel was 70 nm. The microgel was composed of styrene, a bifunctional monomer such as ethylene glycol dimethacrylate (EGDM), macromer A and macromer B. The weight percentage of styrene, EGDM, macromer A and macromer B in microgel was 62.4%, 3.2%, 24% and 10.4%, respectively.

Microgel recipe : (Solid content : 40% : Conversion : 100%)

Styrene/Ethylene glycol dimethacrylate/
Macromer A/Macromer B = 62.4/3.2/24/10.4

Macromer A : Polyoxyethylene group and $-SO_3NH_4$
group containing macromer
Mw=1600

Macromer B :

$$CH_2=C\text{-}C\text{-}O\text{-}(CH_2CH_2O)_{23}CH_3$$

with CH_3 on the carbon and O below the carbonyl.

Mw=1112

2-3 Preparation of paint

The paint composition was prepared by mixing the grind paste of rutile titanium dioxide, emulsion, microgel, two viscosity increasing agents, coalescent solvents and plasticizer. The paint formulation is shown in Table-3. The solid weight ratio of the microgel to emulsion in this experiment was 6 to 35. Weight percentage of microgel of all resins was 14.6% and in the paint film was 11.3%. PVC of the paint is 8.2%. Microgel was not taken into account in the calculation of PVC as both vehicle and pigment. The paint composition was fixed in all samples, except the emulsions particle size, the amount of microgel added and the amount of viscosity increasing agent.

2-4 Measurement of moisture vapor transmission rate (MVTR), moisture vapor transmission coefficient (MVTC), moisture vapor diffusion coefficient (MVDC), CO_2 gas transmission coefficient (CGTC) and CO_2 gas diffusion coefficient (CGDC).

Free films of the paint were used for the measurements of MVTR, MVTC, MVDC, CGTC and CGDC. The paint was applied on a polypropylene sheet to a millimeter wet thickness and the paint film was placed in room temperature for a week. After drying, the film was peeled off the polypropylene sheet. MVTR was measured by ASTM E96-66 (B) method. In MVTR measurements, the average observational error was ±20%. MVTC, MVDC, CGTC and CGDC were measured by the pressure measurement method that is described in Barrer's paper.

Table 3. The Paint Formulation.

	with MG	without MG
Water	1.5	1.5
Ethylene glycol	1.0	1.0
Dispersing Agents	0.8	0.8
Difoaming Agent	0.6	0.3
Titanium dioxige	12.0	12.0
Acrylic emulsion	70.0	70.0
Microgel (emulsion)	15.0	0
Coalescent solvent	1.5	1.5
Plasticizer	3.0	3.0
Thickeners	1.5+α	1.5
28% Ammonia water	0.55	0.55

Properties	Viscosity	:	150cps
	PVC	:	8.2%

2-5   Measurement of film elasticity.

As the parameter for the film elasticity, we measured the film elongation, based on the method of Japanese industrial standard A-6910.

2-6   Estimation of film water resistance.

The paint was applied on a slate panel to a millimeter wet thickness, and placed in room temperature for a week. Then, the test panels were immersed in water at 40°C for a week. Water resistance of the films was estimated by visually observing the change of the film and blistering.

3   Results

The properties of the paint films obtained are shown in Table-4. The particle diameter of the acrylic emulsion used in this study was 180 nm.

Table 4. Properties of the Paint Film.

	Moisture Permeability			CO2 Gas Perm.		Elon-gation (%)	Water Resistance (blister)
	ASTM Method	Pressure Measurement (Barrer's) Method					
	MVTR	MVTC	MVDC	CGTC	CGDC		
Film with MG	230	5.5	10.3	0.05	17	450	Good
Film without MG (Conventional)	20	0.5	1.2	0.03	14	430	Good

MVTR : Moisture Vapor Transmission Rate $(g/m^2 \cdot 24 hours)$

MVTC : Moisture Vapor Transmission Coefficient $(cc(STP) \cdot cm/cm^2 \cdot sec \cdot mmHg)$

MVDC : Moisture Vapor Diffusion Coefficient (cm/g)

CGTC : Carbondioxide Gas Transmission Coefficient $(cc(STP) \cdot cm/cm^2 \cdot sec \cdot mmHg)$

CGDC : Carbondioxide Gas Diffusion Coefficient (cm/g)

Microgel changed the nature of film, by increasing the amount of moisture vapor permeating through the film, but it scarcely increased the amount of carbon dioxide gas going through the film. Elongation and water resistance of the film were not influenced by the addition of microgel. MVTR is the value of the volume of moisture released by the film, which is measured by the ASTM E96-66 (B) method. Different MVTR value would be obtained for the same film composition with a different thickness or under different moisture vapor pressures. Absolute measurements for the transmittance of the gas like moisture vapor or carbon dioxide through the film can be obtained by the pressure measurement method. Both diffusion coefficient (DC) and transmission coefficient (TC) of the gas in the films were measured and Henry's solubility coefficient (S) was calculation by the following equation.

$$TC = DC \times S$$

MVDC value of the film containing microgel is 8.6 times larger than the original film. MVTC value increased 11 times with the addition of microgel. The increase of MVTC was larger than that of MVDC. Microgel increased both DC and S of moisture in the film. On the contrary, microgel had little effect on CGTC and CGDC, and did not change the diffusion and solubility coefficients of carbon dioxide gas in the film. It was expected that the amount of the microgel added would influence MVTR, water resistance and elasticity. At first, MVTR of films containing various amount of the microgel was measured. The results obtained were plotted in Figure-1.

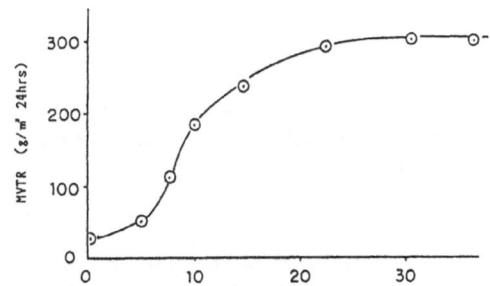

Figure 1. Relationship between MG Amount and MVTR.

MVTR increased as the amounts of the microgel increased from 0 to 20%. From 20 to 36% MVTR saturated to 300g/㎡ 24hours. Figure-2 shows the relationship between the amount of the microgel added and the water resistance of the films.

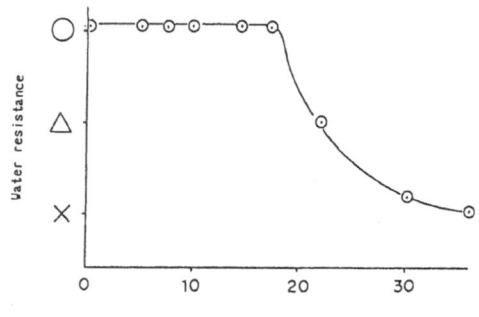

Solid weight percent of MG with emulsion polymer

Figure 2.   Relationship betweem MG Amount and
Water Resistance.

The microgel did not affect the water sensitivity of the films until the amount added reached 18%. Excess addition over 18% decreased the water resistance. Overlapping Figure-1 and Figure-2, we can see that with the addition of 14.6% microgel, the MVTR changed to 230g/㎡ 24hours, this is ten times larger than that of the film without microgel, while water resistance does not decrease.

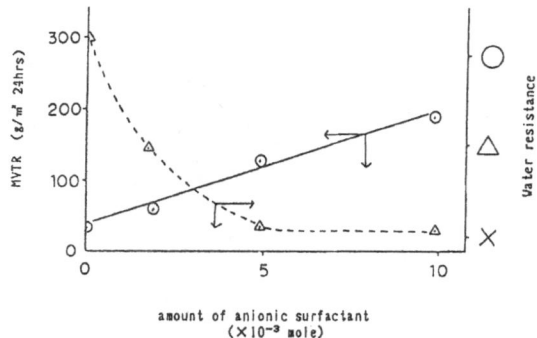

amount of anionic surfactant
($\times 10^{-3}$ mole)

Figure 3.   MVTR and Film Water Resistance Against Anionic
Surfactant Amount.   (Hydrophilic function group:
$-SO_3 NH_4$)

To study these results more precisely, MVTR and water resistance measurement had been carried out with films containing anionic or nonionic surfactants instead of the microgel. The results obtained with an anionic surfactant, sodium dodecylbenzene sulfonate (SDS), are shown in Figure-3. The anionic surfactant significantly lowered the water resistance of the film, while it scarcely improved MVTR.

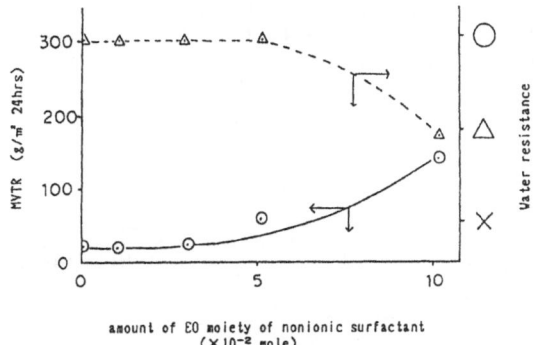

Figure 4. MVTR and Film Water Resistnace Against
           Nonionic Surfactant Amount.

The results obtained with a nonionic surfactant are shown in Figure-4. The nonionic
surfactant is 45 moles ethyleneoxide adduct of nonylphenol, HLB of which is 18. Until
5×10^{-2} mole, the nonionic surfactant does not lower the water resistance, but MVTR
only increases from 20 to 30 g/㎡ 24hour. With the use of a large amount of surfactant beyond 5×10^{-2} mole, water resistance of the film is influenced badly, but MVTR
changes only to 120 g/㎡ 24hours even at 10^{-1} mole addition of surfactant.

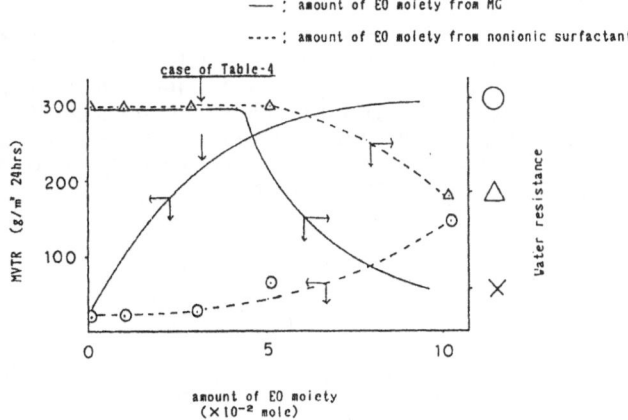

Figure 5. Comparison of MG with Nonionic
           Surfactant.

In case of Table-4, 3.1×10^{-2} mole of the ethylene oxide moiety were introduced to
the film by microgel. Results obtained with the microgel and the surfactants indicate
that the microgel is superior to the surfactants as a moisture transmittable agent.
The surfactants were soluble in the uniformly. The surfactants did not increase MVTR
as much as expected. Large amounts of surfactant damaged water resistance of the film
badly, and did not increase MVTR satisfactorily. On the other hand, the microgel
remarkably increased MVTR to the practical level without decreasing water resistance.
In our experiments, surfactants introduced far more hydrophilic groups to the film
than the microgel. But because surfactants were soluble in the film uniformly, local
concentration of hydrophilic groups was diluted and low. While the hydrophilic groups
introduced by the microgel were limited, but concentrated on the surface of the
microgels. The different behavior of moisture vapor transmission and water resistance

between surfactants and microgel is due to the homogeneous-heterogeneous distribution of hydrophilic groups. Another advantage of the microgel over surfactants was found with the preservation of MVTR after the immersion in water. The results are in Figure-6.

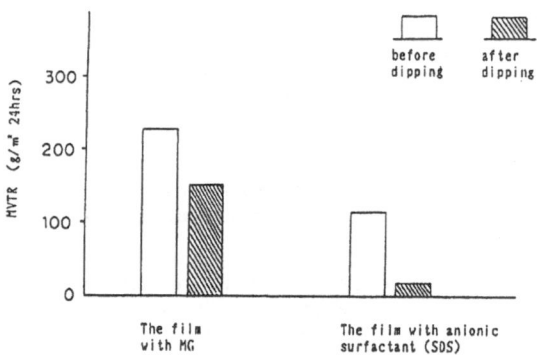

Figure 6. MVTR of Film After Dripping in Water for a Week.

The films containing anionic surfactant (SDS) or the microgel were immersed in water for a week at room temperature. Then the films were taken out of water and allowed to dry for 24 hours at room temperature. The film containing anionic surfactant lost the moisture vapor permeate properties and showed the same MVTR as the film without hydrophilic substance. The film containing microgel was measured to retain 80% of the MVTR before immersion. The results indicated that SDS could easily be extracted from the film but the microgel could be held in the film.

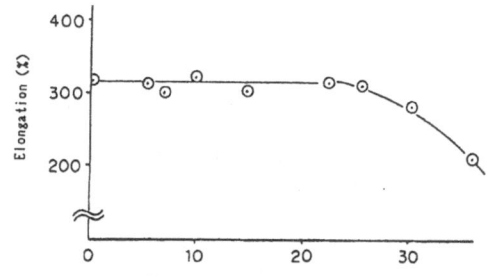

Figure 7. Relationship between MG Amount and Film Elongation.

Figure-7 shows the relationship between amount of the microgel added and the film elongation. Elongation of the film containing the microgel was equal to that of the film without microgel until the amount of microgel added reached to 25%.

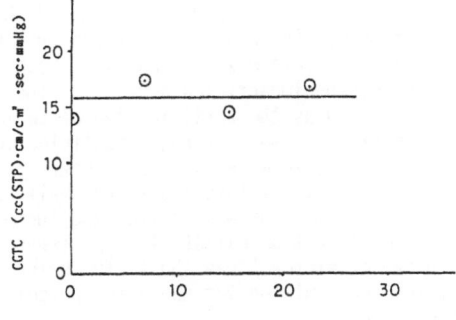

Solid weight percent of MG with emulsion polymer

Figure 8. Relationship between MG Amount and
           CO_2 Gas Transmission Coefficient.

Figure-8 shows the relationship between the amount of the microgel added and CGTC. The CGTC did not increase with addition of the microgel, while the MVTR increases by addition of the microgel as discussed earlier. Surface of the microgel seems to have a strong affinity for moisture, but less affinity for carbon dioxide gas.

4   Discussion

It was assumed that the heterogeneity of the film formed with the emulsion and the microgel would cause the high MVTR without failing good water resistance. The relative particle sizes of the two particles would decide this heterogeneous structure, which would then influence the MVTR. Then, the MVTRs of the films, which contained the microgel and the emulsions with various particle sizes were measured. Particle diameter of the microgel was 70nm, as mentioned above. The emulsions with various particle sizes were composed by the same monomer component and the same amount of emulsifier.

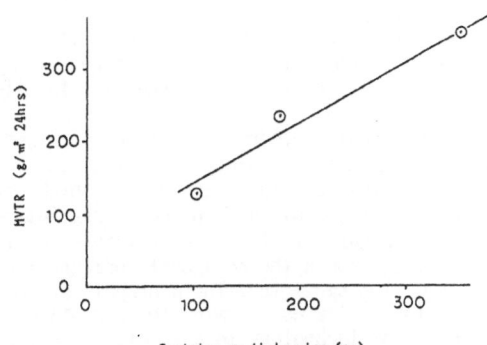

Emulsion particle size (nm)

Figure 9. Relationship between Emulsion Particle
           Diameter and MVTR. (MG/emulsion =
           6/35 (weight ratio))

Figure-9 shows the relationship between MVTR of the film and the particle size of the
emulsion. The paint formulation was the same as described in Table-2. MVTR of the
film increased with the larger particle size of emulsion, because the relative number
of emulsion particles to the microgel particles decreased. The interfacial space
among emulsion particles decreased as the particle size of emulsions increased. With
the use of larger emulsion particles, more microgel particles locate in the interfacial space among emulsion particles to make wider and longer linkages during coalescence, which serve as canals for water moisture but are small enough to block liquid
water. When the particle size of emulsion was smaller and the number of emulsion particles was larger, microgel could not be satisfactorily dispersed around the emulsion
surface and the chain linkage of microgel was short. The heterogeneous structure con-
stituted by the microgel and the emulsion was named the percolator structure. This
structure is shown in Figure-10.

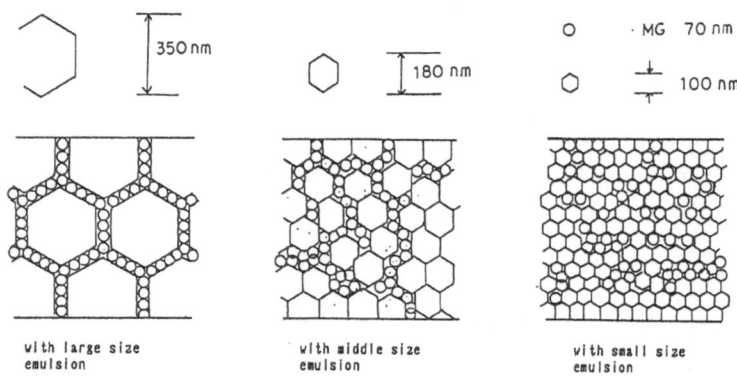

Figure 10.   The Percolator Structure.

The chain linkage of microgels, like a canal, continues from the surface of the film
to the back surface. The canals are wider and longer in the case of the large particle size emulsion. However in case of the small emulsion, the chain linkage would
be thin and short, and the MVTR is lower.
It is expected that glass spheres can be used to make similar structure. But, the
microgel have advantages on balance of MVTR and water resistance of the films to
glass sheres, because microgel can be freely designed its functional groups on its
surface and its diameter.
In order to confirm this hypothesis, observation of the film section by transmission
electron microscope was carried out. The picture obtained is shown in Figure-11. The
model film was made from the emulsion, the microgel and thickener. Pigments were
omitted from the formulation for the picture, because they interfered with the
transparency of the film. In Figure-11, the hexagonal forms that look white are emulsion particles. The black dots around the hexagonal form are the surfactants that
crystallized around the emulsion particles. The microgels are observed as the smaller
gray particles among emulsion particles, forming the percolator like structure. I
have successfully developed the technology to make elastic emulsion paint that are
resistant to water and intercept carbon dioxide gas but transmit moisture ten times
better than conventional paints.

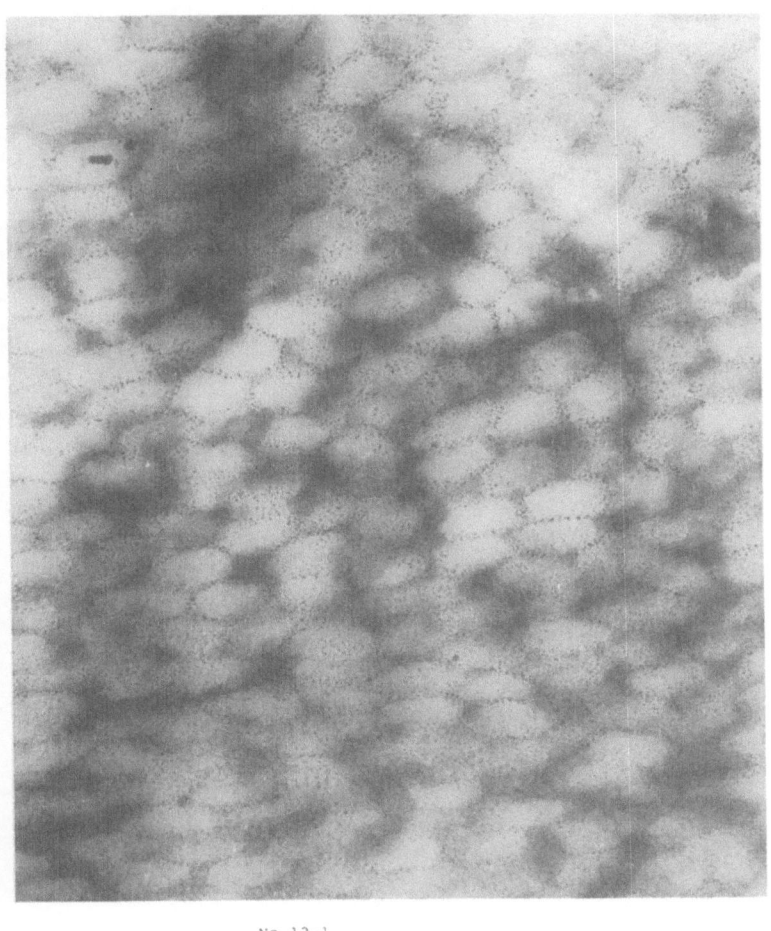

200 nm

No.12-1

80,000   20,000

Figure 11. Transmission Electron Microscope Picture.

5   <u>Reference</u>

1) R.M.Barrer, Trans. Faraday Soc., 35,628,644, (1939)

ALKYD EMULSIONS FOR HIGH GLOSS PAINT SYSTEMS; OLD

PROPERTIES IN NEW PARTICLES

A. Hofland, F.J. Schaap*

DSM Resins BV, Ceintuurbaan 5, 8022 AW
Zwolle, The Netherlands

*DSM Research BV, P.O. Box 18, 6160 MD Geleen,
The Netherlands

Due to environmental legislation, in which the United States
have been a forerunner, more and more coating systems have
been developed, based on water. For the application areas
requiring a gloss higher than 90% under an angle of 60°C
(traditional high-gloss) however, traditional alkyds are
still the binder of choice.

The reason for this lies in the phenomenon of film formation.
Free flowing continuous phase having a Tg of
-60 to -30°C, as is the case in alkyds can be expected to
give a more coherent film than particles having a Tg of +10
to +20°C. MFT's of these acrylic dispersions often have to be
lowered with excessive amounts of coalescing agents.

This paper deals with the parameters concerned with the
emulsification of alkyd resins of moderate viscosity.
Parameters considered are equipment defined (sheer
rate/stress, temperature, etc.) as well as chemistry defined
(nonionic surfactants, anionic surfactants, thickening
agents, etc.).

A very important parameter is the presence of amines, in
order to lower surface tension by neutralization of
carboxylic groups in the resin. Although this well-known
"trick" is very effective and sometimes considered
indispensible, the film properties of alkyds are adversely
affected by it (drying deterioration, yellowing, incomplete
evaporation, environmental problems). This use of amine can
be circumvented in a majority of the cases, as will be shown.

Surface Phenomena and Fine Particles in Water-Based Coatings and Printing Technology
Edited by M.K. Sharma and F.J. Micale, Plenum Press, New York, 1991

INTRODUCTION

Environmental legislation in Europe with regard to the emission of organic solvents is becoming more and more stringent. Although there is no unity between the European countries as yet, the trend is very clear and it can be expected that in the "magic year" 1992 (the European unity) every country from north to south will have some kind of legislation in this field, either to protect the worker (Sweden) or to protect the environment (The Netherlands). Apart from this, public awareness is also growing at an unexpected rate. The reason for this is obvious when we take a look at the next figure:

Interior emulsions:	915,000 tons
Interior gloss	450,000 tons
Exterior trim	185,000 tons
Undercoats	310,000 tons
Woodcare	180,000 tons
Total alkyd based	1,125,000 tons

Average solvent content 30%

337,500 tons solvent

Figure 1. Solvent emission as a result of the use of decorative (trade sales, d.i.y.) paint in Western Europe

Once having made the choice for VOC compliant coatings (as far as there is any choice) there are numerous options, like powder coatings, UV and EB cured coatings, two-component coatings, high solids coatings and water-borne coatings. In the case of alkyd resins, the main use of which is in the decorative field, the last two systems are the more likely ones, anyway for the time being. Whereas in many countries the choice is not very clearly made (Germany e.g.), many Scandinavian countries consider high solids systems only as an "in-between phase" until the water-borne systems can equal or even out-perform the conventional systems.

Once the choice to "go into water" has been made, two options are possible. One can either use dispersions of thermoplastic polymers (styrene- and acrylic type monomers) or, as an alternative, there are the emulsions of oligomeric resins, in the decorative sector being mainly alkyds and polyurethanes. Although neither of these systems can a priori be classified as "good" or "bad", there are some characteristic differences that can make one more suitable for specific applications than the other.

130

Typical advantages of thermoplast dispersions are their low particle size, quick drying, and good outdoor durability. The last two of these advantages can also be regarded as a disadvantage, since the drying often is considered to be too quick when only small amounts (under 5 %) of coalescing agent are used. The nomenclature then changes from "quick drying" to "short open times". The advantage of good outdoor durability is based on (a.o.) a good hydrolysis resistance, meaning that, although it may take a long time, once a coating based on acrylic dispersion has deteriorated, the remainders cannot be removed by the usual alkaline paint stripper. Instead, flakes of coating will come loose and remain in the environment.

Another disadvantage, the seriousness of which is once again depending upon the application area, is that there always has to be some compromise between film formation (low MFT) and good blocking properties (high Tg). Since in alkyd emulsions not all chemistry is "finished" when the film is applied, these emulsions can be expected to outperform the thermoplastic dispersions when it comes to properties that are related with film formation from a low Tg material, such as penetration and inherent adhesion of alkyd emulsions as compared to acrylic dispersion, and gloss. The next figure, a SEM micrograph of some alkyd emulsion particles flowing out and levelling, shows us why this is the case. Instead of being "glued" together with coalescing agent, the particles spread out spontaneously, going from spheres to "pan cakes" and finally to a film. The resulting film will not differ very much from a film originating from organic solvent.

Of course, if alkyd emulsions were perfect in all respects, their use would have been far more widely accepted, the more so since research on these systems has a long history, going some 30 years back in time. Typical problems attached with alkyd emulsions and alkyd solutions are:

*   hydrolysis of the backbone, resulting in loss of drying properties upon storage;
*   complexation of the Cobalt drier with nonionic surfactants of the ethylene oxide type and with neutralizing amines, resulting in deteriorated drying properties;
*   instability of the paint systems.

Some of these problems can be overcome, such as hydrolysis of the backbone. The use of hydrolysis resistant raw materials (isophthalic acid instead of orthophthalic anhydride is a well-known example) does not only yield better retention of molecular weight, but also gives better outdoor durability. Omitting nonionic surfactants with more than 12 ethylene oxide segments as well as omitting amines from the formulation takes care of the complexation problem. The instability of paint systems is inherent to the heterogeneity of the system and phenomena like "surfactant theft". The argument that alkyd emulsions are by definition coarse and thus have bad stability is no longer valid. By careful choice of the surfactant and emulsification equipment, alkyd emulsions that have an average particle diameter of less than

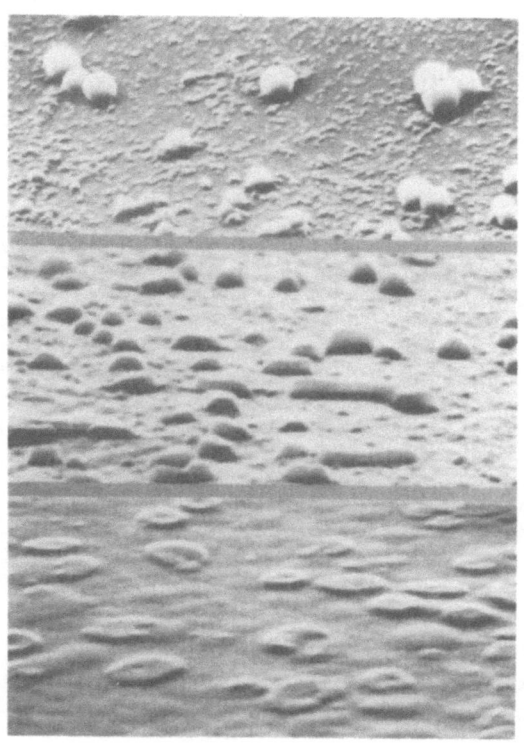

Figure 2. Some SEM micrographs of the spreading and levelling
of alkyd emulsion particles

500 nm can be obtained. Careful paint formulating
nevertheless remains a necessity, as is also the case for
thermoplast dispersions.

PREPARATION AND STABILITY OF ALKYD EMULSIONS

The basic theories

To obtain a stable alkyd emulsion, the knowledge of only
two formulas is of the utmost importance. For one thing,
small particles will have to be made, before being able to
protect them from coalescence [2].

$$ d = \frac{\sigma_i}{\dot{\gamma} \cdot \eta_{cont}} \cdot f \left(\frac{\eta_{disp}}{\eta_{cont}} \right) $$

Figure 3. Creation of small particles

In which:

d = particle size that can be reached
σ_i = interfacial tension
$\dot{\gamma}$ = shear rate
η_{disp} = viscosity of the disperse phase
η_{cont} = viscosity of the continuous phase
f = system dependant function; optimal results are
    obtained with ratio = 1

Once these particles have been obtained, protection from
coalescence can be achieved by ionic as well as by nonionic
means. Ionic stabilization is perfectly described by the
well-known DLVO theory and does not need any extensive
treatment here. Some observations during experiments when
using anionic surfactants will be briefly discussed at the
end of this paper. Nonionic stabilization is mainly based
upon steric repulsion and osmotic forces between the
particles. Specific advantages and disadvantages of this
system, for which DSM Resins has chosen in the past, will be
dealt with later. The need for small particles is very
obvious from Stoke's law [3], represented in figure 4.

Only in a space shuttle large particle size emulsions can
be expected to be stable. Where size is concerned, alkyd
emulsions have a much larger spread in particle size than

$$V_{sed} = \frac{(\rho_{cont} - \rho_{disp}) \times g \times (\bar{d})^2}{18 \times \eta_{cont}} < 1 \text{ mm/day}$$

Figure 4. Prevention of sedimentation

acrylic dispersions. Apart from that, also the average
particle diameter can be much larger. The better the
emulsification process is controlled, the narrower
the particle size distribution. Figure 5, for instance shows
the particle size of an emulsion that has been prepared over
the inversion process [5]. It can be clearly seen that,
although the inversion had been successful in some regions,
it failed in others, leading to multiple emulsions (water-in-
oil-in-water). Although very much desired in margarine, such
large particles are detrimental to the stability of much
lower viscous alkyd emulsions.

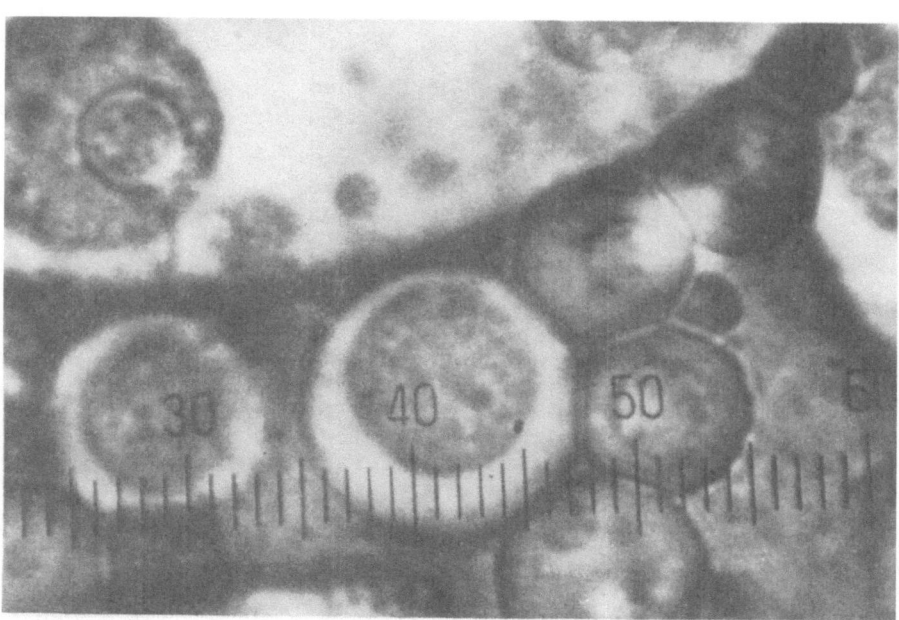

Figure 5. A micrograph of a multiple emulsion, resulting from
badly controlled inversion. One diversion is 800 nm

When the inversion process is not used, the alkyd emulsion can have an average particle diameter of less than 1 micron, but then a very low surface tension is needed, usually obtained by anionic surfactants or neutralization with amine. Also the use of excessive amounts of nonionic surfactants (over 5%, based on resin) is an option.

When the inversion process is used and is kept properly under control (this usually means going slowly over the inversion point at very specific empirically determined temperature and solid content) very small particles can be obtained. To obtain sufficient stability a d50% of less then 750 nm and a d90% of less then 1000 nm is required. This objective can be reached in the majority of the cases. An example is shown in figure 6.

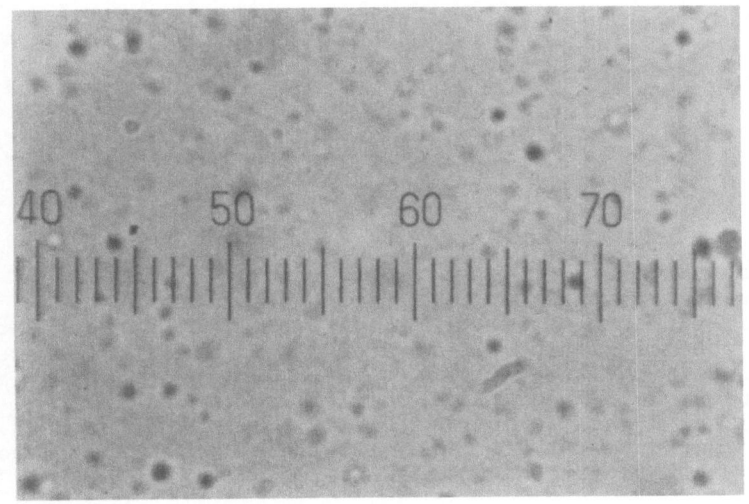

Figure 6. A micrograph of an emulsion, prepared over a properly controlled inversion. The median particle size is 300 nm. One diversion is 800 nm

Comparison of various parameters

Already some of the parameters determining the over-all quality of an alkyd (or indeed any) emulsion have been mentioned, being neutralization and surface tension. Looking at the effect of parameters like these on either particle diminution or stabilization of small particles, it is very helpful to visualize these effects as stated in figure 7.

Increase of:	Effect on particle diminution:	Effect on emulsion stability:
Viscosity of continuous phase	+ +	+
Sheer rate	+ +	o
Viscosity of dispersed phase	– –	o
Temperature	–/+	–
The amount of:		
Neutralizing agents	+ + ($\sigma_i \downarrow$)	+
Anionic surfactants	+ + ($\sigma_i \downarrow$)	+
Cosolvents	+ + (η disp $\downarrow$)	o
Nonionic surfactants	+	+ +
Thickening agents	+ (η cont $\uparrow$)	+ (η cont $\downarrow$)

Figure 7. Comparison of several parameters, associated with alkyd emulsification and alkyd emulsion stabilization

We can see that both conditions as well as chemical constituents that are employed during alkyd resin emulsification can serve either (or even both) purpose. Some typical conditions are: sheer rate, sheer stress (of course interconnected by the viscosity of the continuous phase) and the temperature. Some typical chemical additives are neutralizing agents (amines, alkali), anionic surfactants, nonionic surfactants (cationic surfactants are hardly used in alkyd emulsification) and cosolvents.

One very interesting conclusion can be drawn from this figure, being that amine and shear (at least partly) serve the same purpose, which is particle diminution. Since the use of amine closely resembles Pandora's box when it comes to problems such as drying deterioration, discoloration of white paints and environmental restrictions, it is advantageous for the paint producer that its use can be minimized by applying the optimal shear, i.e. stirring conditions.

NONIONIC, LOW TEMPERATURE SYSTEMS

Main properties of nonionic stabilized systems

Almost all alkyd emulsions of DSM Resins are of the nonionic/neutralizing agent type. These have the advantage

of being stable in the presence of electrolytes (metal soaps
that are used as driers, for instance), although their
thermal stability is poor as compared to ionically stabilized
emulsions. Since some particle charge is already present as a
result of neutralization, anionics are not strictly
necessary.

Furthermore, it can be argued that, since absorption of
surfactants to particle surfaces is a dynamic equilibrium
rather than a static situation, nonionics (being more
lipophilic) tend to replace anionics from the surface. These
anionics, once "kicked" into the aqueous phase have a very
low tendency to re-adhere to the particle. On time average
they are mainly in the aqueous phase, thus giving rise to
foaming and flashrust problems. This was indeed what was
observed in practice.

One extremely important aspect of the nonionic stabilized
system is the Hydrophilic Lipophilic Balance or HLB.

Hydrophilic-Lipophilic balance (HLB) [4]

The HLB, often calculated as the hydrophilic weight
fraction of a nonionic surfactant x 20 or determined
empirically by titration, determines the time average
position of a surfactant with respect to the water-resin
interface, as is depicted in figure 8:

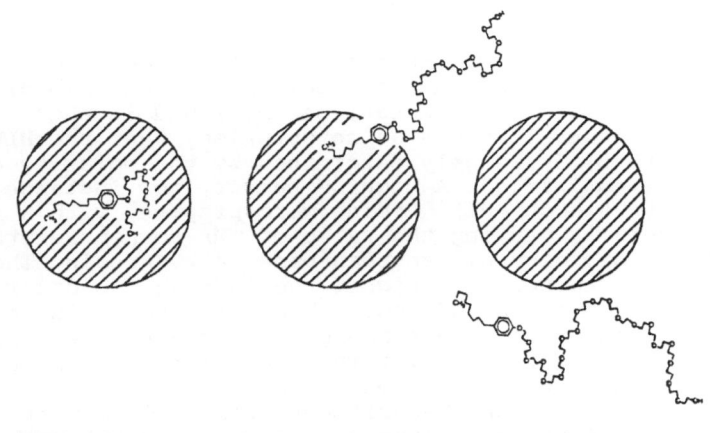

NP 6, HLB =11.6 ;too low   NP 14, HLB = 15.1 = required HLB   NP 20, HLB =16.3 ; too high.

Figure 8. Relative position of a nonionic surfactant in an
alkyd emulsion and its relation with HLB

Of course this phenomenon, like every adsorption
isothermal, is dependent on the concentration of the
surfactant in the solution, according to Langmuir.
The choice of the HLB of the applied surfactant will
determine your emulsion stability and even your emulsion type.

In case of a far too low HLB and a high viscosity resin, the surfactant might never come out of the resin again. A surfactant of too low HLB will lead to water in oil (w/o) emulsion, just where we want the coating to be o/w, for obvious reasons.

Optimal HLB can be determined by visual observation, although also particle size determination (photocorrelation, Mie scattering, Fraunhofer diffraction) can be (and in fact have been) used, which is shown in the next figure:

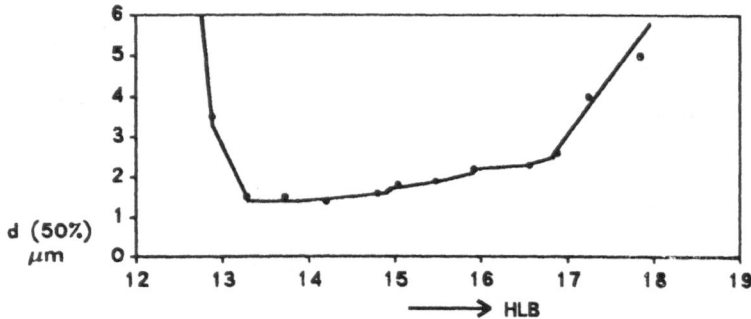

Figure 9. Median particle size for a standard alkyd emulsion (5% w/w surfactant on resin) vs. HLB

Surprisingly we found that even in the case of correctly chosen HLB's (that is, in infinitely stable emulsions) some 25 % of the nonionic surfactant can be found in the continuous phase. It comes as no surprise, that the HLB required (RHLB) is largely determined by the polarity of the resin to be emulsified. A high OH-number, for instance, will give rise to a high RHLB. Resins were prepared with a range of OH-numbers by varying the ratio of TMP to pentaerythritol in a standard medium oil soybean-based alkyd resin. The higher the OH-number, the higher the RHLB as was determined by visual observation. Of course, the same reasoning can be expected to apply to COOH numbers, although this was not fully investigated. A similar relationship has also been found between RHLB and oil length, oil being of course of non-polar contribution. Measurements ranged from pure soybean oil, over two different alkyd resins, to di-octyl phthalate, being an excellent low-viscous model compound for a polyester. The results obtained, indicate that in fact _every_ resin can be emulsified, regardless of OH-value, oil length and acid number, as long as the HLB is adjusted to these variables. The only _real_ problem we have to deal with when emulsifying alkyd resins is the viscosity of the base resin. Finally a similar relationship, fitting perfectly into the scope of this paper, is the one between degree of neutralization and RLHB for one particular resin, this time an oil free alkyd (a polyester, if you wish). The results are shown in the figure 10.

This means that, upon increasing the degree of neutralization, one can not perform these stability tests meaningfully by applying the same nonionic surfactant. It has even been found that upon addition of some amine to an o/w emulsion (either before or after emulsification), (re-)inversion took place and a w/o emulsion was obtained. The reason for this could be the increase in RHLB, rendering the nonionic in use less effective due to a too low HLB. It has already been mentioned that this can very easily lead to an inverse emulsion. Addition of some NP 50 (HLB = 19.5) in the majority of the cases indeed caused the emulsion to (re-)invert to an o/w system. This phenomenon can become very important when the emulsion is prepared via the inversion procedure, i.e. starting out from resins + surfactant (+ neutralizing agent) and then adding the water. Since in this case the inversion <u>has</u> to take place (w/o —> o/w) real trouble appears when the choice of an incorrect HLB causes delay in the inversion (it can be easily envisaged that inversion is facilitated by the presence of the surfactant at the oil/water interface) or it even causes the inversion not to take take place at all (too low HLB).

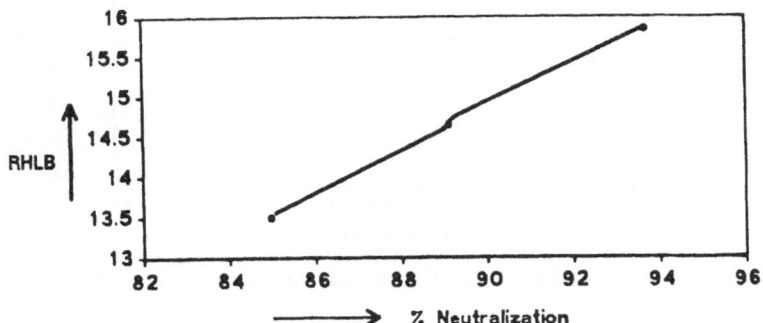

Figure 10. RLHB for a low viscous polyester resin (AV=20, OHV=60) vs. its neutralization grade

The HLB of a mixture of nonionic surfactants can be influenced to a great extent by temperature. Each NP(EO)n has its own cloud point, at which point thermal motion of the ethylene oxide chain causes a decreased hydration. This means a higher Gibbs free energy for the hydrophilic part in water, which in turn means a lower effective HLB. This is represented in figure 11, showing emulsification of our model resin (RHLB = 13.8) with NP 10 and NP 20 at 5 different temperatures.

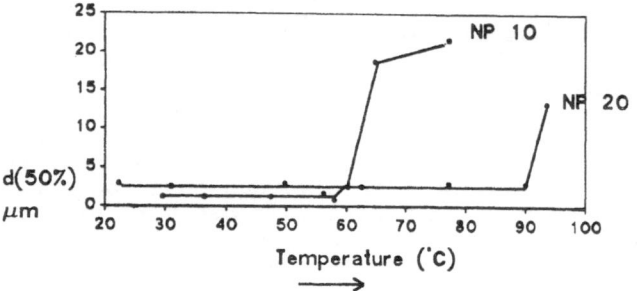

Figure 11. Median particles size for a standard alkyd
emulsion (5% w/w of surfactant on resin) vs.
emulsification temperature

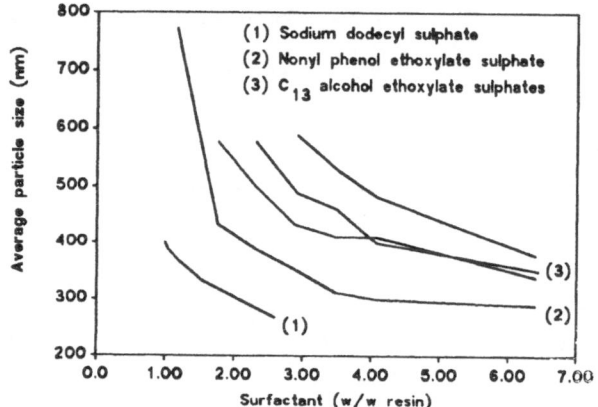

Figure 12. Particle size vs. surfactant concentration

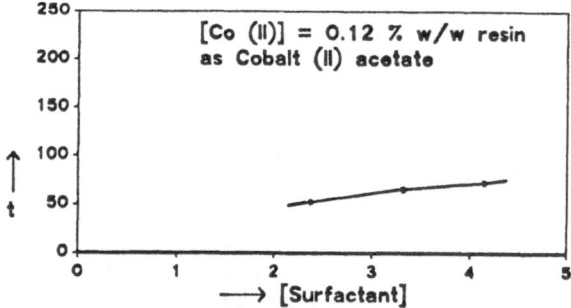

Figure 13. Touch-dry time of a TOFA/IPA/Penta based long oil
alkyd emulsion vs. [surfactant] (Sermul EA 54, %
w/w resin)

Clearly, NP 10 is more or less useless at temperatures above 55°C, being its cloud point. Thus a 1:1 mixture of NP 10 and NP 20 (HLB = 14.9) will behave as though it was a smaller amount of NP 20 at an HLB of 15.9 when emulsification is carried out at e.g. 70°C. This observation need not be measured with sophisticated equipment, but can also be observed very easily.

ANIONIC, HIGHER TEMPERATURE SYSTEMS

As an alternative to the use of nonionic i.e. steric stabilization we can use charge stabilization, brought about by anionic or cationic surfactants, the latter hardly being used in alkyd emulsification. The advantage of anionic stabilization is the relative insensitivity towards high temperatures, which is of particular importance when emulsifying high viscous resins.

Of our base resin, that was shown previously, several alkyd emulsions were made using in-line emulsification at elevated temperatures. Here we found a further advantage, being that, while in the case of steric stabilization ± 5 % of surfactant based upon resin is needed, with charge stabilization an amount of 1.5 to 2 % is sufficient (figure 12).

Interestingly, the well-known and widely used sodium dodecyl sulphate, one of the lowest prized anionic surfactants available, performed the best. It should be noted however, that due to hydrolysis the life time of SDS is limited at higher temperatures, meaning that firstly the residence time in the in-line dispersing unit at high temperatures should be limited to several minutes, and secondly the stability of SDS-emulsions in storage at 50°C (a test, which is in wide-spread use in Europe) is bad. Sedimentation as well as coagulation occurs after as little as 7 days. Better surfactants are the hydrolysis resistant sulphonates like sodium dodecylbenzene sulphonate. Finally it can be seen that ethoxylation does not improve the efficiency of anionic surfactants. It is shown that charge stabilization is far more efficient than steric stabilization.

Since nonionic surfactants of high HLB-values tend to influence the drying properties, it was investigated to what extent this was the case for anionic surfactants. In this case a surfactant possessing ethylene oxide fragments as well as charge (nonylphenol.4EO sulphate) was used. The results are shown in figure 13.

It can be seen that the influence of anionic surfactants upon touch-dry times (and also on final hardness) is negligible. Although in this case the hydrophilic Cobalt acetate was used as a drier, similar results can be obtained when the hydrophobic Cobalt decanate is used. This was in contrast to what was expected. The theory was that when Cobalt is present in the aqueous phase, during film formation it has to diffuse to the resinous phase before it can be active as a drier. Cobalt, present in the resinous phase, is

already in close proximity to the fatty acid chains where the
actual crosslinking takes place. This means that the drying
process with Cobalt in the resinous phase should start
earlier and hence better drying can be expected.
Figure 14 shows that the behaviour of alkyd emulsions is not
very sensitive to the theories one develops by logical
understanding of the systems.

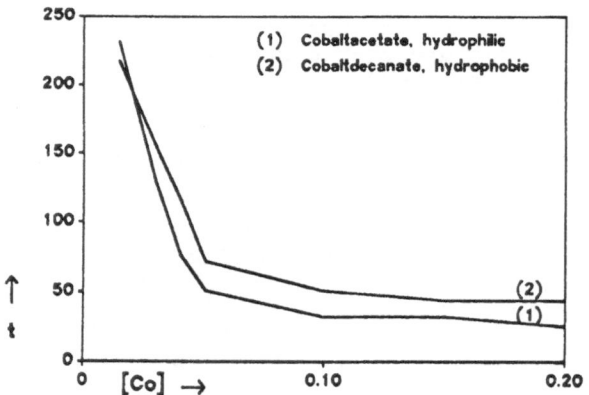

Figure 14. Touch-dry time of a TOFA/IPA/Penta based long oil
        alkyd emulsion vs. [Co(II)], in minutes

THE USE OF AMINE IN ALKYD EMULSIONS

     Earlier in this paper I pointed out that charge and shear
stress are more or less interchangeable. Since shear stress
is limited by the technical standard of available equipment,
omitting a neutralizing agent is also limited to lower
viscous resins. As a rule of thumb we can say that only
resins with a viscosity under 5000 dPa.s can be emulsified
without neutralization or anionic surfactants. Traditionally,
neutralization has been brought about by the use of amines,
because they were believed to evaporate during the drying
process. This evaporation was necessary to prevent charged
loci from staying in the paint film, deteriorating water
resistance. This was also the reason why no alkali like NaOH
or KOH were used. Clearly Na+ and K+ can not be expected to
evaporate. When we studied the behaviour of amines, we found
that of the most widely used amines (triethylamine TEA,
dimethylethanolamine DMEA and aminomethylpropanol AMP) some
25 to 50 % actually still is present in the dried film after
one day. The result of this is clearly noticeable when a
film, drawn from an emulsion of a 47 % soybean oil alkyd is
immersed in water after 1 hr of drying. When neutralization
was accomplished with DMEA (2.1 % on resin), the film turned
white after 15 minutes. When a high sheer emulsion was based
on the same resin and sodium hydroxide (0.4 % on resin), the
same degree of attack by water could be observed after 4
hours. The effect on water sensitivity caused by improved
drying properties on one hand and less charged loci in the
film on the other hand, is clear.

The improvement in drying properties as a result of replacing amines by NaOH can also be visualized by measuring the oxygen uptake, since alkyds dry by reaction of Cobalt-generated radicals with oxygen [6]. The more -oxidative- drying, the more radicals and therefore the more oxygen uptake, as is shown in figure 15.

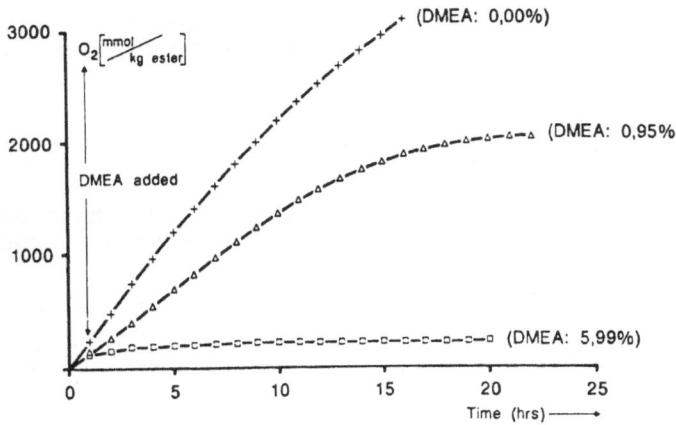

Figure 15. Influence of DMEA on the oxygen uptake of the methylester of safflower fatty acid at 25°C [MeSo] = 10% w/w

One final advantage of this replacement is the reduction of yellowing of white paints. Whereas amine-containing alkyd emulsions tend to give severe yellowing, their NaOH-based counterparts have been found not to show such a behaviour.

CONCLUSIONS

In summary, it can be said that although thermoplast dispersions have reached a high level of sophistication, there is a place in the market for the even more environmentally friendly alkyd emulsions, where high gloss can be obtained without cosolvents, amines and the like. Just water, alkyd, nonionic surfactant and the appropriate equipment is all you need to make a clear varnish with 95% gloss under 60°C, dry in 3-5 hours. It should be always kept in mind that the presence of water alone already diminishes the drying, meaning that alkyd emulsions with exactly the same drying properties as the same alkyd in white spirit do not exist (figure 16).

Nevertheless, we feel that there definitely is a place in the market for alkyd emulsions, next to polymer dispersions.

Copyright for this paper rests with Dr. A. Hofland, employed by DSM Resins BV, Ceintuurbaan 5, Zwolle, The Netherlands.

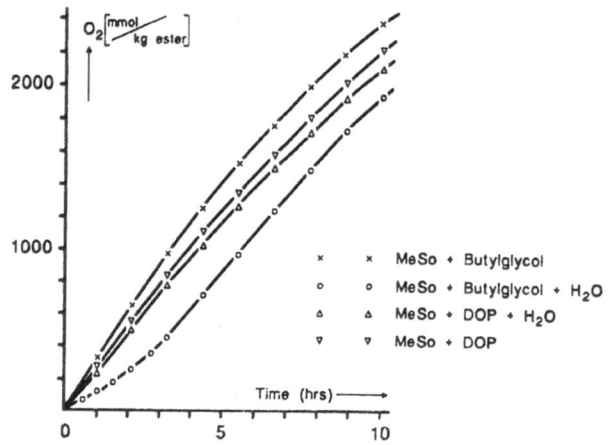

Figure 16. Influence of water on the oxygen uptake of the
methylester of safflower fatty acid at 25°C
[MeSo] = 10% w/w

REFERENCES

[1]   R. Groot, Proceedings of the PRA Conference,
        Frankfurt, 23-25/10/89, p. 153, Paint Research
        Association, Londen.

[2]   S. Wu, Polym. Eng. Sci 27(5), 335 (1987).

[3]   G.G. Stokes, Phil. Mag. 337 (1851).

[4]   W.C. Griffin, J. Soc. Cosmet. Chem. 1, 311 (1949) and
        5, 249 (1954).

[5]   P. Becker, Emulsions-Theory and practice, 2nd ed.
        Reprint, R.E. Krieger, Huntington, N.Y., 1977.

[6]   A. Hulden and K. Eng, Farbe und Lack 28(2), 21
        (1982).

EMULSIFYING EFFICIENCY VERSUS WETTABILITY

AND ADSORBABILITY OF SOME FINELY DIVIDED SOLIDS

HAVING AGRICULTURAL AND PHARMACEUTICAL SIGNIFICANCE

S. N. Srivastava

Agra College
Chemical Laboratories
Agra, India

Most water-based coating formulations contain colloidal suspensions. Therefore, the stability of these dispersions is discussed in this chapter. Fine particles of a solid suspended in a liquid are known to stabilize dispersed droplets of a second liquid against coalescence owing to reasonable adsorption of solid particles and their contact angle at the liquid-liquid interface. The solid particles should also be capable of being wetted by both the liquids, but preferentially by the continuous liquid in which the dispersed solid particles form an acute contact angle. The stability of the emulsions thus promoted depends upon the interaction energy of the dispersed droplets and can be altered by the addition of surfactants, bioamphiphiles and other additives. For stability study, fine particles of the freshly precipitated hydroxides of polyvalent metals, Bordeaux mixtures, clays and drugs have been used to promote stable petroleum ether-in-water emulsions. The stability of the dispersed droplets has been examined in terms of the interaction energy of the emulsion globules whose coalescence necessitates wetting of at least part of one solid particle by the internal phase so as to enable the rupture of the liquid film and this process requires an energy equivalent to several hundreds of kT units. An attempt has also been made to correlate the stability of the emulsions with the interfacial characteristics, adsorption parameters, rheological properties of the liquid film with coated fine particles and their relative wetting by the internal and external phases of the emulsions investigated.

Surface Phenomena and Fine Particles in Water-Based Coatings and Printing Technology
Edited by M.K. Sharma and F.J. Micale, Plenum Press, New York, 1991

INTRODUCTION

The first detailed, suryey of finely divided solids as emulgents was made by Pickering[1] who suggested that the fundamental
condition of the formation of oil-in-water emulsions is that the
solids must be more easily wetted by water than oil. Van der
Minne[2] explained the stability of solid stabilized emulsions in
the light of adsorption of the solid particles at the liquid-
liquid interface and the contact angle against the solid phase.
Later on Briggs[3], Woodman[4], King and Mukherjee[5], Schumlman and
Leja[6], Albers[7], Lucassen-Reynders[8], Vold[9] made significant contributions to the field. However, in recent years detailed in-
vestigations on the subject have been carried out by Srivastava
et al.[10-15], particularly in the light of the interaction energy
of the dispersed droplets.

From the survey of literature it will be seen that the emulsifying efficiency of fine particles of solids is a question of
much difficulty and still much remains to be done by way of
quantitative and systematic study of the mechanism of the formation of the corresponding emulsion and its subsequent stabilization.

Therefore, in the present work, fine particles of the
freshly precipitated hydroxides of polyvalent metals, clays
like kaolinite and fungicides/pesticides like Bordeaux mixtures
and drugs, like sulphapyridine have been used to promote stable
petroleum ether-in-water emulsions. The stability of the emulsion globules has been examined in terms of the interaction energy of the dispersed droplets whose coalescence necessitates
wetting of at least part of the solid particles by the internal
phase so as to enable the rupture of the liquid film and this
process requires an energy equivalent to several hundreds of kT
units. An attempt has also been made to correlate the stability
of the emulsions with the interfacial characteristics, adsorption parameters, rheological properties of the liquid film with
the emulgent coated fine particles, their contact angles and
relative wetting by the internal and external phases of the
emulsions under investigation.

THEORETICAL

Zeta Potential and Interaction Energy

The zeta potential values of the systems under investiga-
tions were calculated from Helmholtz equation:

$$\xi = \frac{4\pi\eta}{\varepsilon x}U$$

For the present systems ξ potential was assumed to be identical with the double layer surface potential, Ψ . Hence, to
evaluate the stability of the systems the interaction energies
were obtained by summing up the repulsion and the attraction
energies using the following equations:

$$V_R = \frac{\varepsilon a\psi_0^2}{2}\ln(1 + e^{-\kappa H})$$

$$V_A = -\frac{Aa}{12H}\left(\frac{\lambda}{\lambda + 3.45H}\right)$$

for H < 15 nm

and

$$v_A = \frac{Aa}{12H}\left[\frac{2.45\lambda}{120H^2} - \frac{\lambda^2}{1045H^3} + \frac{\lambda^3}{5.62x10^4H^4}\right]$$

for H > 15 nm

The thickness ($\frac{1}{\kappa}$) of adsorbed monolayer of emulgent at O/W interface was obtained from the equation[16] given below:

$$\frac{1}{\kappa} = \frac{(4\pi n Z^2 e^2)^{1/2}}{\epsilon \kappa T}$$

where the notations have their usual significance.

Rate of Flocculation

Smoluchowski's theory[17] we know:

$$\frac{1}{n} - \frac{1}{n_0} = 4\pi DRt$$

where n and n_0 denote the number of singlets present at time t and initially, respectively, and D and R (equal to 2a, particle radius) stand for the diffusion coefficient and droplet diameter respectively. ϕ is phase volume ratio and is constant (0.01 by volume) for different systems under examinations. If V_m represents the mean drop volume of n drops at time t, then

$$v_m = \frac{\phi}{n}$$

From equations (6) and (7)

$$v_m = \frac{\phi}{n_0} + 4DR\pi\phi t$$

if ϕ remains constant during coalescence as in our systems, then,

$$\phi = nv_m = n_0 v_{m_0}$$

where V_{m_0} is the mean drop volume of initial drops, hence,

$$v_m = v_{m_0} + 4\pi DR\phi t = \kappa_f t$$

where K_f is Smoluchowski's flocculation rate constant. The plot of V_m against time will be a straight line. The slope of this curve will be equal to $4\pi DR\phi t\, e^{-W}$ 1/kT. The stability factor W_1 was obtained by applying Einstein equation D = kT/6 $\pi \eta$ a, to the above equation.

Rate of Coalescence

The rate of change of specific interface (interfacial area per gm of the dispersed liquid) is proportional to the interfacial area and is given by the equation:

$$-\frac{dS}{dt} = \frac{S}{K} = K_2 S$$

where $K_C = \frac{1}{K}$, being the stability factor and is defined as the reciprocal of the rate of decrease of each sq. cm. of fresh emulsion per unit time.

or

$$\ln S - \ln S_0 = -K_c t$$

or

$$S = S_0 e^{-K_c t}$$

or

$$\log S_0 - \log S = K_c \, t / 2.303$$

This equation conforms to the first order kinetics and coalescence is kinetically of first order. Plot of lnS against t should give a straight line whose slope will give K_C, the rate of coalescence of the investigated systems.

The stability factor is calculated from the slope of each curve and divided by the initial specific area of interface to give the rate of decrease per unit area of interface; this is termed the instability factor $\frac{ds}{dt}$ and is defined as the decrease of each sq. cm. of fresh emulsion per day. The reciprocal of this factor is a direct measure of emulsion stability.

Free Energy and Binding Parameters

The free energy of adsorption and several binding parameters viz., number of the binding sites/cm^2 (N_1), the electrochemical free energy of adsorption per molecule (ΔG^-), the adsorption constants, K_1 and K_2 and the charge density in diffuse layer (σ_s) have been calculated at the isoelectric point of the system using the Stern equation and its different forms. The details are given elsewhere[18,19].

$$\sigma_s = \frac{N_1 e Z}{1 + \dfrac{55.6}{C} \exp(\dfrac{\Delta G^-}{\kappa T})}$$

where C and Z stand for molar concentration and valency of added counter ions.

Equation (14) may be written as

$$\sigma_s = \frac{K_1 C}{K_2 C + 1}$$

which is known as Langmuir adsorption equation, where

$$K_2 = \exp(-\Delta G^- / \kappa T) / 55.6$$

and

$$K_1 = N_1 K_2 e Z$$

The Reversal of Charge and log C- ξ relationship

For large emulsion globules (assumed to be spherical) of the present system, where particle radius is much larger than the double layer thickness ($K a \gg 1$. K and a being respectively Debye-Huckel parameter and radius of globule), the charge density in the diffuse double layer is given by the expression:

$$\sigma_s = \frac{\kappa \varepsilon \Psi \delta}{4\pi} \qquad \text{or} \qquad \sigma_s = \kappa \varepsilon \Delta \Psi \delta / 4\pi$$

where $\Delta \Psi \delta$ is equal to:

$$\Delta \Psi \delta = \Psi \delta^0 - \Psi \delta = \frac{4\pi}{\kappa a} \frac{K_1 C}{1 + K_2 C}$$

which directly relates Ψ_S (Stern potential) with molar concentration C, Ψ_S^0 being constant. With continued adsorption, σ_s may increase till Ψ_S becomes zero and $\Delta \Psi_S^0 = \Delta \Psi_{S_0}$. If the process of binding continues further, $\Delta \Psi_S > \Psi_S^0$ and therefore Ψ has to be opposite in sign thereby reversing the charge of the system.

Successive differentiation of equation (19) provides the relationship in C and Ψ like:

$$\frac{d\Psi \delta}{d \ln C} = -\frac{4\pi}{\kappa \varepsilon} \frac{K_1 C}{(1 + K_2 C)^2}$$

$$\frac{d^2 \Psi \delta}{d \ln C^2} = \frac{4\pi}{\kappa \varepsilon} \frac{K_1 (K_2 C - 1)}{(1 + K_2 C)^3}$$

Since at the zero point of charge from equation 19:

$$\Psi_\delta^0 = \frac{4\pi Z e N_1 C K_2}{\kappa \varepsilon (1 + K_2 C)}$$

for example

$$\frac{1}{C} = \frac{4\pi Z e N_1 C K_2}{\kappa \varepsilon \Psi^0 \delta} - K_2$$

we have

$$\left(\frac{d\Psi \delta}{d \ln C} \right)_{\Psi_\delta^0 = 0} = \frac{\kappa \varepsilon}{4\pi e Z N_1} (\Psi_\delta^0)^2 - \Psi_\delta^0)$$

which shows that at the zero point of charge, the slope of log C- ξ (assuming $\Psi^0 = \xi^0$) curve is independent of K_2 but does depend on Ψ_S^0, the valency of the counter ions adsorbed and N_1.

Wettability of Solid Emulgents

To interpret the results of the present investigation it is also desirable to consider the wetting and the allied parameters of the solid emulsifiers. Wetting is the ability of a liquid to wet a solid surface and is denoted by the Young-Dupre equation as follows:

$$\text{Wetting, } W = \cos \theta = \frac{\sigma_{2,3} - \sigma_{1,3}}{\sigma_{1,2}}$$

where θ is contact angle which is zero for complete wetting (i.e. $W = +1$) or greater and 180° for a hypothetical case of complete non-wetting (i.e. $W = -1$).

The above equation can be derived thermodynamically by im-
agining the movement of a liquid along the solid surface by a
very small distance and having covered the surface area of a
solid equal to $\Delta \xi^1$. Again since specific interfacial energies
are numerically equal to the respective values of the corresponding surface tension, the change in free surface energy,
Δ F is represented as under:

$$\Delta F = \Delta \xi^1 \left(\sigma_{1,3} - \sigma_{2,3} \right) + \sigma_{1,2} \cos(\theta - \Delta\theta)$$

or

$$\frac{\Delta F}{\Delta \xi^1} = \sigma_{1,3} - \sigma_{2,3} + \sigma_{1,2} \cos(\theta - \Delta\theta)$$

But at equilibrium, $\lim \dfrac{\Delta F}{\Delta \xi^{1\to 0}} = 0$

$$\sigma_{1,3} - \sigma_{2,3} - \sigma_{2,3} + \sigma_{1,2} \cos\theta = 0$$

or
$$\cos\theta = \frac{\sigma_{2,3} - \sigma_{1,3}}{\sigma_{1,2}}$$

which is same as eqn. (22) above.

During the wetting process when the liquid spreads, the
total change in free surface energy, Δ F in a system consisting of three contacting phases is given by

$$\Delta F = \left(-\sigma_{2,3} + \sigma_{1,2} + \sigma_{1,3} \right)$$

i.e.

$$\frac{\Delta F}{\Delta \xi^1} = \sigma_{1,3} + \sigma_{1,2} - \sigma_{2,3}$$

But for complete wetting

$$\text{Cos } \theta \quad = \quad 1 \text{ or } > 1$$

From eqn. (22,

$$\frac{\sigma_{2,3} - \sigma_{1,3}}{\sigma_{1,2}} \rangle 1$$

or
$$\sigma_{2,3} \rangle \sigma_{1,3} + \sigma_{1,2}$$
or
$$\sigma_{1,3} + \sigma_{1,2} - \sigma_{2,3} \langle 0$$

On comparison of eqns. (23) and (24) one may conclude that a
decrease in the free surface energy takes place when the liquid spreads and wets a surface and this decrease can be taken
as a measure of the corresponding wetting. Hence the decrease
in surface free energy should result in the liberation of heat
during the wetting process and this heat per sq. cm. of the
surface is known as the heat of immersion or heat of wetting.

For the present emulsion systems instead of the solid-liquid
-gas contact angle we have solid-water-oil contact angle be-

cause two immiscible liquids oil and water are involved. Accordingly, if θ is the solid, S (emulgent)-water, W-oil, O contact angle, the same equations are applicable and equation (22) can be written as

$$\sigma_{S,O} = \sigma_{S,W} + \sigma_{W,O} Cos\theta_{S,W,O}$$

where the contact angle $\theta_{S,W,O}$ is measured in the water phase.

Therefore, the above mathematical formulations can be represented in a slightly different manner given below.

When a lump of solid is dropped into a liquid, or when a liquid is poured into a tube containing the solid, the liquid penetrates the solid with the resulting disappearance of the solid surface and the formation of a new liquid solid interface. If the initial surface tension of the solid, σ and surface area, ξ^1 after immersion become $\sigma_{S,L}$ and ξ^1 respectively, then according to Razouk[21] the heat of immersion of one gram of the solid will be:

$$-\frac{H}{\xi^1} = \left(\sigma_S - T\frac{d\sigma_S}{dT}\right) - \left(\sigma_{S,L} - T\frac{d\sigma_{S,L}}{dT}\right)$$

With solid adsorbents where the swelling is of small order of magnitude, the surface area is not appreciably altered by the wetting process and so ξ may be assumed to be equal. Hence the above equation reduces to

$$-\frac{H}{\xi^1} = \left(\sigma_S - T\frac{d\sigma_S}{dT}\right) - \left(\sigma_{S,L} - T\frac{d\sigma_{S,L}}{dT}\right)$$

$$= \left(\sigma_S - \sigma_{S,L}\right) - T\left(\frac{d\sigma_S}{dT} - \frac{d\sigma_{S,L}}{dT}\right)$$

If the variation of surface tension with temperature is also neglected, the above equation becomes

$$-H = \xi\left(\sigma_S - \sigma_{S,L}\right) = \xi A(S-L)$$

where A (S-L) is the adhesion tension of the liquid, L against the solid, S, and amounts to the work required for separating the unit surface area of the solid from the liquid and is equal to (σ_S - $\sigma_{S,L}$) 0 by definition. From the above equation it is clear that on wetting the adsorbent with any liquid an amount of heat is set free which is equivalent to the decrease in surface tension. The greater, therefore, the heat of wetting the greater is the lowering of surface tension produced by the solid emulgent and greater is its adsorption at oil-water interface.

EXPERIMENTAL

All the chemicals used for the present investigation were of BDH analar grade. The details of preparation of the hydroxides of polyvalent metals used as emulgents is reported in earlier publication[10].

The emulsions were prepared by dispersing 1.0% by volume of petroleum ether (boiling range, 120-160° and a density of

$0.756 \mathrm{cm}^{-3}$ purified through alumina column) in 1% aqueous solution of the corresponding hydroxide. The mixture was then homogenized in a Fischer homogenizer with stainless steel stirrer and pyrex glass container. The emulsions were than sampled under identical conditions.

The flocculation of the emulsion was the followed haemocytometrically using an improved neubauer model of Haemocytometer and Fischer Tally counter under Leitz microscope with 15x95 magnification. The pH adjustments were done by Cambridge pH meter. The zeta potentials were calculated from electrophoretic mobilities which were determined in a rectangular Northup and Kunitz flat type microelectrophoretic cell attached to a Carl Zeiss Jena microscope, the details being described elsewhere[19].

The rheological parameters were determined by the oscillating bob technique. For this a torsion pendulum based on the apparatus of Biswas and Haydon[23] and Criddle and Meader[24] was used and shear modulii (elasticity G_f and interfacial viscosity η_f) of the emulgent film were calculated from the equation:

$$G_f = \pi I \left(\frac{1}{r_1^2} - \frac{1}{r_2^2} \right) \left(\frac{1}{T^2} - \frac{1}{T_0^2} \right)$$

$$\eta_f = \frac{1.15 I}{\pi} \left(\frac{1}{r_1^2} - \frac{1}{r_2^2} \right) \left(\frac{\Delta f}{T_f} - \frac{\Delta o}{T_o} \right)$$

where I is the moment of inertia of the suspension device and T_o and T are oscillation periods of the clean interface and the emulsion respectively. Δ_f and Δ_o are the logarithmic decrements in the presence and absence of the film respectively.

The interfacial tensions between the dispersed petroleum ether and the dispersion medium (aqueous solution of emulgent) were measured by a modified drop volume method using an 'Agla' micrometer syringe which can deliver liquids accurately upto $\pm$ 0.001 ml petroleum ether was taken in a tall tube and the aqueous solution, (0.01% protein + 0.01M KCL + a certain concentration of a bioamphiphile/surfactant in the syringe).

The tip of the syringe was dipped inside petroleum ether and the drops were developed at a very slow speed of about 2-3 minutes. All the system, Agla micrometer and the tall tube, was kept in an air thermostat at 25°C. The volume of a single drop was averaged from five to six measurements. The densities of the phases were determined with the help of stoppered pyknometer.

With the help of the mean volume of one drop the interfacial tension was calculated by the following equation[27]:

$$V = \frac{v(d_2 - d_1) g f}{r}$$

where d_1 and d_2 and the densities of the disperse phase and dispersion medium respectively, r is the radius of the edge of the capillary, V is the Volume of one drop and f is a factor depending upon the value of V/r^3 being taken from the standard tables.

The viscosities of the emulsions were determined by a Rheoviscometer (according to Hoppler-Eastern German-Patent No. 210) using a constant shear stress at $25^{\circ}C \pm 0.5^{\circ}C$.

To estimate the solid-oil-water contact angles the method of Bartell and Osterhoff[25] was used which is based on the measurement of the pressure set up by the liquid entering a highly compressed plug of the powder of the finely divided solid emulgent. The powder is compressed in a brass cell, connected with a monometer and the liquid is allowed to enter from one end. Accordingly, the pressure required to prevent the forward movement of the liquid varies directly with its surface tension and the cosine of the contact angle and inversely with the radius r of the capillary. Hence the pressure P is given by

$$P = \frac{2}{r}\cos\theta$$

The effective capillary radius r for equivalent, cylindrical uniform bores is determined by mesasuring the pressure set up by a standard liquid which wets the powder completely.

HEAT OF WETTING

The calorimeter used for the measurement of heat of wetting (immersion) was made of a Dewar's flask tightly stoppered with a cork through which passed the thermometer (Beckmann), the stirrer and a tube containing the solid. The heater consisted of a manganin coil about 4 cm long wound around a closed end of glass tube and soldered to thick copper leads. The resistance of the heater was about 4 ohm and was determined at the end of each experiment.

The calorimeter was fitted up and when the equilibrium conditions were established, the substance was allowed to fall from a thin walled bulb containing it by breaking the bulb. The rise of temperature was determined by measuring the rise of temperature owing to the passage of known amount of current through the heating coil for a definite time. The current was measured by a millimeter. The formulae used are:

$$q = \frac{RI^2t}{4.183} = \frac{EIt}{4.183}$$

where q is the heat dissipated in categories, S the time in seconds, I the current in amperes, R the resistance dhm of the heating element, E, the voltage drop across the heating coil; and

$$q = dt\ (W + w)$$

where W is the water equivalent of the calorimeter, the weight of water within the calorimeter and dt the corrected temperature rise.

RESULTS AND DISCUSSIONS

Quite a few hydroxides of polyvalent metals, fungicides/pesticides (Bordeaux mixtures), a clay and a drug were used to promote reasonably stable emulsions of petroleum ether in water. Their initial characteristics are listed in Table I and a typical microphotograph of CR $(OH)_3$ stabilized emulsion is given in Figure 1. It can be seen that these emulsions though coarse, their globule size being greater than 1 um, exhibit considerable stability despite rather low zeta potentials. It is also clear that their stability is neither dependent on the degree of dispersion nor it is appreciably pH dependent. Again the plots of the droplet concentration n against time t exhibit an eventual levelling off presumably owing to the existence of a less degenerate quasi equilibrium which might have arisen from the coexistence of singlets and doublets of the emulsion droplets. This also implies that the flocculation of the present emulsion systems occurs in the secondary minimum of the energy profiles and is to a major extent reversible flocculation, its rate constant being of a low order of 10^{-15} cm^3 sec. This in its turn can be ascribed to energy barriers which inhibit the flocculation and change even its nature. The energy barriers assumed to be proportional to the stability factor, W_1 could be estimated from the plots V_m against t (Cf. eqn. 10) and are reflected in Figure 2. This can be further checked by constructing the energy profiles and obtaining the corresponding secondary minima which are about 6 kT deep and can trap the emulsion droplets and cause reversible or secondary minimum flocculation because of the inter-droplets interaction.

Table I. Initial attributes of the emulsions stabilize by fine solid particles.

Sample No.	Emulgent	pH	ξ (mV)	Radius a for emulsion droplet	$Kx10^{-7}$	Stability factor W_1 $(10)^3$
1	Al(OH)$_3$	3.9	34	2.4	0.2	2.7
2	Cd(OH)$_2$	7.0	24	2.8	0.3	1.1
3	Cr(OH)$_2$	5.8	32	2.3	2.1	2.4
4	Th(OH)$_4$	3.8	38	2.6	0.4	6.2
5	Ti(OH)$_3$	3.3	46	2.4	0.1	1.3
6	La(OH)$_3$	8.2	17	2.5	3.7	4.6
7	Kaolinite	4.5	35	3.1	0.1	4.6
8	Acid Bordeaux	7.5	19	2.9	0.6	3.8
9	Alkaline Bordeaux	6.9	25	2.7	0.1	4.2
10	Neutral Bordeaux	7.4	27	2.6	0.6	1.3
11	Sulpha-pyridine	9.5	19	1.2	2.7	4.6

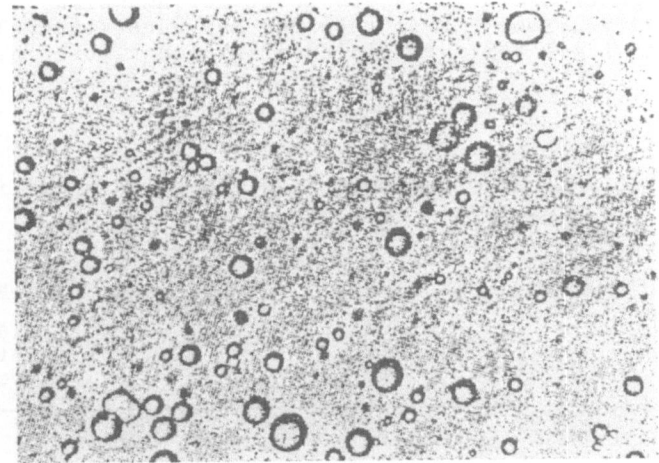

Fig. 1. Microphotograph of the partially
flocculated Cr (OH)$_3$ sol stabilized
emulsion X 600 times magnified.

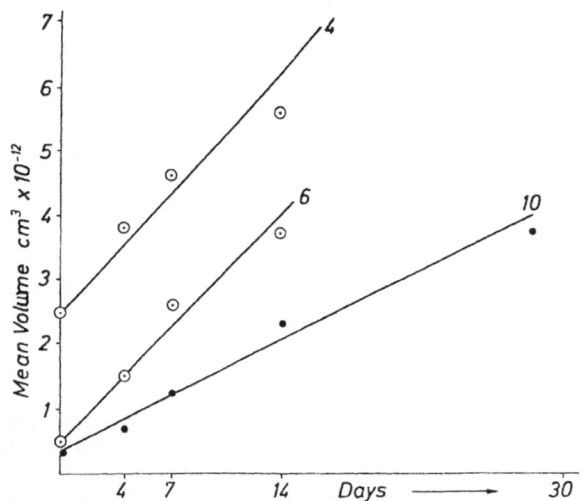

Fig. 2. Plot of mean volume versus time.

Coagulation of an emulsion globules comprises flocculation followed by coalescence. The latter was determined using equation (12) and its order of magnitude was found to be 10^{-7} sec.$^{-1}$[26] which is much slower than the rates obtained elsewhere[26] because of the considerable resistance offered by the solid emulgents films which are about 250 nm which conforms to colloidal dimensions. From this one may guess that clusters of the individual solid particles might constitute the film.

Table II. Kinetic parameters of the solid stabilized emulsions.

Sample No.	Emulgent	$K_f \times 10^{15}$ cm^3 Sec^{-1}	$\dfrac{K_o}{K_f} \times 10^3$	$K_c \times 10^7$	Stability (10^{-3}) from interfacial area
1	$Al(OH)_3$	3.2	3.1	1.2	3.8
2	$Cd(OH)_2$	5.1	1.9	4.6	3.2
3	$Cr(OH)_3$	3.4	2.9	1.3	5.6
4	$Th(OH)_4$	2.1	4.7	1.5	5.7
5	$Ti(OH)_3$	2.9	3.4	2.0	4.7
6	$La(OH)_3$	2.5	4.0	1.6	2.4
7	Kaolinite	2.8	3.6	1.5	5.3
8	Acid Bordeaux	4.8	2.0	2.6	2.5
9	Alkaline Bordeaux	5.2	1.7	3.2	4.3
10	Neutral Bordeaux	4.7	2.1	4.1	2.6
11	Sulpha-pyridine	2.2	4.8	1.8	6.3

Accordingly, it is likely that steric hindrance might be hindering coalescence which should perhaps be caused by the removal of the solid particles at least from a fraction of two dispersed droplets thereby leading to the high stability (except in the case of the drug emulgent) of the systems as reflected from the high stability factors (derived from the dampened rate of decrease of interfacial area) of Table II.

Influence of Cationic and Non-ionic Surfactants and Bioamphiphiles on the Stability

The cationic surfactants used to affect the change in zeta potential and eventual stability of the emulsions under study are LPC, CPC, CPB and CTAB. These have been found to affect the coagulation of the negatively charged emulsions of the present system more easily.

Binding parameters and surface energetics were calculated with the help of the flocculating concentration of the surfactants and zeta potential-long C curves. The results are reflected in Table III. From Table III, it is clear that the value of adsorption constant K_2 are in the order of LPC<CPC<CPB<CTAB. This order clearly indicates that the adsorption of these surfactants depends on their chain length. The value of number of binding sites available, N_1 decreases with increasing chain length of the surfactants which shows that the N_1 would depend upon its capacity for charge reversal.

Table III. Adsorption parameters for LPC, CPC, CPB and CTAB.

Surfactants	Flocculating concentration M/1	N_1 Groups/ $cm^2 x 10^{-1}$	K_1 $(x10^{-8})$	K_2 (10^{-3})	$-G_A$ Kcal/ More	K_1/K_2 $(x10^{-4})$
LPC	$4.1x10^{-7}$	8.5	2.6	7.6	0.89	3.3
CPC	$3.4x10^{-5}$	7.3	2.2	6.1	7.52	3.4
CPB	$2.7x10^{-5}$	7.7	2.3	6.2	7.64	3.7
CTAB	$3.2x10^{-6}$	1.1	1.1	3.1	9.81	3.5

The high value of $-\Delta G$ obtained supports the view that the adsorbed ions are tightly bound in the stern layer. The value of $-\Delta G$ was found to increase with increasing chain length of the surfactants which is indicative of the influence of chain length on their flocculating power because smaller the free energy of adsorption greater is the amount of surfactants required to reach the critical value of charge reversal concentration.

From the changes of interfacial tension in the presence of surfactants/bioamphiphiles on estimation of area of adsorbed molecules was made by calculating interfacial tensions in the presence of various concentrations of surfactants and plotting those values against log molar concentrations of the surfactants. With the help of interfacial data, the surface pressure, number of molecules adsorbed and their area were calculated using the following relation[27,28].

$$\pi = \gamma_o - \gamma_s$$

and

$$n = \frac{1}{A} = \frac{1}{-KT}\left(\frac{d}{d\ln c}\right)_T = -\frac{c}{KT}\left(\frac{d}{dc}\right)_T$$

where π is the surface pressure, γ_o and γ_s are the interfacial tensions of clean surface and in the presence of emulgent and 0.01MkCl respectively, n is the number of molecules adsorbed per cm^2 and A is the area per adsorbed molecule. The value of ($\frac{d\gamma}{d\ln c}$) has been estimated by tangent method. The values obtained are recorded in Table IV. The influence of some nonionic detergents of Triton X series was also examined. These are the alkyl phenyl ethers of polyethylene glycol having the formula:

Triton X - 114 = C_8H_{17} C_6 H_5 $(CH_2CH_2O)_{7.5}$

Triton X - 100 = C_8H_{17} C_6 H_5 $(CH_2CH_2O)_{9.7}$

Triton X - 102 = C_9H_{19} C_6 H_5 $(CH_2CH_2O)12.3$

Triton N - 101 = C_9H_{19} C_6 H_5 $(CH_2CH_2O)_{9.7}$

The above detergents lower the zeta potential by virtue of adsorption but are unable to affect the charge reversal, the trend being depicted by zeta-log c curves of Figure 3, and efficiency being in the order of:

Triton X-102 > Triton N-101 > Triton X-100 > Triton X-114

The effect of some bioamphiphiles like, BSA, RNA and DNA on the zeta potential and stability of the present emulsions has also been studied. The results obtained are given in Table V and shown in Figure 4.

Wettability of Solid Emulgents Vis-A-Vis Emulsion Stability

To correlate the wettability of the solid emulsifying agents with the stability of the corresponding emulsions promoted, the heats of wetting and sedimentation of various solid emulsifiers in the oil (petroleum ether) and water phase were determined and related with their emulsifying efficiency. The data obtained are listed in Table VI. It is implicit that the solids having higher heats of immersion have promoted stable emulsions with high initial specific areas as well as high stability factors. Since low interfacial tension is favorable to emulsification and a preliminary lowering of surface tension ensures small particle size and a stable emulsion hence higher values of heats of wetting (i.e. greater lowering of surface tension and greater dispersibility) should correspond to finer and more stable emulsions with high specific areas

Table IV. Variation of Interfacial Tension and Area in the Presence of Cationic Surfactants.

Bioamphile Moles/1 concentration	Dynes/cm-1 γ	Dynes/cm-1 π	Area/ Molecule A^2	No. of Molecules adsorbed/ cm^2x10^{-1}
LPC				
$1.0x10^{-6}$	10.1	5.9	169	5.90
$1.0x10^{-5}$	16.3	8.7	146	6.83
$1.0x10^{-4}$	13.3	11.7	102	9.80
$1.0x10^{-3}$	7.0	18.0	40	2.46
CTAB				
$5.0x10^{-6}$	13.1	11.9	75	1.32
$1.0x10^{-5}$	10.5	14.6	56	1.78
$5.0x10^{-5}$	4.7	20.4	48	2.07
$1.0x10^{-4}$	2.0	23.0	29	3.40
CPC				
$1.0x10^{-5}$	11.0	14.0	98	1.02
$1.0x10^{-4}$	4.7	20.4	79	1.26
$1.0x10^{-3}$	0.6	24.4	73	1.36
CPB				
$5.0x10^{-6}$	11.1	14.3	78.2	1.3
$1.0x10^{-5}$	9.1	15.8	82.3	1.1
$5.0x10^{-5}$	4.2	18.9	76.1	1.2
$1.0x10^{-4}$	2.6	21.1	66.1	1.2

Table V. Influence of bioamphiphiles on solid stabilized emulsions.

Sample No.	Emulgent	Bioamphi-philes	Molar Concentration					
			5×10^{-6}		5×10^{-5}		5×10^{-4}	
			Zeta Pot. mV	Stabil-ity $x10^{-3}$	Zeta Pot. mV	Stabil-ity	Zeta Pot.	Stabil-ity $x10^3$
1	Al(OH)$_3$	RNA	34	2.6	15	2.1	14	2.2
1	Al(OH)$_3$	DNA	31	3.1	14	1.3	12	1.4
1	Al(OH)$_3$	BSA	36	3.7	19	3.8	18	3.9
4	Th(OH)$_4$	RNA	35	3.1	28	1.6	26	1.5
4	Th(OH)$_4$	DNA	33	2.8	26	1.8	25	1.4
4	Th(OH)$_4$	BSA	39	3.9	26	2.1	24	1.7
10	Kaolinite	RNA	28	2.5	25	2.6	24	2.7
10	Kaolinite	DNA	29	2.2	23	1.8	22	1.8
10	Kaolinite	BSA	31	2.7	29	2.8	30	3.0

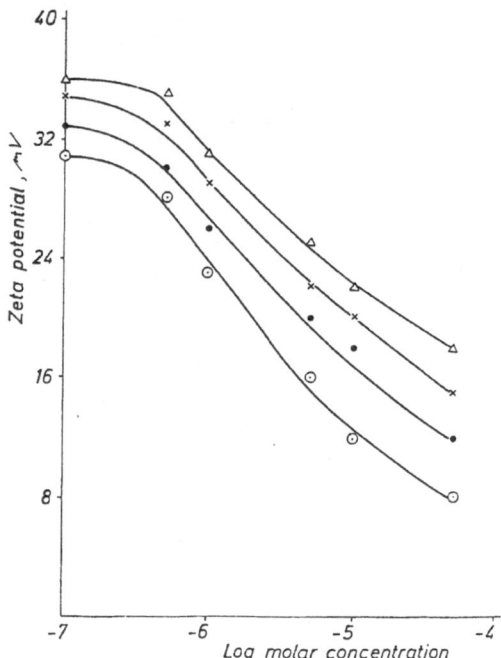

Fig. 3. Influence of non-ionic
surfactants on zeta
potential, ⊙ Triton X-102,
● Triton N-101, X Triton
X Hundred, △ Triton X-114

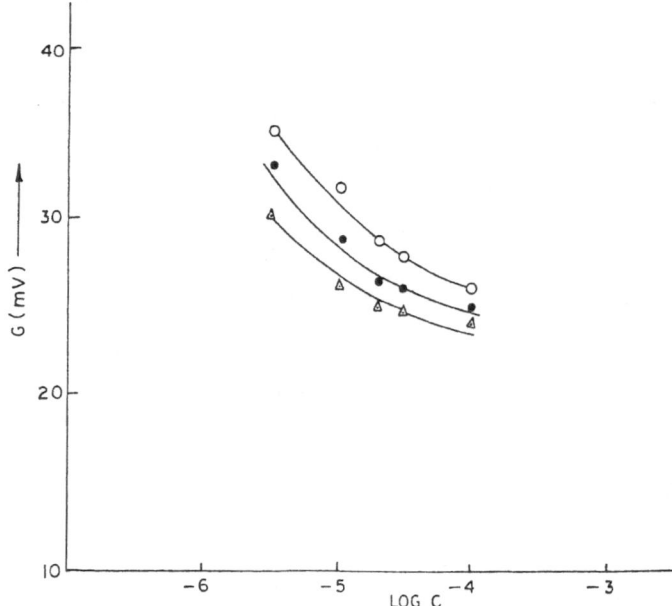

Fig. 4. Plot of G versus log C for the
effect of RNAO, DNA ●, BSA △ on
Th(OH)$_4$ stabilized emulsion.

and stability factors. Solids would not adsorb/accumulate at the interface, particularly on the water side, unless this arrangement results in a decrease of the overall free energy of the system and consequent lowering of the interfacial tension and a quick film formation which gives protection to the globules against coalescence.

Again since heat of wetting is directly proportional to the adhesion of the solid against the liquid, from the values of the heats of immersion of various solids the relative stability of the suspensions of the solids in the internal (oil) and outer phase (water) of the emulsions can be predicted i.e. stability being greater in the liquid with higher adhesion tension for the solid. As these solids do not generate measurable quantity of heat when immersed in petroleum ether whereas appreciable heat of wetting is noticeable in water, the affinity of these solids for water which is the continuous phase is greater than for petroleum ether. This also explains the formation of o/W emulsions with these solid emulgents.

The above-mentioned experimental results were corroborated by measuring the corresponding contact angles and calculating therefrom the heats of wetting which are given in the penultimate column of Table VI. One can see that the two sets of the results are consistent and lead to the conclusions discussed above.

Rheological Parameters

The presence of an interfacial film is likely to affect emulsion viscosity owing to its effect on the internal circulation of the emulsion globules. The interfacial and rheological characteristics of the adsorbed film originate from the presence of the emulgent itself. Hence the viscosity of the emulsion depends on its type (i.e. W/O or O/W). pH, droplet size and concentration of the emulsifier used. In general, the basic six factors governing the rheological properties of an emulsion as pointed out by Sherman[20] are: (a) η_o, viscosity of the outer phase; (b) ϕ, volume concentration of the disperse phase; (c) η_i, viscosity of the internal phase; (d) nature of the emulgent and the resulting interfacial film; (e) electroviscous effect and (f) size frequency of the emulsion droplets.

Some of the above-mentioned rheological characteristics of the emulsion samples under investigation are listed in Table VII and are consistent with the theoretical expectations. The effect of rate of shear and homogenization on the viscosity of the solid stabilized emulsions has also been studied. The results are reflected in the typical graphs of Figure 5 where viscosity is found to decrease with the increase in the rate of shear and seems to vary with interfacial area. This is similar to the work of Richardson[30] who inferred that the emulsion viscosity varies inversely with the mean droplet diameter at constant con-centration of the emulgent.

Finally an attempt was made to measure the film elasticity, G_f and viscosity η_f of the solid emulgents at the oil/water interface. The variation of G_f with the rate of coalescence is given in Figure 6 and data for η_f are given in the penultimate column of Table VII.

Table VI. Heats of wetting, specific areas, contact angles of solid emulgents.

Sample No.	Emulgents	Experiments Heats of Wetting in cals/sq.cm	Initial Specific Areas in sq. cm/ cm^3	Contact Angle in Degree	Calcu- lated Heats of Wetting in cals/ sq.cm	Stability of the Emulsion Promoted x10^3
1	Al(OH)$_3$	5.1×10^{-4}	5392	53.9	2.9×10^{-4}	2.7
2	Cd(OH)$_2$	3.1×10^{-4}	3563	49.5	2.2×10^{-4}	1.1
3	Cr(OH)$_3$	3.2×10^{-4}	3933	51.2	2.7×10^{-4}	2.4
4	Th(OH)$_4$	4.9×10^{-4}	3568	49.8	2.1×10^{-4}	3.8
5	Ti(OH)$_3$	2.8×10^{-4}	2505	61.3	3.4×10^{-4}	6.2
6	La(OH)$_3$	1.6×10^{-4}	3367	58.0	2.6×10^{-4}	1.3
7	Kaolinite	4.7×10^{-4}	3933	64.5	2.2×10^{-4}	4.6

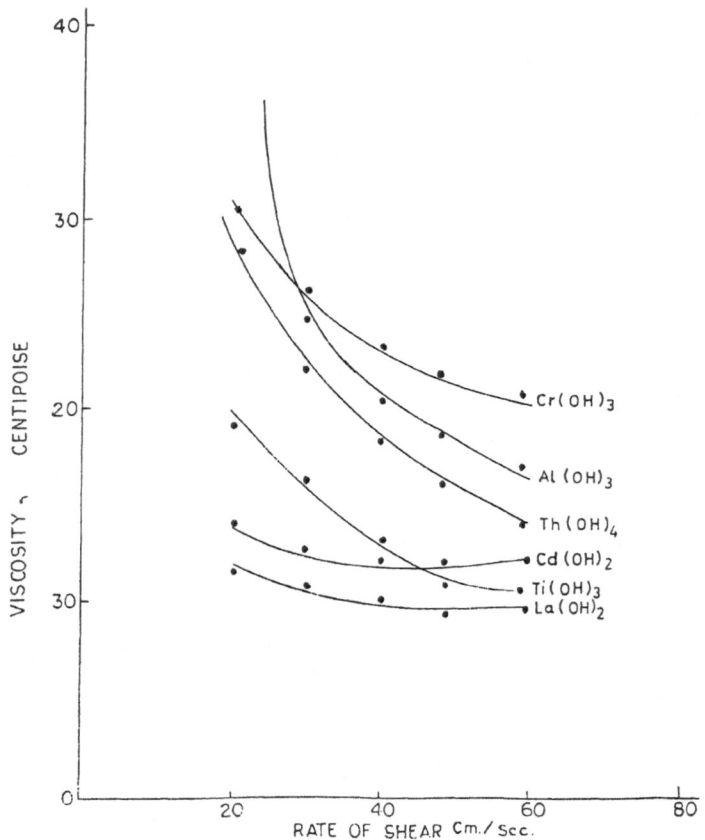

Fig. 5. Viscosity of emulsion stabilized by hydroxide of metal as a function of rate of shear at 25°C.

Table VII. Rheological parameters of the solid-stabilized petroleum ether in water emulsions with volume concentration of the disperse phase = 0.05.

Sample No.	Emulgents	Droplet Diameter μ	$K_c \times 10^7$ sec^{-1}	n at $25^{\circ}C$ Centipoises	Film Elasticity G_f dyne/cm^2	Electro-Viscous Effect
1	Al(OH)$_3$	2.35	1.2	4.7	15.4	4.3
2	Cd(OH)$_2$	2.75	4.6	2.8	11.3	3.9
3	Cr(OH)$_3$	2.38	1.3	4.3	13.6	3.8
4	Th(OH)$_3$	2.62	1.5	5.2	19.2	2.8
5	Ti(OH)$_3$	2.76	2.0	3.3	12.2	3.5
6	La(OH)$_2$	2.82	1.6	3.2	6.8	3.9
7	Kaolinite	2.39	1.5	4.1	8.2	5.4
8	Acid Bordeaux	2.58	2.6	3.8	16.6	6.5
9	Alkaline Bordeaux	2.86	3.2	3.7	11.2	6.9
10	Neutral Bordeaux	2.72	4.1	3.7	14.8	6.1
11	Sulpha-pyridine	1.20	1.8	2.1	5.3	3.6

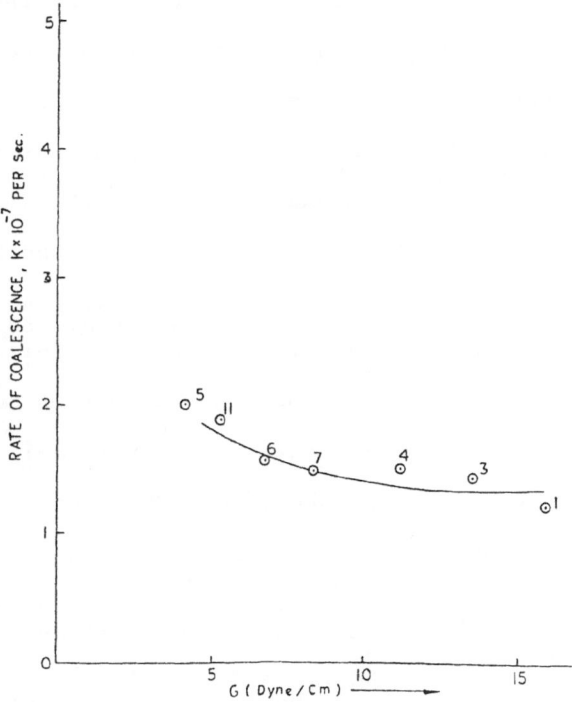

Fig. 6 Rate of coalescence of different samples of emulsions as a function of the film, viscosity, G.

CONCLUSION

Fine particles of the freshly precipitated hydroxides of polyvalent metals, Bordeaux mixtures, a clay and a drug have been used to promote stable petroleum ether-in-water emulsions. The stability of the dispersed droplets has been examined in terms of the interaction energy of the emulsion globules whose coalescence necessitates wetting of at least part of one solid particle by the internal phase so as to enable the rupture of the liquid film and this process requires an energy equivalent to several hundreds of kT units. An attempt has also been made to correlate the stability of the emulsions with the interfacial characteristics, adsorption[31] parameter/surface energies, rheological properties of the liquid film with coated fine particles and their relative wettability[32] estimated from the corresponding heats of immersion and contact angles. In general, the stability of emulsions examined is attributed to the strength of the compact interfacial film of the adsorbed finely divided solid emulgent.

REFERENCES

1. S. U. Pickering, J. Chem. Soc., 91:2001 (1907).
2. J. L. Van der Minne, Over Emulsions, Amsterdam, 66 (1928).
3. T. R. Briggs, Ind. Eng. Chem., 13:1008 (1921).
4. R. M. Woodman, J. Phys. Chem., 34:299 (1930).
5. A. King, and L. N. Mukerjee, J. Soc. Chem. Ind., 58:243 (1939).
6. J. H. Schulman and L. Leja, Trans. Faraday Soc. 50:598 (1954).
7. W. Albers and J. Th. G. Overbeek, J Coll. Sci., 14:509 (1959).
8. E. H. Lucassen-Reynders, Stabilization of water-in-oil emulsions, Utrecht, (1962).
9. M. J. Vold and D. V. Rathnama, J. Phys. Chem., 64:1619 (1960).
10. L. N. Mukerjee, S. N. Srivastava, Kolloid and Zeit, 147:146 (1956).
11. L. N. Mukerjee and S. N. Srivastava, ibid, 149:35 (1956).
12. L. N. Mukerjee and S. N. Srivastava, ibid, 150:144 (1957).
13. S. N. Srivastava and S. P. Jain, Kolloid Zeit, 235:1230 (1969).
14. S. N. Srivastava, Progr. Colloid and Polymer Sci., 63:41 (1978).
15. S. P. Jain and S. N. Srivastava, J. Surf. Sci. Tech., 3(1), 39 (1987).
16. B. Derjaguin and M. Kussakov, Acta Phys. Chem. URSS, 10:25 (1939).
17. M. V. Smoluchowski, Zeit, Physik. Chem., 92:129 (1917).
18. O. Stern, Z. Electrochemie, 30:568 (1924).
19. R. H. Ottewill et al, Trans. Faraday Soc., 56:854 (1960).
20. A. W. Adamson, Physical Chemistry of Surfaces, Intersciences, 267 (1960).
21. J. Razouk, J. Phys. Chem., 45:180 (1941).
22. I. Langmuir, J. Am. Chem. Soc., 38:2221 (1916).
23. B. Biswas and D. A. Haydon, Proc. Roy. Soc., A271:296 (1963).
24. D. W. Criddle and A. L. Meader, J. Appl. Phys., 26:838 (1955).

25. F. E. Bartell and H. J. Osterhoff, Colloid Symposium Monograph, The Chemical Catalog Co.,Inc., New York, 113 (1928).
26. V. K. Sharma and S. N. Srivastava, in Macro and Micro Emulsions, Theory and Practice, ACS Symposium Series, 272:398 (1985).
27. J. H. Brooks and A. E. Alexander, Proc. 3rd Int. Congr. Surf. Chem., Cologne, 535 (1960).
28. D. K. Chattoraj, J. Phys. Chem., 70:3743 (1966).
29. P. Sherman, Research (London), 8:396 (1955).
30. E. G. Richardson, J. Coll, Sci., 5:404 (1950, 8:396 (1955).
31. M. J. Schwuger and W. Rybinski on "Adsorption and Wetting" in "Nonionic Surfactants: Physical Chemistry", M. J. Schick, 23:45 (1987).
32. M. J. Hey, J. W. MacTaggart, and C. H. Rochester, J. Chem. Soc. Faraday Trans., I80:699 (1984).

LATEX CHARACTERIZATION TO

DETERMINE ADHESION FAILURE

Peter J. Palackdharry

Dexter Corporation
Packaging Products Division
1-7 East Water Street
Waukegan, Illinois 60085

Acrylic latices of similar compositions are being commercially used as adhesives to laminate metallized polyester to paperboard. The laminated sheets are then converted into trays to pack microwaveable pizza products. Adhesion failure on laminates was observed with a particular latex composition and only marginally acceptable adhesion with another. These two latices, together with one which exhibited quite acceptable adhesion, were characterized for particle size distribution, FTIR spectroscopy for compositional variations, conductometric titration to locate carboxylic acid functionalities on the particle and DSC for Tg ranges. In addition, wet samples were characterized for viscosity and also potentiometrically titrated to establish the total ammonia base content and the amount involved in carboxylic acid neutralization.

The significant features apparently required to perform acceptably as a laminating adhesive is a lower Tg range coupled with a particle size distribution containing a higher level of the larger sized particles produced in the polymerization process. Also, the relative amount of carboxyl functionalities readily available for bonding to the respective substrates is certainly a contributory factor. The extent of neutralization of surface carboxyl groups contributed significantly to the viscosity achieved in a particular latex batch. There were indications that when the amount of base involved in latex acid neutralization was greater than 20% of the total contained in the formulation, the resulting product would meet all performance requirements.

INTRODUCTION

This paper deals with the analytical characterization of three different batches of an acrylic latex composition which exhibited significant differences in adhesion as laminating adhesives. Since the monomer compositions were identical, the physical characteristics of these latex particles together with their respective internal and surface structures seemed very critical to the eventual performance requirements as adhesives. One batch failed completely (A), another gave only marginal adhesive strength (B), while the third (C) was totally acceptable for bond strength. The end use was to laminate metallized polyester to paperboard from which trays were fabricated to microwave pizza products.

Surface Phenomena and Fine Particles in Water-Based Coatings and Printing Technology
Edited by M.K. Sharma and F.J. Micale, Plenum Press, New York, 1991

167

THEORETICAL

The monomer composition of this commercial latex is of styrene, butyl acrylate, methyl methacrylate, methacrylic acid, and a wet adhesion monomer. The product is synthesized without the use of surfactants or emulsifiers. Once the polymerization process is completed, aqueous ammonia is utilized to provide particle dispersion stability through neutralization of surface carboxyl groups. This procedure is also essential to increase the viscosity which is dependent on the extent of neutralization. A specific amount of ammonia is added to reach a specified pH.

Theoretically, once the polymer is synthesized and neutralized within the range of the specification, the adhesive property for lamination should not vary significantly from the standard bond peel strength requirement. However, once this variation is manifested as in the current case being discussed, the reason for this behavior must be discovered. For the product to be commercially successful, the quality in performance must be maintained at all times.

Quality Control Data on Latex Batches

Latex	Date Manufactured	pH	Viscosity (cps)*	Peel Strength** (g/in.)
A	12/09/88	6.38	115	148 (poor)
B	01/10/89	6.45	180	290 (marginal)
C	02/07/89	6.49	260	630 (excellent)

*RVT   – #4 Spindle – 20 rpm
**Instron – 5 inches per minute

EXPERIMENTAL RESULTS AND DISCUSSION

1. Particle Size Distribution

This analysis was executed by transmission electron microscopy (TEM) in which a cold stage attachment was used to freeze the relatively soft latex particles to maintain their original shape and to minimize electron beam damage. In addition, a phosphotungstic acid staining procedure provided enhanced resolution of particles. Approximately 500 particles were measured altogether on a series of micrographs for each latex sample at magnifications of 17,000X, 36,000X and 100,000X. A computer program was then used to calculate the respective particle size distribution and to generate plots of these distributions. The calculated values are listed in Table 1.

There may be a correlation between particle size and the relative degree of adhesion between these latex samples. Sample C with the best overall adhesion has small particles distribution of significantly greater diameters than sample A which exhibited poor adhesion. Although sample C has practically the same small particles diameter distribution as sample B (marginal adhesion), the agglomerates of sample C has a significantly lower diameter than sample B. Blends of this latex composition made at different particle sizes, have provided substantial proof that adhesive strength improved as the overall particles diameter distribution approached that of sample C.

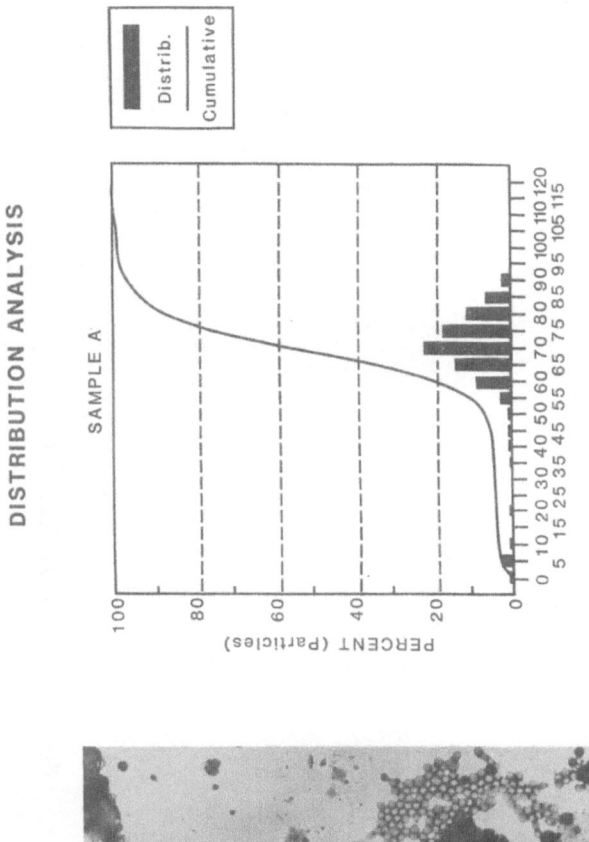

DISTRIBUTION ANALYSIS

SAMPLE A

PERCENT (Particles)

DIAMETER (nm)

Distrib.

Cumulative

Figure 1A. Sample A - 17Kx

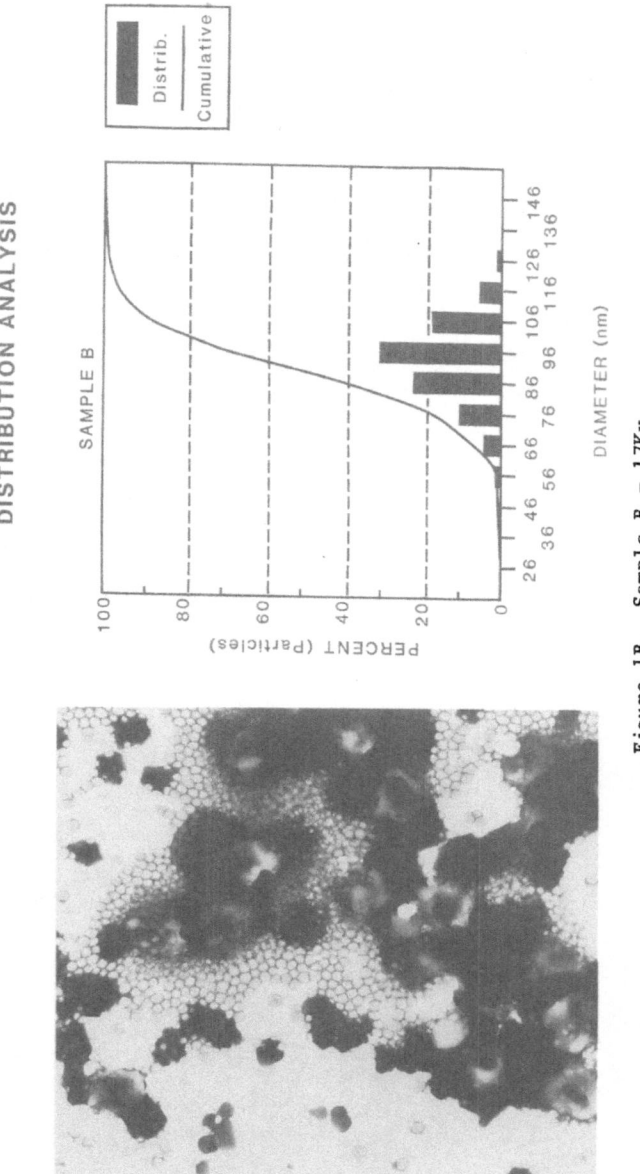

Figure 1B. Sample B – 17Kx

170

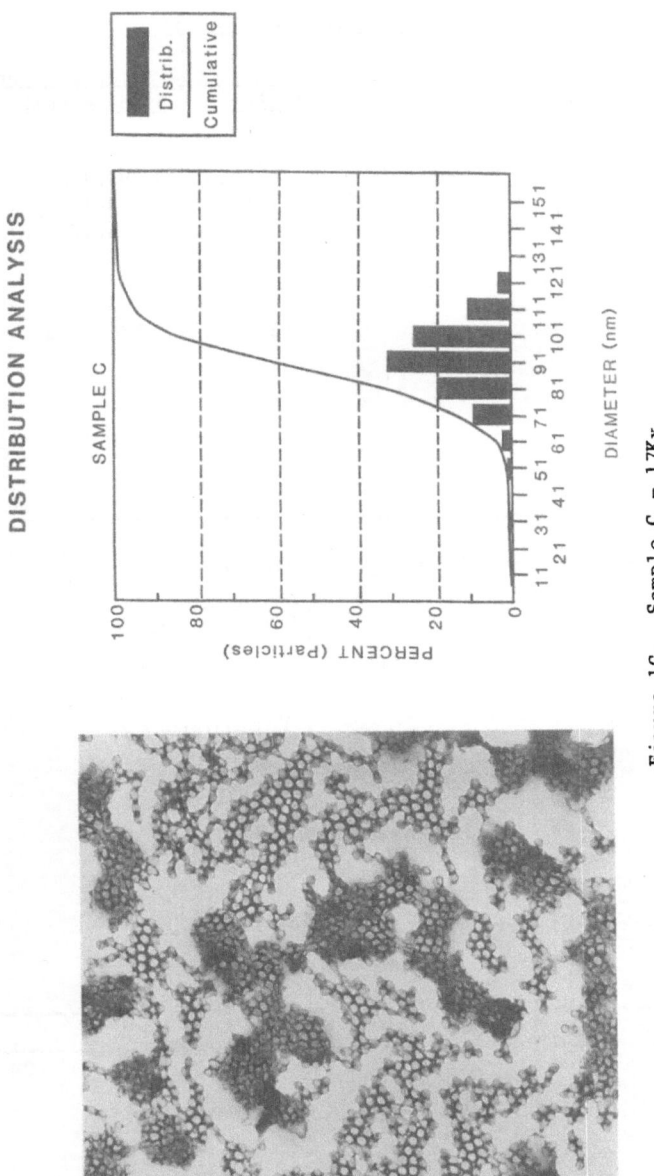

Figure 1C. Sample C - 17Kx

171

Table 1
Particle Diameter Calculation

| | Samples Characterized for Adhesion | | |
Small Particles	A (Poor)	B (Marginal)	C (Good)
Number Average Particle Diameter (Dn)	69.5 nm	93.9 nm	92.2 nm
Weight Average Particle Diameter (Dw)	77.1	100.2	98.8
Polydispersity Index (PDI)	1.11	1.07	1.07
Agglomerates			
Average Diameter	362.0 nm	536.0 nm	362.0 nm

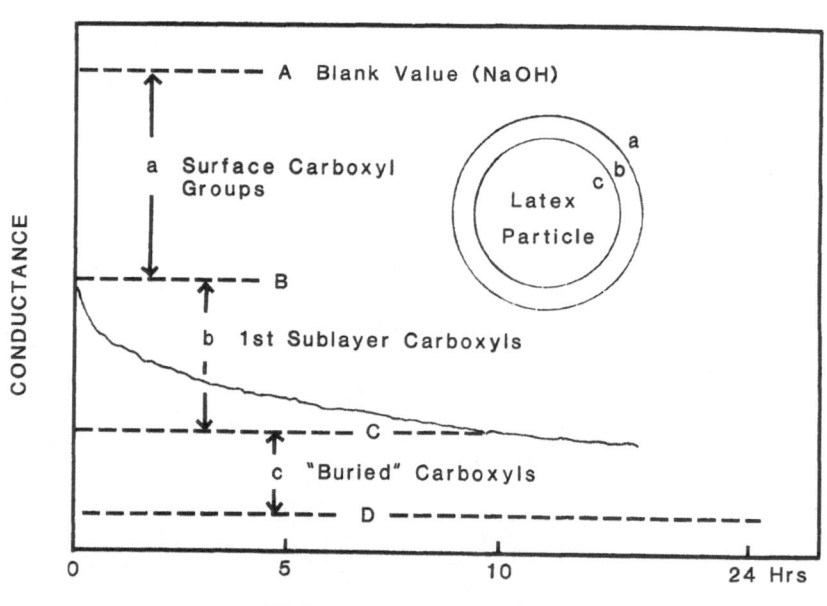

Figure 2A

Schematic representation of Conductometric Titration Procedure to
Determine the Carboxyl Group Distribution in Latex Particles.

The TEM micrographs of each sample at 17Kx are shown in Figures 1A, 1B and 1C, respectively.

2.   Conductometric Titration

Before this titration method could be effectively carried out, the samples had to be thoroughly cleaned to remove any adsorbed substances. This was accomplished by the use of ion exchange resin in the appropriate ionic form and mixed to construct a neutral ion exchange bed.   Each latex sample was diluted to 2% solids for this cleaning procedure.   For titration, weighed amounts of each cleaned sample were diluted with distilled deionized water in a borosilicate jar through which nitrogen was bubbled to remove any dissolved carbon dioxide gas.   Titration of carboxyl groups was carried out with a 0.02N sodium hydroxide solution and the change in conductivity with time over a ten minute period was recorded for each latex.   These neutralized samples were then sealed and stored over nitrogen for twenty-four hours.   The samples were then back-titrated with 0.02N hydrochloric acid to determine the amount of carboxyl groups neutralized over the twenty-four hour period.

A schematic representation of this type of conductometric titration is depicted in Figure 2A.   Line 'A' represents a blank value in which the selected amount of alkali hydroxide base was added only to the distilled deionized water.   The distance 'a' represents the immediate drop in conductance as the surface carboxyl groups are neutralized.   Distance 'b' the lowering of conductivity over a ten minute period as carboxyl groups

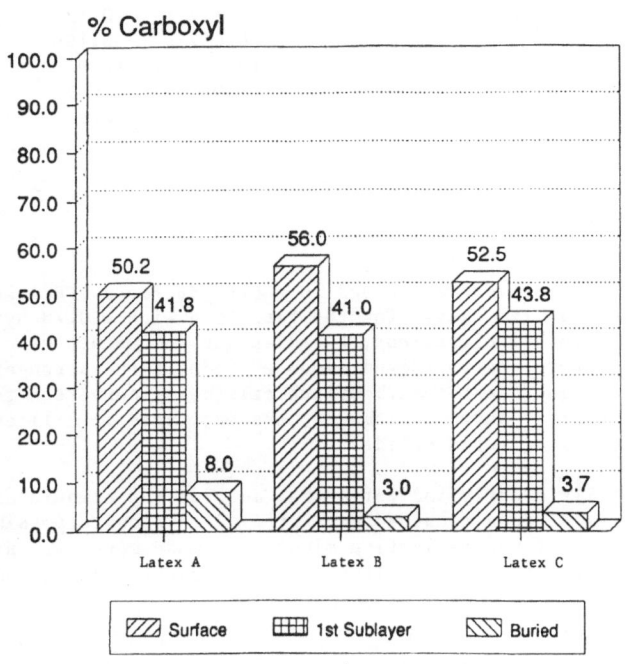

Figure 2B

Carboxyl Group Distribution

in the first sub-layer are being neutralized. Distance 'c' represents the neutralization over a twenty-four hour period, of the carboxyl group buried within the latex. The distribution of carboxyl groups within the surface, first sub-layer, and particle interior, for each latex sample is shown in Figure 2B.

This analysis did indicate that the distribution of the acid functionalities on the surface and sub-layers of the three different batches of this latex composition were within accpetable variation limits. It was difficult to judge whether the slightly higher amount of acid lie buried in sample A would cause such a "dramatic" decrease in adhesive bond strength.

3.   Potentiometric Titration

This particular titration was done in order to establish the total amount of ammonia which was contained in each latex sample added to promote particle stability and to achieve the specified viscosity. In addition, the titration was designed to determine the amount of unreacted 'free' ammonia and the amount involved in carboxylic acid neutralization.

Table 2

Potentiometric Titration for Base Content

	Latex A	Latex B	Latex C
Total Meq Ammonia/g Latex Solid	0.112	0.122	0.144
Meq/g Unreacted Ammonia	0.092	0.100	0.110
% Ammonia Reacted	17.9	18.0	23.6
Meq Carboxylic Acid/g Solid	1.09	1.07	1.07
% Neutralization	1.8	2.3	3.2

Each latex sample was dissolved in tetrahydrofuran (THF) and diluted with distilled deionized water for the titration with a 0.1N hydrochloric acid solution. In all titrations, two distinct end points were recorded. The first was due to the unreacted ammonia which is the stronger base and the second end point was the carboxylate salt being converted to the acid form. These results, shown in Table 2, are expressed in milliequivalents (meq) per gram (g) of resin solids.

These results clearly indicated that most of the ammonia added to the respective latex batches remained unreacted. It is quite possible that the higher degree of neutralization might have made more carboxyl groups available for adhesion of the polymer to the substrate being coated.

4.   FTIR Analysis

Fourier Transform Infrared analyses of air cast films showed very close overlapping of peaks from the spectrum of each sample. There were no obvious differences in polymer structure or composition. An example of the spectra is shown in Figure 3.

174

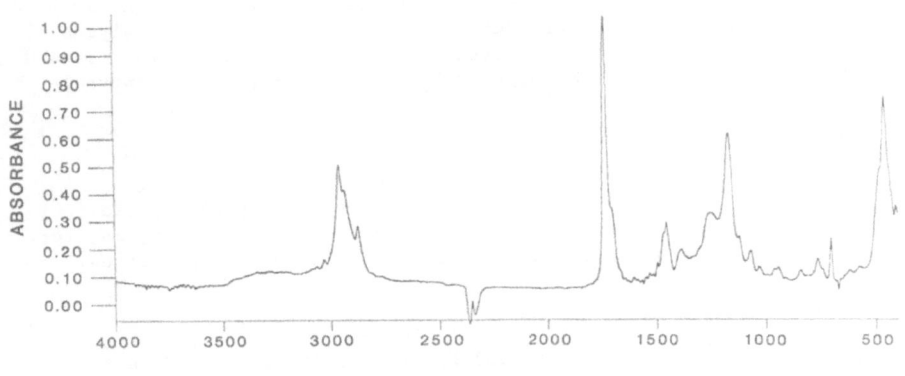

WAVENUMBERS

Figure 3

FTIR Spectrum

5. DSC Analysis

A Mettler TA3000 instrument, using the DSC module, was used to determine the glass transition temperatures (Tg) for each latex sample. Air cast films, stored in a dessicator, were used for these analyses. Each film was scanned at a rate of 10°C per minute over a temperature range extending from –100°C to 100°C. Below, Table 3 lists three]alculated glass transition temperature values, derived from the exothermal DSC scan, for each latex sample.

The first temperature shows the onset of Tg, the second the mid-range value, and the third the ending Tg temperature. The mid-range value is the most appropriate for comparison. From these results, the lowest Tg (37.55°C) latex provided the best adhesive properties. The highest Tg (47.86°C) latex exhibited relatively poor bond peel strength characteristics.

Table 3

DSC Results for T_g

	Latex A	Latex B	Latex C
Initial Tg in °C	43.20	35.61	27.90
Middle Tg	47.86	41.77	37.55
Final Tg	52.51	48.10	47.38

CONCLUSION

Significant differences between three lots of latex adhesives of similar composition were observed by analyses. Latex C, which exhibited quite acceptable adhesive properties, was composed of relatively large particles compared to latex A with poor adhesion, and smaller sized agglomerates compared to latex B with intermediate adhesive strength. Conductometric titration revealed that a greater percentage of carboxyl groups lie "buried" within latex A. Potentiometric titration showed that the lowest level of carboxyl neutralization was with latex A with sample C containing almost twice as much neutralized acid groups. DSC analyses of the glass transition temperature revealed the lowest Tg value for latex C with the best adhesion and the highest Tg for latex A with unacceptable bond strength properties.

ACKNOWLEDGEMENT

The author wholeheartedly thanks the Analytical Department staff of the Emulsion Polymers Institute at Lehigh University, Bethlehem, PA, for performing much of the analytical work reported in this presentation.

REFERENCES

1.   Buckley, D.H., "Role of Material Properties In Adhesion," NASA Technology Memo 83792, 1984.

2.   Eagland, D., "Adhesion and Adhesive Performance," Endeavour, Volume 12 (4), 1988, pp. 179-184.

3.   Kim, J., Kim, K.S., Kim, Y.H., "Mechanical Effects In Peel Strength,"Journal of Adhesion Science and Technology," Volume 3 (3), 1989, pp. 175-189.

4.   Smith, N.C., "Water-Based Coatings Problems and Solutions," Tappi Journal, Volume 71 (7), 1988.

5.   Wicks, Z.W. Jr., "Film Formation," FSCT Series on Coatings Technology.

HIGH TEMPERATURE INORGANIC ANTI-CORROSIVE COATINGS USING ORGANOMETALLIC TITANATE AND ZIRCONATE QUAT BLENDS (QB)

S. J. Monte and G. Sugerman

Kenrich Petrochemicals, Inc.
Bayonne, N.J. 07002-0032

Organosoluble and hydrolytically stable liquid pyrophosphato functional titanate and zirconate organometallic coupling agents are made into water soluble salts via quaternization with certain methacrylic and acrylic functional amines. The resultant individual quats may be combined to form Quat Blends (QB's) having new and novel synergistic performance benefits in unfilled and particulate filled inks, coatings and films.

Previous work by the authors is referenced to gain a historical perspective of evolved chemistry and performance efficacy. Data shows QB additive containing coatings to have superior corrosion rate (mils/year) performance when compared against controls without and with two and one-half times as much strontium chromate in various water based lithium silicate, sodium methyl silicate and tris (hydroxyethyl) ammonium silicate, zinc rich conversion coatings spray applied to carbon steel, aluminum and zinc substrates exposed to various conditions such as repeated cycles of dipping in liquid nitrogen and 200°C Dowtherm A, followed by extraction (to constant weight) with 5% potassium tartrate in boiling triglyme containing 0.5% weight crown 12 phase transfer agent.

INTRODUCTION

During the past decade a substantial body of literature (Ref. 1 to 112) has been developed which collectively demonstrated the rather substantial efficacy of organotitanate and zirconate coupling agents as alternatives to conventional heavy metal based anti-corrosive pigmentary materials in high performance anti-corrosion organic coatings. The authors' efforts are references 1 through 8, 29, 111 and 112. This work plus other references 9 through 110 respectively have variously demonstrated the utility of these organometallic additives as anti-corrosives in vehicles ranging from alkyds through acrylics, epoxies, polyamides, polyurethanes, in both aqueous and solvent based vehicles, applied to metals such as aluminum, copper, carbon and stainless steel, tin plate and zinc, as well as substrates as varied as thermoset polyester and carbon fiber. Applications have varied from enhanced

Surface Phenomena and Fine Particles in Water-Based Coatings and Printing Technology
Edited by M.K. Sharma and F.J. Micale, Plenum Press, New York, 1991

177

Table 1

KEN-REACT QUAT BLEND SELECTION TABLE FOR WATERBORNE SYSTEMS

RESIN	HARDENER	SUBSTRATES(s)	PRIMARY EFFECTS	QUAT BLEND[a]
Acrylic	None	Metals	Adhesion/Anticorrsion	QB 310
Acrylic	None	P.O.,Styrenics	Adhesion	QB 310
Acrylic	None	Concrete	Adhesion/Anticorrosion	QB 012
Acrylic	Melamine	Metals	Adhesion/Anticorrosion	QB 012
Alkyd(S.O.)	Melamine	Metals	Adhesion/Anticorrosion/Gloss	QB 310
Alkyd(L.O.)	None	Metals/Eng.Plastics	Adhesion/Anticorrosion/Gloss	QB 521
Epoxy	None	Metals	Adhesion/Impact/Anticorrosion	QB 310
Epoxy	None	Metals	Intercoat Adhesion	QB 521
Epoxy	None	Concrete/Silicate	Adhesion/Gloss	QB 012
Epoxy	Amine	Metals	Anticorrosion	QB 310
				(Add to Hardener)
Epoxy	Melamine	Metals	Flow Impact/Anticorrosion	QB 012
Polyester	Melamine	Metals	Adhesion/Anticorrosion	QB 012
Urethane	None	Metals/Plastics	Adhesion/Anticorrosion	QB 310
Urethane	None	Metals	Anticorrosion/Primers	QB 521
Urethane	None	Concrete	Anticorrosion/Primers	QB 012
Urethane	Blocked	Metals/Concrete	Anticorrosion/Adhesion	QB 521

[a]Notes:

1.

Quat Blend Components

Coupling Agent	Function
KR 238M	Provides quick dry and water resistance.
LICA 38J	Acts as hydrotrope for water solubility.
NZ 38J	Accelerates cure of KR 238M and LICA 38J and increases chemical resistance.

2.

Quat Blend Part Ratio

Designation	KR 238M	:LICA 38J	:NZ 38J
QB 310	3 :	1 :	0
QB 012	0 :	1 :	2
QB 521	5 :	2 :	1

3.   Water Solubility

a) KR 238M & QB 521 are insoluble in H_2O; at 1% concentration are soluble in H_2O containing 36% organics; at 0.2% concentration are soluble in H_2O containing 12% organics.

b) LICA 38J is soluble in H_2O.

c) NZ 38J is soluble in H_2O at concentrations equal to, or less than 1%.

d) QB 310 is emulsifiable directly into H_2O at concentrations equal to, or less than 1%. Emulsification is facilitated when 12% organics are present in the water.

e) QB 012 is soluble in water.

scrubability of latex flat wall paint to exotic EMI shielding coatings.

The mechanism or mechanisms by which these materials provide the indicated benefits are incompletely known. However, various attributions have certainly been contributory. These include: massive improvements in coating to substrate and coating to internal particulate adhesion (Ref. 9-30); anti-corrosive participation by the additive as a consequence of surface pacification (Ref. 1-63); dispersion enhancement of anti-corrosive pigments (Ref. 64-85); enhanced stability of pigmentary dispersions and/or other particulates (Ref. 86-100); and most controversially -- catalytic modification of resin matrix chemistry through repolymerization and/or cross-link not normally associated with conventional esterification and/or adhesion promoters (Ref. 101-109 respectively). The aforementioned summary represents an ultra concise and conservative assessment of literature claimed-for-benefits obtained with the usage of organometallic coupling agents, specifically those primarily derived from titanium and/or zirconium organopyrophosphates, at levels of the order of 1 to 5 parts per thousand on formulation solids in place of, or as a supplement to, conventional dosages of anti-corrosive pigments. However, common to the bulk of the cited references are that the organic resin derived and pigmented coatings discussed have thermal limitations which preclude their usage at temperatures much in excess of approximately 600°F, which temperature is becoming more frequently a temperature performance base for usage requirements in stack applications, and in high technology systems such as compressors and aerospace equipment. In contra-distinction, their application to inorganic resin coatings has only infrequently appeared in the literature (Ref. 59, 61 and 63). In reference 58, usage of organotitanates in conjunction with borosilane and borosiloxane formulations were shown to provide substantial improvement in adhesion and moisture resistance, whereas in references 59 and 61, usage of organotitanates were shown to drastically improve salt spray resistance in zinc rich sodium silicate based coatings. This paper represents an extension of the art with respect the usage of organotitanium and zirconium based organopyrophosphates as anti-corrosives in clear coat silicates and in zinc rich silicate systems. Table I provides a summary of the recommended applications of various Quat Blends (QB) designated by the ratio of the various component quats: KR® 238M (Titanium IV ethylenediolato bis(dioctyl)pyrophosphato-O, bis(dialkyl) amino alkyl-2-methyl propenoate); LICA® 38J (Titanium IV 2,2(bis 2-propenolatomethyl) butanolato, tris (dioctyl) pyrophosphato-O); and NZ® 38J (Zirconium IV 2,2-dimethyl 1,3 propanediolato, bis (dioctyl) pyrophosphato-O, (adduct) 2 moles N,N-dimethylamino-alkyl propenoamide). Since KR® 238M, LICA® 38J, and NZ® 38J vary greatly as to their solubility in water and water/organic based grind vehicles, it is important to check the solubility of each of the Quat Blends (QB) in water and water/organic vehicles before proceeding with the formulation of the coating. Please see the notes at the bottom of Table I for detailed comments as to the solubility of the individual Quats and their Quat Blends (QB).

EXPERIMENTAL

The sample preparation and methods are described in the appropriate tables.

RESULTS AND DISCUSSION

Based on data developed in this study (cf. Tables II through VI), it will appear that the primary mechanism for beneficiation of silica

Table II

ORGANOMETALLIC COUPLING AGENTS AS CORROSION INHIBITORS
IN A LITHIUM SILICATE CONVERSION COATING ON METAL

Experimental

Sample Preparation:
Five per cent aqueous lithium silicate[1] solutions containing the
indicated proportions of stated anticorrosive(s) (PPT as shown)
were spray applied to metallic panels. The coatings were tunnel
dried at air temperatures of 350°F for 5 minutes.

Salt Spray Corrosion Tests:
Panels were next subjected to attack by oxygen saturated 10 wt.
per cent aqueous salt spray at cycle temperatures of 25 to 150°F
(thermocycle 5°F/Min). Sample panels were stripped of coating
and accumulated oxide by repeated (5 cycle) dipping in liquid
nitrogen and 200°C Dowtherm A, followed by extraction (to
constant weight) with 5% potassium tartrate in boiling triglyme
containing 0.5% wt. crown 12 phase transfer agent. Panel weight
loss was calculated in Mils/Year (8766 Hr.) by extrapolation.

Results

Carbon Steel

Additive(s)	PPT	Corrosion Rates (Mils/Year)		
		500 Hr.	1000 Hr.	2000 Hr.
None	--	3.1	6.0	41.2
Strontium Chromate	9	0.4	2.0	6.3
QB 310	8	0.1	0.1	0.7
QB 012	8	0.2	0.4	0.9
QB 521	8	0.1	0.1	0.2

Aircraft Aluminum

None	--	6.7	14.4	107.8
Strontium Chromate	9	0.9	4.7	21.2
QB 310	8	0.2	0.3	0.4
QB 012	8	0.2	0.3	0.4
QB 521	8	0.1	0.2	0.3

Zinc

None	--	4.9	10.2	67.3
Strontium Chromate	9	1.0	4.9	7.9
QB 310	8	0.1	0.2	0.4
QB 012	8	0.1	0.1	0.4
QB 521	8	0.1	0.1	0.2

1 - DuPont Polysilicate 48

Table III

OXIDATIVE STABILIZATION OF CARBON STEEL, ZINC
AND AIRCRAFT ALUMINUM BY COUPLING AGENT
MODIFIED LITHIUM SILICATE CONVERSION COATINGS

Experimental

Sample Preparation:
Five per cent aqueous lithium silicate solutions containing the
indicated proportions of stated anticorrosive(s) were spray
applied to metallic panels. The coatings were tunnel dried at
air temperatures of 350°F for 5 minutes.

Oxidation Stability Corrosion Rate Evaluation Test:
Air Thermocycle 0 to 600 °F at 20 °F/min. (oven temp.) of 5%
water (w/w) in air stream at 45° angle to panels. Sample panels
were stripped of coating and accumulated oxide by repeated (5
cycle) dipping in liquid nitrogen and 200 °C Dowtherm A, followed
by extraction (to constant weight) with 5% potassium tartrate in
boiling triglyme containing 0.5% wt. crown 12 phase transfer
agent. Panel weight loss was calculated in Mils/Year (8766 Hr.)
by extrapolation.

Results

Carbon Steel

Additive(s)	PPT	Corrosion Rates (Mils/Year)		
		500 Hr.	1000 Hr.	2000 Hr.
None	--	3.2	3.0	3.4
Strontium Chromate	9	1.9	2.7	2.8
QB 310	8	2.3	2.4	2.5
QB 012	8	1.4	1.4	1.9
QB 521	8	2.0	2.0	2.6

Aircraft Aluminum

None	--	1.0	0.9	0.8
Strontium Chromate	9	0.4	0.7	0.9
QB 310	8	0.6	0.7	0.8
QB 012	8	0.2	0.3	0.6
QB 521	8	0.4	0.6	0.7

Zinc

None	--	4.7	5.9	7.6
Strontium Chromate	9	0.6	0.8	4.2
QB 310	8	0.5	0.8	4.7
QB 012	8	0.6	0.7	2.1
QB 521	8	0.6	0.8	2.6

1 - Lithium Corp - Lithium Silicate 6

Table IV

EFFICACY OF ORGANOMETALLIC COUPLING AGENTS AS ANTICORROSIVES
IN A HIGH TEMPERATURE SODIUM METHYL SILICATE COATING

Experimental

Sample Preparation:

Aqueous (10% solids @ 400 °F) sodium methyl silicate solution
containing the indicated proportions of stated additives was
spray applied in two one mil (wet) passes with 20 minute 200 °F
(air temp) dry.   The resultant coatings were baked at 500 °F
(air temp) for 10 minutes.

Corrosion Test:

Baked coatings were subjected to attack by air saturated 10%
aqueous acetic acid containing 5% dissolved salt at 200 °F.

Results

Carbon Steel

Additive(s)	PPT	Corrosion Rates (Mils/Year)		
		500 Hr.	1000 Hr.	2000 Hr.
None	--	4.4	7.9	31.4
Strontium Chromate	20	0.9	2.1	6.3
QB 310	8	0.7	0.9	1.4
QB 012	8	0.4	0.4	0.5
QB 521	8	0.6	0.9	1.2

Aircraft Aluminum

Additive(s)	PPT	Corrosion Rates (Mils/Year)		
		500 Hr.	1000 Hr.	2000 Hr.
None	--	11.2	21.7	42.9
Strontium Chromate	20	1.4	1.9	2.8
QB 310	8	1.2	1.4	1.5
QB 012	8	0.7	0.7	0.9
QB 012	16	0.5	0.6	0.8
QB 521	8	0.9	0.9	1.2

Table V

OXIDATIVE STABILIZATION OF A SODIUM SILICATE ANTICORROSIVE
COATING BY ORGANOMETALLIC COUPLING AGENTS

Experimental

Sample Preparation:

Aqueous (10% solids @ 400 °F) sodium methyl silicate solution
containing the indicated proportions of stated additives was
spray applied in two one mil (wet) passes with 20 minute 200 °F
(air temp) dry. The resultant coatings were baked at 500 °F (air
temp) for 10 minutes.

Corrosion Test:

100 to 900 °F thermocycle @ 20 °F/min. utilizing 5% SO_3 and 10%
water (w/w) in air as the corrosives (Note: condensation observed
below 280 °F). Sample panels were stripped of coating and
accumulated oxide by repeated (5 cycle) dipping in liquid
nitrogen and 200 °C Dowtherm A, followed by extraction (to
constant weight) with 5% potasium tartrate in boiling triglyme
containing 0.5% wt. crown 12 phase transfer agent. Panel weight
loss was calculated in Mils/Year (8766 Hr.) by extrapolation.

Results

Carbon Steel

Additives	PPT	Corrosion Rates (Mils/Year)		
		500 Hr.	1000 Hr.	2000 Hr.
None	--	8.2	12.7	17.1
Strontium Chromate	20	3.8	4.9	12.4
QB 310	8	2.9	4.1	9.6
QB 012	8	1.6	2.2	2.3
QB 521	8	1.9	2.4	3.1

Aircraft Aluminum

Additive(s)	PPT	Corrosion Rates (Mils/Year)		
		500 Hr.	1000 Hr.	2000 Hr.
None	--	9.4	10.2	29.6
Strontium Chromate	20	2.7	3.6	17.9
QB 310	8	2.9	4.9	6.3
QB 012	8	1.4	1.5	1.5
QB 521	8	2.6	3.8	3.9

Table VI

EFFICACY OF ORGANOMETALLICS AS ANTICORROSIVES
IN A ZINC RICH SILICATE COATING

Experimental

Sample Preparation:

Dispersions containing 20 weight percent of zinc dust (-400 mesh)
in 10% aqueous tris (hydroxyethyl) ammonium silicate containing
the indicated additives(s) concentrations, were spray applied (2
mils. wet) to metal panels. The coatings were air dried at 250
°F for one hour prior to evaluation.

Corrosion Test

Panels were subjected to attack by oxygen saturated 5% aqueous
caustic at cycle temperatures of 25 to 150 °F (thermocycle 5
°F/Min. Sample panels were stripped of coating and accumulated
oxide by repeated (5 cycle) dipping in liquid nitrogen and 200 °F
Dowtherm A, followed by extraction (to constant weight) with 5%
potassium tartrate in boiling triglyme containing 0.5% wt. crown
12 phase transfer agent. Panel weight loss was calculated in
Mils/Year (8766 Hr.) by extrapolation.

Results

Carbon Steel

Additive(s)	PPT	Corrosion Rates (Mils/Year)		
		500 Hr.	1000 Hr.	2000 Hr.
None	--	0.9	1.4	12.3
Strontium Chromate	20	0.9	1.3	7.4
QB 310	8	0.9	1.0	1.6
QB 012	8	1.4	1.7	2.2
QB 521	8	0.4	0.5	0.5

Aircraft Aluminum

None	--	2.9	5.7	31.7
Strontium Chromate	20	1.8	2.9	16.3
QB 310	8	2.4	2.6	8.2
QB 012	8	1.9	2.1	7.4
QB 521	8	1.4	1.5	1.9

adhesion of silicate based coatings in formulations containing such additives also undoubtedly play a significant role in improving performance.

Combinations of organotitanates and zirconates were found to be uniformly more efficient than strontium chromate as anti-corrosives in aqueous lithium silicate systems subjected to either salt spray at cycle temperatures of 25 to 150°F (cf. Table II) or 0 to 600°F, high humidity, air thermal cycling (cf. Table III). These same combinations of organometallics were shown to provide significant improvements in corrosion resistance to attack by air saturated 10% aqueous acetic acid containing 5% dissolved salt at 200°F (cf. Table IV), and 100 − 900°F air thermocycles employing 5% sulfur trioxide and 10% water as the corrosive medium (cf. Table V). Similarly combinations of organotitanium pyrophosphates and organozirconium pyrophosphates were shown to provide substantial performance upgrade with respect to the resistance of an ammonium silicate based zinc rich coating on carbon steel and aircraft aluminum against oxygen saturated, 5% caustic in a 25 − 150°F thermocycle (cf. Table VI). It should be noted that benefits observed by the usage of indicated combinations of organometallics are measured on a comparative basis against equal or greater proportions by weight of strontium chromate and simultaneously against no additive control systems subjected to comparable environments under identical conditions. Further details regarding the specific tests performed are given in Tables II through VI.

CONCLUSION

Usage of appropriate blends of organotitanates and zirconates has been shown to provide an efficacious environmentally sound means of upgrading the performance of inorganic silicate coatings against both high temperature oxidated air attack and that of corrosives such as: salt spray, acid and alkali. Further, these benefits are directly available as a vehicle soluble, colorless, environmentally acceptable additive system (QB) which exceeds the performance of comparable formulations employing like, or somewhat higher, proportions of strontium chromate as an alternative anti-corrosive with much less environmental acceptability. Formulations and procedure specifics with respect to Tables II through VI are available to interested parties.

REFERENCES

1. Monte, S.J. and G. Sugerman, Kenrich Petrochemicals, Inc.: S.M. Gabayson and W.E. Chitwood, General Dynamics, "Enhanced Bonding of Fiber Reinforcements to Thermoset Resins", 33rd International SAMPE Symposium, Anaheim, CA March 7-10, 1988

2. Monte, S.J. and G. Sugerman, "Heavy Metal Free, High Solids Anticorrosive Baked Coatings"

3. Gabayson, S.M., Ph.D., General Dynamics and G. Sugerman, Ph.D. and S.J. Monte, Kenrich Petrochemicals, Inc., "Role of Coupling Agents In Aerospace Composite Failure"

4. Monte, S.J. and G. Sugerman, "Corrosion Resistant One Hundred Percent Solids, Environmentally Sound Coatings", Water-Borne & Higher Solids Coatings Symposium, New Orleans, LA Feb. 3-5, 1988

5. Monte, S.J. and G. Sugerman, Ph.D., "Alkoxy Titanates and Zirconates as Corrosion Inhibitors in Clear Coats and Unfilled Polymers", Corrosion/88, NACE, St. Louis, MO, March 21-25, 1988

6. Sugerman, G., Ph.D. and S.J. Monte, "Very High Solids and Waterborne Anticorrosive Coatings", Western Coatings Societies' 19th Biennial Symposium and Show, Anaheim, CA, March 14-16, 1989

7. Monte, S.J. and G. Sugerman, Ph.D., "The Usage of Organometallic Reagents as Catalysts and Adhesion Promoters in Reinforced Composites", Second International Conference on Composite Interfaces(ICCI-II), Case Western Reserve University, Cleveland, Ohio, June 13-17, 1988

8. Eur. Patent Appl. 270, 271; 08 June 1988; G.S., S.J.M.

9. Ger. Offer DE 3,800,889; 28 July 1988 Kohm T.S. (Kollmorgan Corp)

10. Kryszafkiewicz A.J., "Modification of waste silicas by silanes and titanates, and their use as reinforcing fillers in elastomer composites", Adhes. Sci Tech. 1988 2(3)203-213 (Polytech Inst Pozman Pol.)

11. Application of Titanium Based Coupling Agents To Thermosets; Nabeau T., Netsu Koskasei Jushi 1988 9(3)149-161

12. Polymerization on the Surface of Submacro Particles; Caris C.H.M., Van Herk A., Verfteronieke 1988 61(11) 482-4

13. Jpn. Patent JP 60/34841 A2, Kawasaki Steel Corp, 22 Feb 1985

14. CA 37(9) 2645-59, "Coconut Fiber Reinforced Rubber Composites" Arumugam N. et al, J. Appl. Polymer Science (1989)

15. Vazirani, H.N., "White ink for video jet printing", Bell Labs, SAMPE 1978, 835-44

16. Jpn. Patent JP 58,81467, Nihon Shashin, 16 May 1983 (Insatsu KK)

17. Ger. Offen DE 3,525,910, Girot M.C.J., 23 Jan 1986 (Tioxide Group PLC)

18. CA 101(22):193220G, Lawrence Livermore Natl. Lab, 1984

19. Jpn. Patent JP 58,171,364, Dainippon Printing Co. Ltd., (08 Oct 1983)

20. Jpn. Patent JP 60,249,145, Konishiroku Photo Industry Co., Ltd., (09 Dec 1985)

21. Jpn. Patent JP 59/123863, Cannon K. K., (17 Jul 1984)

22. CA 108/133427q, Adhesion promoters, (At Weapons Res. Establ. Aldermaston, UK) 1987

23. Jpn. Patent JP 60 88,027, Toyobo Co. Ltd., (17 May 1985)

24. Eur. Patent Appl. EP 244,952 (Toray Silicone Co. Ltd) 11 Nov 1987

25. Jpn. Patent JP 010184, Tokyo Shibaura Elec Ltd., (05 08 82)

26. Jpn. Patent JP 59 66,427, Mitsubishi Gas Chemical Co. Inc.,
    (14 Apr 1984)

27. Jpn. Patent JP 59 24,760, Pentel Co. Ltd., (08 Feb 1984)

28. Jpn. Patent JP 62 33,781, Nisshin Steel Co., Ltd., (13 Feb 1987)

29. Monte, S.J. and G. Sugerman, KPI; A. Damusis and P. Patel,
    Polymer Institute, University of Detroit, "Application of
    Titanate Coupling Agents in Mineral and Glass Fiber Filled
    RIM Urethane Systems," SPI Urethane Division 26th Annual
    Technical Conference (November 1981).

30. Jpn. Patent JP 61,106,618, Nippon Soda Co., Ltd. (24 May 1986)

31. FR. Patent FR 2,513,545, Sion, Edouard Jean, (01 Apr 1983)

32. Addison Jr., R.C., M.W. Kendig and S.J. Jeanjaquet, "In-Situ
    Measurement of Cathodic Disbonding of Polybutadiene",
    Rockwell International, CA

33. Jpn. Patent JP 81 99,266, Dainippon Toryo Co., Ltd., (10 Aug 1981)

34. Jpn. Patent JP 80,120,649, Hitachi Chemical Co., Ltd.,(17 Sep 1980)

35. Jpn. Patent JP 63/65087 A2, Mori, Taizo; Tsugawa, Shunic (1988)

36. Jpn. Patent JP 61,108,724, Kawasaki Steel Corp, (27 May 1986)

37. CA 93: 151713F, Functional improvement and economy of water
    durable paint, 1980

38. Jpn. Patent JP 62, 07,784, Nissan Motor Co., Ltd., (14 Jan 1987)

39. Jpn. Patent JP 61,279,687, Nisshin Steel Co., Ltd., (10 Dec 1986)

40. Jpn. Patent JP 61/99680 A2, Nisshin Steel Co., Ltd., (17 May 1986)

41. Jpn. Patent JP 61/99679, Nisshin Steel Co., Ltd., 17 May 1986

42. Jpn. Patent JP 61,231,177, Nisshin Steel Co., Ltd., (15 Oct 1986)

43. Jpn. Patent JP 61/23768 A2, Nisshin Steel Co., Ltd., (1 Feb 1986)

44. Jpn. Patent JP 61/91368 A2, Pentel Co., Ltd., (9 May 1986)

45. Jpn. Patent JP 62 54,772, Shinto Paint Co., Ltd., (10 Mar 1987)

46. Jpn. Patent JP 81,165,393, Sumitomo Electric Ind., Ltd.,
    (18 Dec 1981)

47. Jpn. Patent JP 62,280,271, Sunstar Engineering Co., Ltd.,
    (05 Dec 1987)

48. Jpn. Patent JP 60,231,769, Showa Electric Wire and Cable Co., Ltd.,
    (18 Nov 1985)

49. U.S. 4,382,981, Method for Shielding Electronic Equipment by Coating with Copper Containing Composition, Stoetzer, S.R., Robert E. Wiley, MI

50. Jpn. Patent JP 58,179,642, Matsushita Electric Works, Ltd., (20 Oct 1983)

51. U.S. 4,413,047, Mita Industrial Co., Ltd.

52. Jpn. Patent JP 60 30,200, Mitsui Mining and Smelting Co., Ltd., (15 Feb 1985)

53. Jpn. Patent JP 60 94,419, Mitsui Petrochemicals Ind., Ltd., 27 May 1985

54. Jpn. Patent JP 60,120,767, Toshiba Chemical Corp., (28 Jun 1985)

55. Jpn. Patent JP 60,120,768, Toshiba Chemical Corp., (28 Jun 1985)

56. U.S. US 4,560,716, Toyota Central Research and Development Lab., Inc. (24 Dec 1985)

57. CA 92: 43315y, Aoki, Junjiro, 1979

58. U.S. Patent 4,656,097, Claffey et al., (Apr 7, 1987)

59. U.S. Patent 4,555,450, Goodyear Aerospace Corp., (26 Nov 1985)

60. Span. Patent ES 535,240, Colores Hispania S.A., (16 Jun 1985)

61. Jpn. Patent JP 60 75,366, Dainippon Toryo Co., Ltd., (27 Apr 1985)

62. Jpn. Patent JP 80,152,759, Dainippon Toryo Co., Ltd., (28 Nov 1980)

63. Jpn. Patent JP 60 64,667, Dainippon Toryo Co., Ltd., (13 Apr 1985)

64. U.S. Patent 4,560,716, Shigeyuki Sato, Toyota; Mitsumasa Matsuskita

65. Jpn. Patent JP 80 62,975, Kansai Paint Co., Ltd., (12 May 1980)

66. Jpn. Patent JP 62/27436 A2, Kawasaki Steel Corp, (5 Feb 1987)

67. Jpn. Patent JP 58,152,065, Toyo Rubber Ind., Co., Ltd., (09 Sep 1983)

68. Jpn. Patent JP 61 23,659, Kawakami Paint Mfg., Co., Ltd., (01 Feb 1986)

69. Jpn. Patent JP 62,265,366, Deer Island Ind., Ltd., (18 Nov 1987)

70. Jpn. Patent JP 80 62,977, Daicel Chemical Ind., Ltd., (12 May 1980)

71. U.S. Patent US 4,490,282, Corboy, Thomas A; Joseph Philipson, (18 Feb 1983)

72. U.S. Patent 4,469,637, Trakka Corporation, New Jersey

73. Jpn. Patent JP 60,120,768, Toshiba Chemcial Corp., (28 Jun 1985)

74. Jpn. Patent JP 59/12443 A2, Minolta Camera Co., Ltd., (23 Jan 1984)

75. Jpn. Patent JP 60,143,820, Nippon Soda Co., Ltd., (30 Jul 1985)

76. Ger. Patent DE 3339244 A1, Henkel, (29.10.82)

77. CA 103: 54840n, Inst. Plast, Caucho, Spain, (1985)

78. CA 101: 192914z, Inst. Plast, Caucho, Spain, (1984)

79. UK Patent GB 2,147,592, BASF Wyandotte Corp., (15 May 1985)

80. Jpn. Patent JP 82 55,007, Matsushita Electric Ind., Co., Ltd.,
    (01 Apr 1982)

81. Jpn. Patent JP 60 71,625, Nippon Solda Co., Ltd., (23 Apr 1985)

82. Jpn. Patent JP 57,207,651, Dainichiseika Color and Chemicals
    Mfg. Co., Ltd., (20 Dec 1982)

83. Jpn. Patent JP 57,151,616, Matsushita Electric Works, Ltd,
    (18 Sep 1982)

84. Jpn. Patent JP 62 84,175, Nissan Motor Co., Ltd., (17 Apr 1987)

85. Jpn. Patent JP 61,228,073, Kawasaki Steel Corp., (11 Oct 1986)

86. Jpn. Patent JP 57, 151,617, Matsushita Electric Works, Ltd.,
    (18 Sep 1982)

87. Ger. Patent 2,758,112, BASF Wyandotte Corp., (13 Jul 1978)

88. CA 107: 116324r, Wuhun Ind. Univ., Wuhan, China, (1986)

89. Jpn. Patent JP 60,233,170, Mitsubishi Pencil Co., Ltd.,
    (07 May 1984)

90. Jpn. Patent JP 60,233,171, Mitsubishi Pencil Co., Ltd.,
    (19 Nov 1985)

91. Jpn. Patent JP 60,120,766, Tokyo Printing Ink Mfg., Co., Ltd.,
    (28 Jun 1985)

92. Jpn. Patent JP 62/172372, Minolta Camera Co., Ltd., (29 Jul 1987)

93. Jpn. Patent JP 81,155,240, Matsushita Electric Works, Ltd.,
    (01 Dec 1981)

94. Jpn. Patent JP 59,155,989, Sumitomo Metal Mining Co., Ltd.,
    (25 Feb 1983)

95. U.S. Patent 4,363,887, Seymour of Sycamore, Inc., (14 Dec 1982)

96. Jpn. Patent JP 58 74,764, Pentel Co., Ltd., (06 May 1983)

97. Jpn. Patent JP 60 42,467, Totoku Toryo K.K., 06 Mar 1985)

98. U.S. Navy, Silicone Alkyd Water-Displacing Paint, Hegedus
    Charles R. and Kenneth G. Clark, (Code 302) 30 July 1981

99. CA 103: 7766h, Wartusch, Johann; Kurz, Rolf, (21 Feb 1985)

100. EP 0 028 880 A1, Dow Corning Corporation, Michigan

101. CA 98: 180344v, Inst. Colloid Interf. Sci, Sci. Univ. Tokyo, (1983)

102. CA 104: 169441w, Natl. Ind. Res. Inst., S. Korea, (1984)

103. Jpn. Patent JP 81,147,843, Matsushita Electric Ind., Co., Ltd., (17 Nov 1981)

104. Ger. Patent DE 3,224,258, Matsumoto Seiyaku Kogyo Co., Ltd., (29 Dec 1983)

105. Jpn. Patent JP 82 96,057, Showa Electric Wire and Cable Co., Ltd., (15 Jun 1982)

106. EP 0 104 814 A1, Occidental Chemical Corp., Nigara Falls, NY

107. Jpn. Patent JP 82 14,651, Matsumoto Seiyaku Kogyo Co., Ltd., (25 Jan 1982)

108. U.S. Patent 4,397,983, Aerofoam Ind. Proprietary Ltd., South Africa

109. Varma, D.S., Manika Varma and I.K. Varma, "Coir Fibers: IV - Effect of Isopropyl Triisostearoyl Titanate on Fibers Properties, Oxford & IBH Publishing Co. Pvt. Ltd.

110. Journal of Polymer Material, 101-108 (3), 1986

111. U.S. Patent 4,816,522, Kenrich Petrochemicals, Inc., (28 Mar 1989)

112. Monte, S.J. and G. Sugerman, KPI, "Titanate Coupling Agents - 1985 Urethane Applications," SPI Urethane Division 29th Annual Technical/Marketing Conference (October 1985).

190

STABILIZATION OF EMULSIONS WITH STYRENE-ETHYLENE OXIDE BLOCK

COPOLYMERS

P. Bahadur

Department of Chemistry
South Gujarat University
Surat-395 007, India

Several coatings and printing ink formulations involve
oil-in water emulsions (o/w) preferably stabilized by
polymers used as a binder. The stabilization of o/w
emulsions by two and three block styrene-ethylene oxide
copolymers (PS-PEO and PEO-PS-PEO) were investigated.
It was observed that the stability of emulsions depends
largely on the nature of the dispersed phase, phase volume
ratio, type and molecular characteristics (total molecular
weight and % PEO) of the copolymer. Though, the block
copolymers particularly of high molecular weight were not
effective in reducing the surface/interfacial tension
significantly, they exhibit good emulsifying characteristics
due to steric stabilization. A particular type of emulsion
(o/w or w/o) can be stabilized at a desired percentage of
dispersed phase by properly selecting the copolymer.

INTRODUCTION

Block copolymers exhibit micellar and adsorption phenomena in solution
in analogy to the conventional low molecular weight surfactants. Colloidal
aspects of block copolymer solutions are reviewed by Riess, Bahadur
and Hurtrez[1]. We have previously examined the micellar behavior of
styrene-ethylene oxide block copolymers in water[2-4]. Depending on the
type (two, three and star block) and molecular characteristics (total
molecular weight and % polyethylene oxide), the polymer remains singly
dissolved or aggregates to form micelles which vary in size or even behave
as an insoluble surfactant. The micellar characteristics were found to be
depend largely on the molecular weight and block composition of the
copolymer and the presence of additives like salt, alcohol etc. in water.
This paper describes various surface and bulk properties of both w/o and
o/w emulsions stabilized by styrene-ethylene oxide block copolymers.

EXPERIMENTAL

The styrene-ethylene oxide block copolymers used were synthesised in
the laboratory of Prof. G. Riess in Mulhouse (France). These copolymers

Surface Phenomena and Fine Particles in Water-Based Coatings and Printing Technology
Edited by M.K. Sharma and F.J. Micale, Plenum Press, New York, 1991

191

Table 1. Molecular characteristics of styrene-ethylene oxide and ethylene oxide-styrene-ethylene oxide block copolymers.

Copolymer	% PS	M_n PS	M_n Total
Diblock			
33	10.5	11,800	1,12,000
39	10.8	4,200	38,000
41	15.6	3,700	23,600
Triblock			
42	40.5	2,400	6,000
45	18.4	1,600	8,800
49	21.0	1,800	8,600
51	9.1	1,800	19,800

were characterized for molecular weight and % PEO. The specifications of the block copolymers used are given in Table 1.

The oil-in-water type emulsions were prepared by mixing water (containing styrene-ethylene oxide block copolymer and KCl) and the oil phase, to get the emulsion of the desired phase volume ratio. The final concentration of emulsifier was 1 wt.%. Potassium chloride (0.01 M) was kept constant to maintain the ionic strength.

The emulsions were examined haemocytometrically by monitoring the decrease in droplet number with time. To study the phase inversion of emulsion, viscosity and the dye test methods were used. Surface tensions of aqueous block copolymer solutions were measured. Conductivity meter, viscometer and Tensiometer were used to measure resistance, viscosity and surface tension, respectively.

RESULTS AND DISCUSSION

Using block copolymer as emulgents has several advantages from applied point of view. The hydrophilic-lipophilic balance (HLB) that determines the effectiveness can be optimized by carefully selecting a copolymer. Living anionic polymerization leads to a tailor made block copolymer, the molecular characteristics of which can be changed in different ways. Different types of block copolymers (two-, three-, multi- and star block) can be obtained (Figure 1). Further, a carefully controlled polymerization can yield several variations in molecular characteristics by changing the molecular weight or block composition or both (Figure 2). Several two (PS-PEO) and three (PEO-PS-PEO) block copolymers were used as emulsifiers. The emulsions with high block copolymer concentrations may be beneficial for coating and printing processes.

The surface tension as a function of concentration for the block copolymers in water is shown in figure 3. It is evident from the figure that the block copolymers are less surface active than the conventional nonionic surfactants. The minimum surface tension value was 56.5 dyne/cm. The values of surface tension remain almost constant with the copolymer concentration. Moreover, no inflection could be observed in the plots, signifying very low CMC values for block copolymers in aqueous solutions, which are too difficult to be determined precisely. The small variation in surface tension values with copolymer concentration and very low CMC values can be accounted for by the fact that the hydrophobic moiety of the block copolymers has a high molecular weight and even a trace amount of copolymer is sufficient to saturate the surface, resulting very low CMC values.

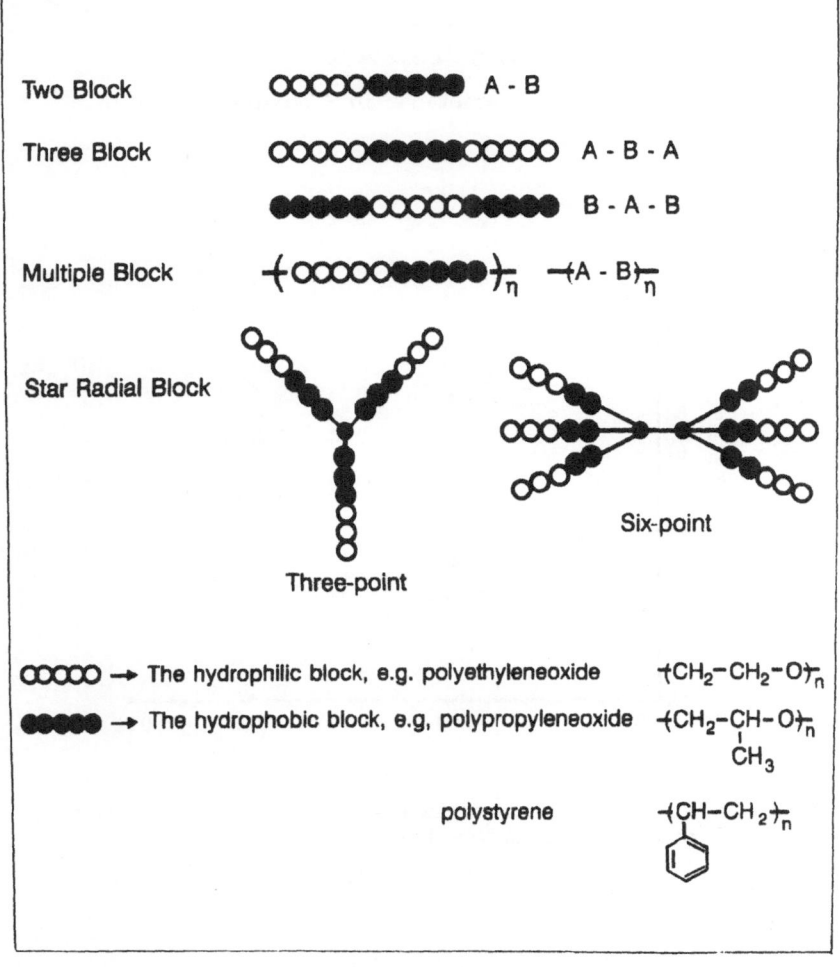

Figure 1. A Schematic Presentation of Different Block Copolymers.

2. Variation in % composition (mol. wt. constant)

3. Variation in % composition of one block
(mol. wt. of other block constant)

Figure 2. Possible Variations in Molecular Structure.

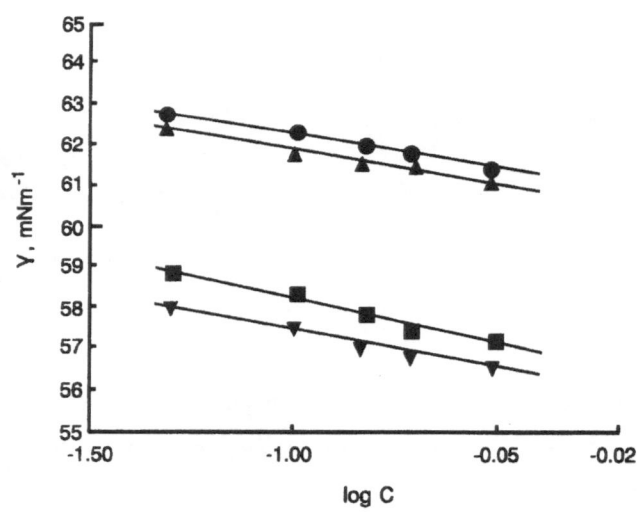

Figure 3. Surface tension versus log concentration for different
copolymers in water at 25°C.
(●) Cop 41; (▲) Cop 39; (■) Cop 45; (▼) Cop 49.

Despite their inefficiencies in reducing surface tension, the styrene–ethylene oxide block copolymers yielded stable emulsions. The stability of emulsions was examined haemocytometrically (particle counting method) by monitoring the change in droplet number with time. Stable benzene-in-water and styrene-in-water emulsions were obtained by emulsifying 5 wt% of nonaqueous phase with aqueous salt (0.01 M) solution containing 1 wt% of the block copolymer. The size of emulsion droplets depends on several factors such as phase volume ratio, concentration and molecular characteristics of the block copolymer used and the homogenization time. Microscopic examination showed the homogenized emulsion, after skimming off the cream, had both individual droplets and flocs. The initial droplet diameter was about 2 micron. However, the total number of dispersed droplets (both individual and flocculated) remains almost the same with time implying that although flocculation occurred in the system at significant level, the emulsions were stable against coalescence. The thick film of the block copolymer is formed by the adsorption of the hydrophobic polystyrene moiety to the nonpolar dispersed oil droplets and anchoring polyethylene oxide chains toward an aqueous phase.

Due to the combined effect of all monomeric units adsorbed, the decrease in free energy upon adsorption per polymer molecule becomes very high. The steric stabilization of emulsions and other colloidal dispersions by polymeric materials, of which block copolymers are most efficient, has been discussed in several monographs[5-8]. Table 2 shows the data on benzene-in-water emulsion stabilized by different block copolymers.

Table 2. Benzene-in-water emulsions (phase volume ratio=0.05) stabilized by 1.0% block copolymer.

Copolymer	Average droplet m	Total droplet number x 10^{-8}	
		at t=o	at t = 2 hrs
33	1.05	1.52	1.38
39	0.90	1.60	1.28
41	0.90	1.48	1.12
42	1.00	1.55	1.17
45	1.05	1.62	1.34
49	1.05	1.72	1.36
51	1.10	1.40	1.10

An interesting observation that could be found was that the boundaries of emulsion formation can be varied. Large amount (even more than 50%) of one liquid could be dispersed into other by carefully selecting the emulsifier. A general feature of emulsion inversion is presented in figure 4. More hydrophobic block copolymer could stabilize water-in-benzene emulsion which contained even upto about 60% of water. The emulsion inverted to benzene-in-water type when water content increased above 60%. Similar phase inversion behavior was observed for benzene (oil-in-water emulsion to water-in-oil emulsion) in the presence of block copolymers which had high content of the hydrophilic PEO moiety (>50%).

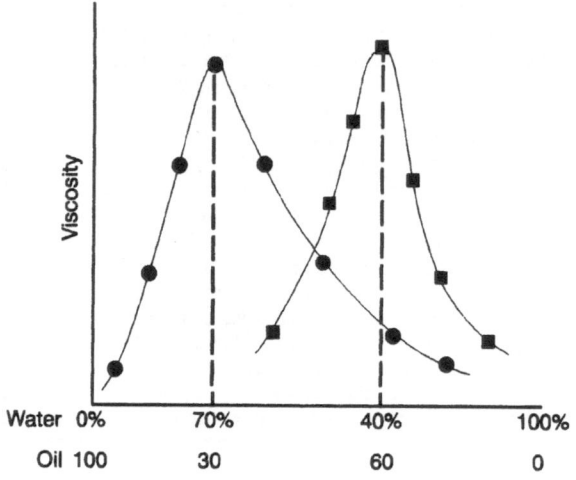

Fig.4. Effect of molecular parameters of block copolymer on
emulsion type.
● high PEO content,
■ high PS content.

ACKNOWLEDGEMENT

The block copolymers samples were gifted by Prof. G. Riese, Mulhouse,
France.

REFERENCES

1. Riess, G., Bahadur, P. and Hurtrez, G., "Block Copolymers" In
    Encyclopedia of polymer Science and Engineering (second Edition),
    Wiley, N. Y.,Vol 2 (1985),PP 324-434.

2. Bahadur, P. and Sastry, N.V. and Riess, G., In Surfactant in Solution,
    vol. 7 (Ed. K.L. Mittal) ,Plenum, N.Y. (1989).

3. Bahadur, P. and Sastry, N.V., Eur. Polym. J., 24(3), 285 (1988).

4. Bahadur, P., Sastry, N.V., Rao. Y.K. and Riess, G., Colloids and
    Surfaces 29, 343 (1988).

5. Sato, T. and Ruch, R., Stabilization of Colloidal Dispersion by
    Polymer Adsorption, Surfactant Science Series (Vol. 9), Marcel
    Dekker Inc., New York (1983).

6. Napper. D.H., Polymer Stabilization of Colloidal Dispersions, Academic
    Press, N.Y. (1982).

7. Tadros, Th.F., The Effect of Polymers on Dispersion Properties,
    Academic Press, N.Y. (1982).

8. Goddard, E.D. and Vincent, B., Polymer Adsorption and Dispersion
    Stability, Royal Soc.Chem., London (1984).

PROBING EXHUDATION OF SURFACTANT MOLECULES IN

LATICES BY SURFACE FT-IR SPECTROSCOPY

Kevin W. Evanson and Marek W. Urban[*]

Department of Polymers and Coatings
NORTH DAKOTA STATE UNIVERSITY
Fargo, ND 58105

This chapter covers some of the recent
advances and applications of the surface Fourier
transform infrared (FT-IR) spectroscopy in
characterizing latex systems. A particular
emphasis is given to the recent studies on the
behavior of surfactants in latex films as a result
from the externally applied stresses and surface
tension differences. Photoacoustic (PA) and
attenuated total reflectance (ATR) Fourier
transform infrared (FT-IR) spectroscopy are used
to monitor and characterize sodium
dioctyl-sulfosuccinate/ ethyl acrylate/methacrylic
acid (SDOSS/EA/MAA) copolymer latex films at both
the film-air and film-substrate interfaces. The
distribution of surfactant at the interfaces
depends upon the water flux out of the film in the
early stages of film formation and the surface
tension of a substrate. Mechanical elongation of
the latex films also influences the surfactant
distribution across the film. To our best
knowledge, this is the first spectroscopic study
illustrating the factors affecting surfactant
exhudation from the latex films.

INTRODUCTION

An intentional addition of small molecules to polymeric
systems has been recognized many years ago as a means to
improve such properties as processibility of polymeric
materials. For example, plasticizers are often introduced to
a polymer matrix in an effort to lower the temperature of
processibility by effectively lowering overall T_g of the
system. These monomeric species, however, have a significant
effect on adhesion to plastic substrates since plasticizers
tend to migrate to the film-air interface and thus may
diminish adhesion. While the use of plasticizers in

---

[*] - the author to whom all correspondence should be addressed.

Surface Phenomena and Fine Particles in Water-Based Coatings and Printing Technology
Edited by M.K. Sharma and F.J. Micale, Plenum Press, New York, 1991

197

polymeric films may be avoided, in the case of latex films, the presence of surfactants during the latex preparation is an absolute necessity governed by the formation of surfactant micelles responsible for the latex particle formation. It is, therefore, appropriate to address the question as to how the presence of these monomeric species may influence the latex film properties? This issue is of a particular importance in the systems in which surfactants exhibit lack of compatibility with the polymer network. Of course, it is also important to remember that starting with a compatible system does not guarantee that the surfactant retains its chemical composition since it may undergo hydrolysis and other chemical changes during latex preparation resulting in incompatibility with a polymer system. As a result, one may expect phase separation, diffusion and increased mobility, as these small molecules are free to move within the network. Thus, the behavior of small molecules added to various polymeric systems is one of the critical factors determining many properties of coatings. One can speculate, for example, that if the surfactant is capable of migrating in some preferential direction across the film, the direction of its propagation may affect such film properties as adhesion or durability. This is because many surfactants are water soluble and if their concentration at the film-air interface is increased, it may increase free volume allowing penetration of water and other species into the film. On the other hand, the increased concentration of surfactant at the film-substrate interface will most likely effect adhesion. Although some of these issues had been recognized earlier with the pioneering work at Lehigh University, we would like to further understand the origin of forces governing the behavior of additives in latex films.

Although polymer latices and their technology have been established many years ago, an extensive effort is given to develop novel modes of preparation of latices, techniques for cleaning and methods for determining their surface or bulk characteristics. A particular interest has been given to the potential applications of polymer latices as model colloids.[1] In spite of the fact that typical formulations of styrene consisting of monomer, water, water soluble initiator and emulsifier result in a polydisperse latex,[2] several approaches including seed growth[2], controlled emulsifier concentrations[3,4] and combinations of ionic and non-ionic surfactants[5] have been employed to achieve monodispersity. Such monodisperse latices can be made reproducibly, but the detailed surface characterization remains[6] to be an elusive goal. Following the definition of La Mer[6], latices are considered to be "monodispersive" when the variation of the mean diameter is less than 10%. The techniques which may be employed to determine average particle diameters include electron microscopy[7,8], light scattering[9,10,11], ultra-centrifugation[10], small angle X-ray scattering[12] soap titration[13] and photon correlation spectroscopy[14]. Transmission electron microscopy is the technique most frequently utilized since it allows a visual assessment of the size, shape and the size distribution of particles. The capability of visual assessments has enabled the observation of interparticle bridge formation[15], anomalous particle morphology[16] and the presence of a

198

fraction of small particles resulting from secondary nucleation[17].

While monodispersity and the particle size diameter are essential factors determining film properties, it is also important to establish interactions between individual latex components. In this case, spectroscopic techniques must be employed since information on the molecular level is sought.

Among spectroscopic techniques, Fourier transform infrared spectroscopy (FT-IR) has found only limited applications to the latex systems. This was mainly attributed to the strong absorbance of water in the infrared region. Only recently, Zhao et. al.[18] have demonstrated the kinetics of surfactant exhudation in latex films containing anionic surfactants by ATR FT-IR. Surfactant enrichment at both the film-air and film-substrate interface was observed. The enrichment was dependant upon the nature of the surfactant, the interface involved, the global concentration of surfactant and the coalescence time. The concentration of surfactant at the film-air interface was higher than at the substrate interface and the distribution was mostly established after three hours. However, analysis of surfactant exhudation in the early stages of coalescence as well as interactions between polymer and other components of the latex system have not been reported. Moreover, the effect of the substrate surface tension as well as the effect of external forces applied to the films have not been addressed in the past.

Since the properties of latex films may be inherently influenced by the surface activity of surfactants, determination of the distribution of surfactant throughout the latex film is essential and, as mentioned earlier, plays a significant role in the film properties such as adhesion and film durability. Therefore, the latex composition at the film-substrate and the film-air interfaces is of a particular interest. These reasons motivated us to further investigate the distribution of surfactants throughout latex films during film formation.

The present study is concerned with a SDOSS/EA/MAA (sodium dioctyl sulfo-succinate/ethyl acrylate/methacrylic acid) latex system in which the polarity of the system was modified by the addition of a small fraction of acid monomer[19]. In most latices, a small percentage of acid monomer is incorporated to improve adhesion, increase stability of the latex particles, control viscosity and provide crosslinking sites for interparticle thermosetting reactions. ATR, Circle ATR and photoacoustic FT-IR techniques will be employed to the analysis of latex films at the film-substrate and film-air interfaces. Circle ATR was originally developed for the analysis of aqueous solutions which are difficult to analyze due to strong water absorbance in transmission measurements[20]. Using Circle ATR, the distribution of surfactant at the substrate interface can be monitored immediately after application and throughout the early stages of film formation. Hence, it is well suited for the analysis of film composition near the substrate with significantly improved signal-to-noise ratio over that of the traditional ATR technique[21]. Photoacoustic

FT-IR allows[22,23,24,25,26] the detection of the air-film interface during the early stages of film formation and thus both techniques compliment each other.

EXPERIMENTAL

Latex Preparation

Ethyl acrylate and methacrylic acid monomers (Aldrich Chemical Co.) were copolymerized by a semicontinuous emulsion polymerization process in a glass four neck break-away reaction vessel equipped with a mechanical stirrer, addition funnel, thermometer, a condenser, and a nitrogen inlet tube. A typical recipe based upon total formulation weight included: DDI water, 53.4%; ethyl acrylate, 42.1% (95.8% based upon total monomer wt.); methacrylic acid, 1.9% (4.2% based upon total monomer wt.); sodium dioctyl sulfo-succinate (SDOSS, Aerosol OT; American Cyanamide), 2.3% (4.0% based on total monomer wt.); and ammonium persulfate (Aldrich Chemical Co.), 0.26% (0.60% based on total monomer wt.). The monomers were mixed together thoroughly and placed in the addition funnel. The reaction vessel was purged for 20 minutes with nitrogen then all the water, SDOSS, and initiator were added and allowed to stir for 5 minutes while heating. At 40°C, 20% of the monomer mixture was slowly added with stirring over a 5 minute period. At this point the temperature was raised to 70°C and held constant until a white milky emulsion was observed, followed by slow monomer addition over a 3 hour period. The temperature was then raised to 85°C and held for 3 hours or until no monomer odor could be detected. The final product was 46% solids. The preparation procedure of the pure copolymer was reported in the literature.[27]

Film Preparation and Properties

All films were prepared to maintain the film thickness between 75 and 100 μm. Upon deposition on a substrate, the films were all air dried for 72 hours (unless otherwise specified) at 23°C. For the PAS measurements, the films were prepared by direct deposition of a pre-determined amount of latex on an aluminum photoacoustic sample cup. A typical thickness of dry films was approximately 100 μm. The films studied by Circle ATR were applied directly to the ATR crystal to yield films approximately 75 μm thick. The films prepared on other substrates such as polytetrafluoroethylene (PTFE) and liquid mercury were prepared in a similar manner.

The glass transition temperature of the latex films (-5°C) was determined from the average of three runs on a DuPont Instruments 910 differential scanning calorimeter equipped with a DuPont Thermal Analyzer 2000. The temperature range was from -50 to 250°C at a heating rate of 20°C/min.

Spectral Measurements

Transmission, attenuated total reflectance (ATR), Circle ATR, and photoacoustic (PA) FT-IR techniques were used to monitor chemical composition of the latex film.

Transmission, Circle ATR, and rectangular ATR FT-IR spectra were collected on a Mattson Cygnus 25 instrument equipped with a single beam spectrometer (Sirius 100). In a typical experiment, 200 scans at a resolution of 4 cm^{-1} were collected. Circle ATR spectra were obtained using a cylindrical internal reflection accessory with a ZnSe circular crystal (Spectra Tech Inc.). In order to obtain the film-substrate interface spectra, the latex film was deposited directly on the crystal and allowed to coalesce. Figure 1 schematically depicts the cylindrical ATR attachment. The rectangular ATR attachment (Mattson Instruments) was equipped with a KRS-5 crystal aligned to give an incident beam angle of 45 degrees.

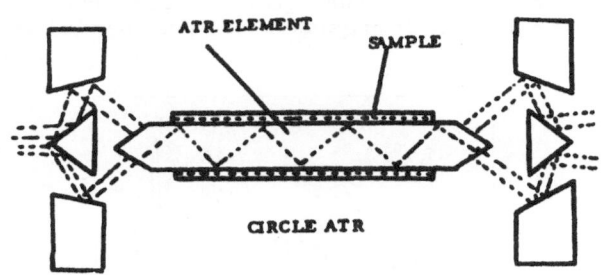

Figure 1. A schematic diagram of the Circle ATR setup: A combination of mirrors bring the energy of the FT-IR spectrometer into focus at an average angle of 45° on the surface of ATR element. The light propagates through the crystal and exits through the other end, where it is redirected by another set of mirrors toward the IR detector. Circle ATR for aqueous solutions would contain the solution in a boat surrounding the element. For the latex studies, one layer of film was deposited on the ATR element, as shown by the area marked sample. Other applications of this technique can be found in ref. ([28]).

Photoacoustic FT-IR spectra were recorded on a Digilab FTS-10M equipped with a photoacoustic cell (Digilab). 400 single beam spectra were signal averaged at a resolution of 4 cm^{-1} and ratioed against a carbon black reference. Each sample was purged with helium for 5 minutes prior to the spectral collection. Both FT-IR instruments were purged continuously with purified air (Balston filter system).

Spectral Analysis

All spectra were transferred to an AT compatible computer for further spectral analysis utilizing Spectra Calc software (Galactic Ind.).

RESULTS AND DISCUSSION

Before we begin the analysis of the surface infrared spectra of ethyl acrylate-methacrylic acid copolymer latex, let us set the stage by defining the relevant features in the infrared spectra of all individual components of the

latex. Figure 2 illustrates photoaccoustic FT-IR spectra of
sodium dioctyl sulfo-succinate (trace A), ethyl
acrylate/methacrylic acid copolymer only (trace B), and the
corresponding ethyl acrylate-methacrylic acid copolymer
latex (trace C). While all three spectra exhibit a strong
carbonyl band due to alkyl ester groups centered at 1735
cm^{-1}, the shoulder at 1703 cm^{-1} appears only in the spectra
of the solution copolymer acrylic (trace B) and the latex
(trace C). The 1703 cm^{-1} band is characteristic of
carboxylic acid dimers which have a tendency to form
intermolecular hydrogen bonding.[29] The presence of SO_3^-
groups in the surfactant is manifested by the strong band at
1050 cm^{-1} (trace A), attributed to the symmetric[30] S-O
stretching vibrations of the sodium sufonate groups. The
asymmetric S-O stretching mode appears at 1210 cm^{-1}. A
closer examination of the latex spectrum (trace C)
indicates, however, the presence of two bands at 1056 and
1046 cm^{-1}.

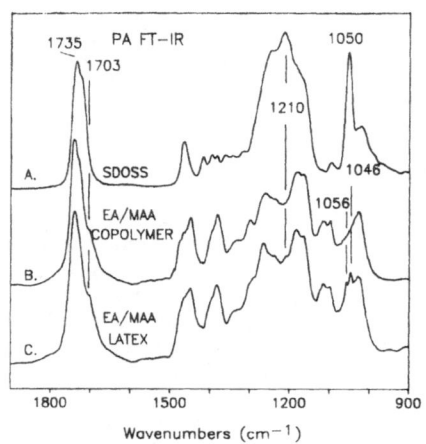

Figure 2. PA FT-IR spectra in the 1900 to 900 cm^{-1} region: A
- SDOSS (surfactant only); B - EA/MAA copolymer; C - EA/MAA
latex.

     Postponing temporarily the evaluation of the
differences between these three spectra until later on, let
us focus on the C-H stretching region of the surfactant
(trace A), EA/MAA copolymer only (trace B), and the latex
(trace C). They are shown in Figure 3. The spectra of the
latex and the copolymer exhibit the band at 2981 cm^{-1},
attributed to the asymmetric C-H stretching normal
vibrations of the CH_3 units. The bands at 2960 and 2934 cm^{-1}
are due to the asymmetric and symmetric stretching
vibrations of the CH_2 groups, respectively. Relative to the
spectrum of the EA/MAA copolymer only (trace B), the
spectrum of the latex (trace C) indicates a marked increase
of the intensity of the bands at 2960 and 2934 cm^{-1}. Since
photoacoustic FT-IR spectroscopy is a surface sensitive
technique, and these bands are attributed to the C-H normal
vibrational modes of the surfactant, these observations
suggest a greater concentration of the surfactant (SDOSS) on
the latex surface. This in turn, would indicate that the
surfactant exhudates to the film-air interface as a result
of coalescence of the latex film.

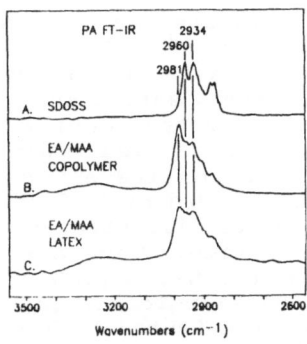

Figure 3. PA FT-IR spectra in the 3550 to 2550 cm^{-1} region: A - SDOSS; B - EA/MAA copolymer; C - EA/MAA latex.

In an effort to establish if indeed the exhudation process occurs, the latex film infrared spectra at the film-air and film-substrate interfaces were recorded using Circle ATR FT-IR spectroscopy. The experimental section provides all the necessary details relevant to the use of Circle ATR. Figure 4 illustrates a series of ATR FT-IR spectra recorded at the film-air (trace A) and film-substrate (trace B) interfaces. Similarly to the PA FT-IR results, the spectrum of the film-air interface (trace A) indicates the presence of the bands at 1056 and 1046 cm^{-1}. Changing the configuration of the sample with respect to the Circle ATR crystal allows us to detect the film-substrate interface. Trace B of Figure 3 illustrates the results of this analysis. It appears that both the 1056 and 1046 cm^{-1} bands are not observed indicating that during coalescence the surfactant exhudates to the film-air interface. In order to further substantiate the observed phenomenon, and taking advantage of the fact that the surfactant is water soluble, the film-air interface was washed with DDI water-MeOH (80:20-v/v). Such a surface modified sample was analyzed at the film-air interface and the resulting spectrum is shown in Figure 4, trace C. It is apparent that the bands at 1056 and 1046 cm^{-1} are absent, such as that seen in spectrum for the film-substrate interface (trace B). These observations again indicate a greater concentration of surfactant at the film-air interface.

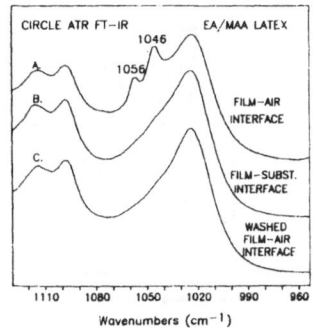

Figure 4. Circle ATR FT-IR spectra of EA/MAA latex in the 1130 to 950 cm^{-1} region: A - film-air interface; B - film-substrate interface; C - film-air interface washed with MeOH/DDI H$_2$O solution.

While the above studies were conducted on fully coalesced film, it is also of interest to establish the stage of coalescence at which the two bands begin to appear. Since exhudation of the surfactant away from the film-substrate interface occurs during the latex film formation, the process can be also monitored using Circle ATR by depositing wet latex on the ATR cylindrical element (shown in Figure 1) and recording infrared spectra during its coalescence. Figure 5a illustrates infrared spectra recorded as a function of time. Trace A exhibits a shoulder at 1046 cm^{-1} assigned to the S-O asymmetric stretching mode of the SO$_3^-$ groups. As the film formation proceeds, the 1046 cm^{-1} band diminishes as illustrated by traces A (5 min.), B (30 min.), and C (4 hours). After 4 hours, the film is completely transparent, and the band at 1046 cm^{-1} is no longer present. While the concentration gradient of the

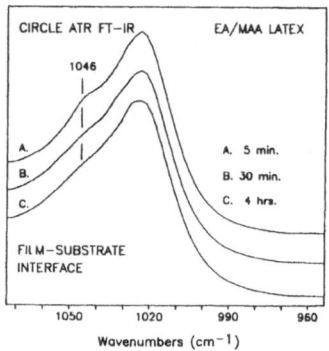

Figure 5a. Circle ATR FT-IR spectra in the 1070 to 960 cm^{-1} region collected as a function of time for the EA/MAA latex as it coalesces: A - 5 min.; B - 30 min.; C - 4 hrs.

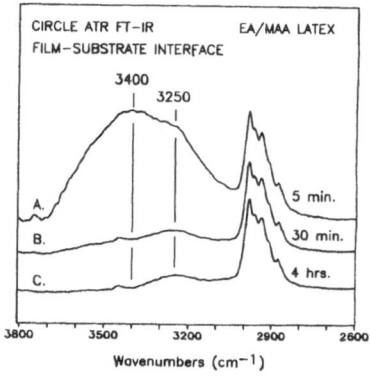

Figure 5b. Circle ATR FT-IR spectra in the 3800 to 2600 cm^{-1} region collected as a function of time for the EA/MAA latex as it coalesces: A - 5 min.; B - 30 min.; C - 4 hrs.

surfactant across the latex film has been reported earlier[18], it is apparent that during the early stages only one band at 1046 cm^{-1} is detected, whereas the film-air interface exhibits an additional band at 1056 cm^{-1}. Thus, the observed phenomenon is the interface specific.

Recently, Vanderhoff[31] reported that surfactants may become insoluble in the polymer film and are then forced to the film-air interface due to incompatibility. Moreover, as a result of the surface tension and surface tension differentials, surfactants may accumulate at the interface in order to lower the high surface tension present between the polymer and the air. Such water soluble surfactants[18] may be carried to the film-air interface during coalescence by the water flux that passes between latex particles during film formation. As seen in Figure 5a, this effect is most pronounced in the very early stages of film formation when the concentration of water is the highest. As the water diffuses out of the film and away from the substrate, one would expect to observe the bands due to surfactant diminish. Additionally, spectra collected at the film-air interface for the case where a second layer of latex was prepared over a pre-existing latex film resulted in very pronounced increases in the bands due to surfactant (not shown), indicating that the excess surfactant may quite easily move with the water flux through the film.

The transient intensity changes of the 1046 cm^{-1} band are associated with and parallel the disappearance of a broad band around 3400 cm^{-1} attributed to water evaporating from the film. This is demonstrated in Figure 5b. Because the surfactant is water soluble, during the initial stages of film formation, the concentration of the surfactant diminishes at the film-substrate interface as water evaporates from the system. It should also be noted that the spectrum (not shown) obtained by subtracting out contributions of water from a 0.05M aqueous solution of SDOSS reveals the spectrum of SDOSS with a strong band centered at 1046 cm^{-1}. This observation may indicate that the band at 1046 cm^{-1} of the latex spectrum (Figure 2, trace C) is a result of hydration of the sulfonate group of the surfactant since these groups tend to be highly hydrophilic.

In an effort to understand the origin of the exhudation process, we will focus on the two bands at 1056 and 1046 cm^{-1} (Figs. 2 and 4). It is, therefore, appropriate to address the question as to why the two bands are observed only for the surface-air interface. Both bands originate from the asymmetric S-O stretching vibrations of the SO_3^- groups, but at this point their origin is uncertain and may account for surfactant-copolymer or surfactant-initiator interactions followed by the formation of the ordered surfactant surface layers. Since the copolymer contains carboxylic acid groups, hydrogen bonding between the carboxylic acid groups and the sulfonate groups or ionic interactions could be likely responsible for the occurrence of the two infrared bands in this region. The nature of these interactions and their origin will be addressed in the upcoming publication.[32] Here, the primary focus will be on the factors affecting surfactant exhudation.

During the course of this study it was noted that both the surface energy of the substrate and mechanical stretching of the latex film may affect the intensity of the bands at 1056 and 1046 cm^{-1}. Therefore, it is appropriate to address the question as to the effect of surface tension of the substrate and the effect of film removal from the substrate on interactions between latex components. In an effort to gain further knowledge about these effects, a rectangular ATR cell was used as the most convenient means of analysis. Figure 6, trace A, illustrates the film-air interface spectrum, whereas trace B is the film-substrate interface obtained for a 75 μm film cast on a poly(tetrafluoroethylene) (PTFE) substrate and air dried at room temperature for 72 hours.

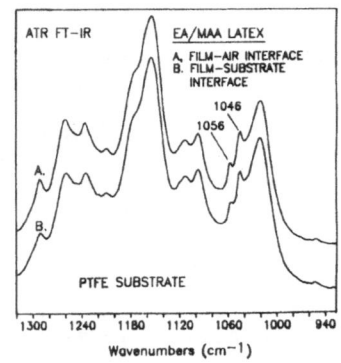

Figure 6. Rectangular ATR FT-IR spectra in the 1320 to 940 cm^{-1} region of EA/MAA latex prepared on a PTFE substrate: A – film-air interface; B – film-substrate interface.

Similarly to the Circle ATR measurements, the spectrum of the film-air interface shows the presence of two bands at 1056 and 1046 cm^{-1}. Surprisingly however, the spectrum of the film-substrate interface, (trace B), reveals the presence of the bands at 1056 and 1046 cm^{-1}, not present in the spectrum recorded from the film-substrate interface by cylindrical ATR. The only difference between the two measurements is that in order to obtain the film-substrate spectrum, the sample must be removed from the PTFE. Thus, it was our hypothesis that the presence of surfactant at the film-substrate interface detected in the rectangular ATR mode of detection may result from the increased surface tension brought on by mechanical stretching of the film during film removal from the substrate.

In order to verify this hypothesis, the latex films were prepared in the same manner as before. The excess of surfactant was removed from the film-substrate and the film-air interfaces by washing it with a DDI/MeOH solution after which the films were allowed to air dry for 5 minutes. Next, the films were elongated (20%) in a manner which closely approximated that encountered during the initial removal from the PTFE substrate. Figure 7 illustrates the spectral changes occurring at the film-substrate interface recorded as a function of time. The initial spectrum trace A, recorded 5 minutes after washing and prior to elongation does not exhibit the bands at 1056 and 1046 cm^{-1}. However,

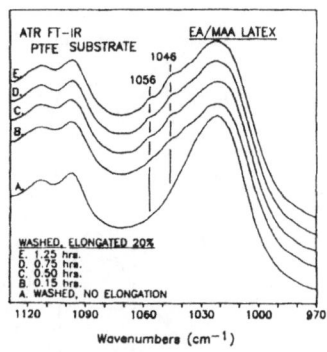

Figure 7. Rectangular ATR FT-IR spectra in the 1320 - 940 cm^{-1} region of EA/MAA latex prepared on a PTFE substrate (the sample was washed with DDI/MeOH solution and elongated 20%; A - no elongation; B - 0.15 hrs.; C - 0.50 hrs.; D - 0.75 hrs.; E - 1.25 hrs.

after elongation (traces B-E), the presence of the 1056 and 1046 cm^{-1} bands becomes more pronounced with time. These results indicate that mechanical stretching of the film influences the distribution of surfactant throughout the latex film. By stretching the film, the surface area of the latex film is increased, which results in a lower concentration of surfactant at the surface and increased surface tension. In order to lower this increased surface tension, more surfactant migrates to the interface. This process is schematically depicted in Figure 8, the ramifications of which may be important to adhesion for cases where latex films are used to protect substrates subject to expansion and contraction. Another possible reason for the increased concentration of surfactant at the film-substrate interface is the high surface energy of water (72 mN/m) which is lowered by the surfactant in order to obtain a continuous latex film (the surface energy of PTFE is 18.5 mN m^{-1}, 20°C)[33].

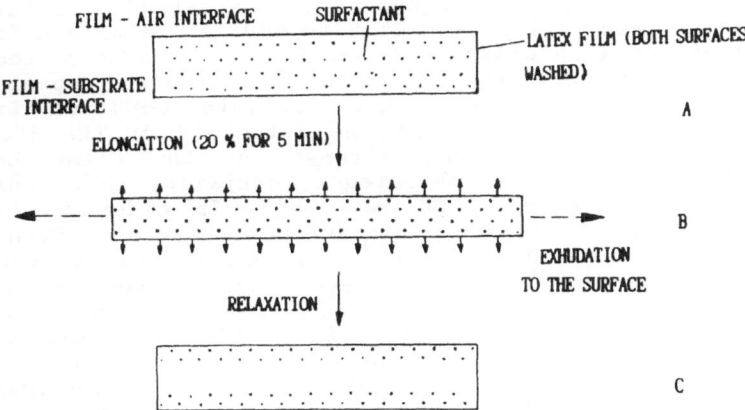

Figure 8. Proposed mechanism of exhudation induced by elongation.

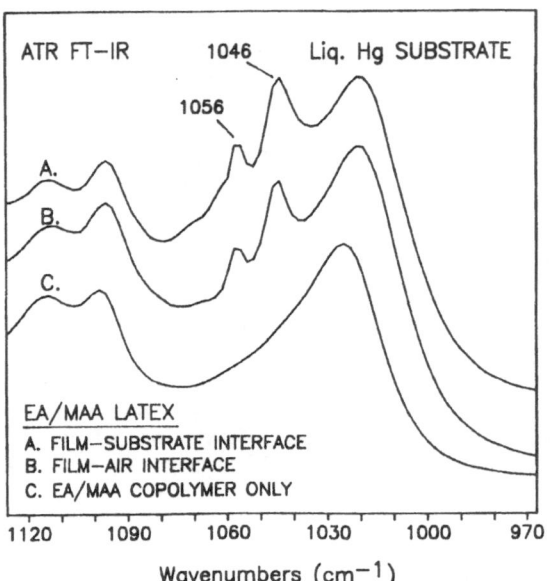

Figure 9. Rectangular ATR FT-IR spectra in the 1120 to 970 cm^{-1} region of EA/MAA latex prepared on liquid mercury: A - film-substrate interface; B - film-air interface; C - EA/MAA copolymer only.

While elongation experiments reveal the nature of surfactant exhudation upon stretching (Fig.8), at this point it is appropriate to address the issue of the effect of substrate surface tension on the surfactant distribution across the latex films. For that reason the latex films were prepared on liquid mercury. This enables the easy removal of the films from the substrate with minimal mechanical disturbance to the film. Furthermore, such procedure should provide further information as to the effect of the substrate surface tension (Hg, 415 mN/m; 20°C)[34] along with the ability of monitoring the film-substrate and the film-air interfaces. Figure 9 illustrates a series of spectra in the region from 1130 to 970 cm^{-1} obtained for the films prepared on a layer of Hg. Traces A and B reflect the film-substrate and the film-air interfaces, respectively, while trace C is the spectrum of the latex copolymer free of surfactant. It appears that the intensity of the 1056 and 1046 cm^{-1} bands are much stronger at the film-substrate interface which is demonstrated by comparing their relative intensities with the 1025 cm^{-1} band. In order to further substantiate the effect of surface tension, Figure 10 illustrates the 820 - 480 cm^{-1} spectral region for the same samples. The spectra of the latex film at both interfaces reveals several bands at 677, 652, 598, 581, 535, and 500 cm^{-1} not present in the spectra of the copolymer only, trace C. Although the origin of these changes in still under investigation, similarly to the bands at 1056 and 1046 cm^{-1} (Figure 9, A and B), the intensities of all bands in the spectrum of the film-substrate interface are stronger, indicating a greater concentration of surfactant or its

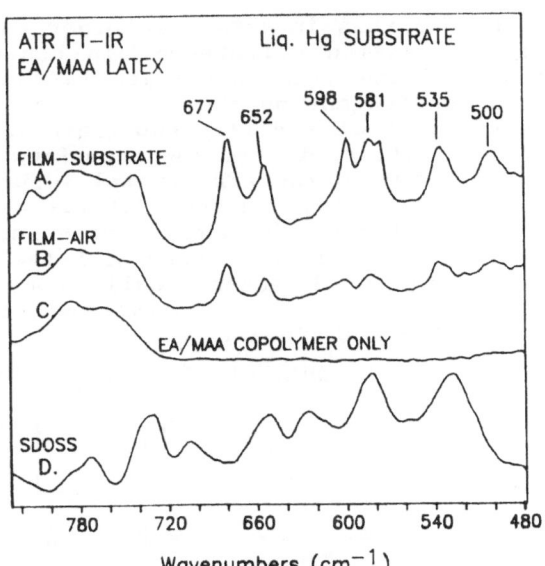

Figure 10. Rectangular ATR FT-IR in the 820 to 480 cm^{-1} region of EA/MAA latex prepared on liquid Hg: A - film-substrate interface; B - film-air interface; C - EA/MAA copolymer only; D - SDOSS only.

derivative at the film-substrate interface. It should be noted that these bands do not directly correspond with those detected in the surfactant spectrum (trace D). They may, however, arise from interactions with the copolymer or from surface phenomena that results in orientation effects on the surface.

Although these results demonstrate the complexity of the surfactant interactions with other latex components, it is apparent that the exhudation of surfactant is a function of the water flux out of the film as well as the substrate surface tension. In a view of the above considerations, it is important to reveal why no surfactant was found at the film-substrate interface with Circle ATR whereas it was present for the films prepared on PTFE and mercury.

Using a simplistic view it is known that for two different surfaces which come into contact with each other, the surface with the lower surface tension will wet the one of higher surface tension.[33] The films prepared on the PTFE substrate with a low surface tension (18.5 mN/m), exhibit an excess of surfactant at the film-substrate interface. Most polymeric materials have surface tension in the range of 25-60 mN/m. Hence, they do not wet the PTFE surface. In order to lower the interfacial-surface tension difference, the surfactant is forced to the film-substrate interface. In the case of the films prepared on liquid mercury, the initial interfacial boundary is between two liquids. Due to the high surface tension of mercury (416 mN/m), the mercury is wetted by the latex. However, as coalescence proceeds and water leaves the system, a solid polymeric film is formed which now acts as the substrate with a much lower surface tension compared to the liquid mercury. Therefore, the surfactant migrates to this interface in order to lower the interfacial surface tension difference. In contrast, when latex is deposited on the glass substrate (ATR crystal), a greater content of surfactant on the latex film surface and no detectable surfactant at the film-substrate interface is observed. This is attributed to the lower surface tension of the polymer with respect to the ATR crystal (appr. 70 mN/m) which is able to wet the glass surface. Therefore, there is no driving force to maintain the water soluble surfactant at the interface and it thus exhudates away with the water flux out of the film. Figure 11 schematically summarizes the described phenomena and arrows point a preferential direction of the surfactant exhudation.

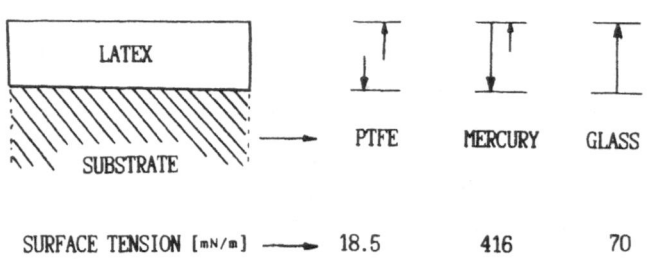

Figure 11. A direction of the surfactant exhudation in the latex film as a function of substrate surface tension.

It is now possible to provide a better picture of the coalescence process and the factors affecting it. During the initial stages of coalescence, water evaporates from the film. As a result, the surfactant, as being incompatible with the copolymer, is carried out of the film by the water flux. This is evidenced by the results obtained using the Circle ATR cell. As the film coalesces, an excess of surfactant is detected at the film-air interface in all the latex films in this study. However, the direction of surfactant exhudation may change when there is a large surface tension difference between the substrate and the latex film. This process is governed by interfacial surface tension differentials and is clearly observed in the case of films prepared on mercury where a higher concentration of surfactant at the film-substrate interface is detected.

The elongation studies on the latex films reveal two important points. First, surfactant exists not only at the film interfaces, but also throughout the continuous phase within the latex network. Secondly, the surfactant within the continuous phase may be forced to exhudate to either interface by mechanical extension of the film which results in increased surface area and thus increased surface tension.

CONCLUSIONS

These recently initiated studies demonstrate the utility of FT-IR spectroscopy for the analysis of the film-air and the film-substrate interfaces along with the distribution of surfactant throughout latex films. The results indicate that surfactant may exhudate to both the film-air and film-substrate interfaces during film formation and depends upon the film-substrate surface tension difference. This process competes with the water flux which evaporates from the film during its coalescence. Although it is quite obvious that more research is needed in order to establish the interactions between various latex components, this study demonstrates the capability of surface FT-IR spectroscopy as a useful tool for establishing the molecular level interactions in multicomponent systems and the factors affecting diffusion of small molecules in water-borne systems.

REFERENCES

1. Vanderhoff,J.W., ACS Div. Org. Coatings and Plastics, Preprints, 1964, 24, 223.

2. Wilson, E.A., Miller,J.R., and Rowe,E.H., J.Phys.Coll. Chem., 1969, 53, 357.

3. Vanderhoff,B.M.E., J.Polym.Sci., 1958, 33, 487.

4. Roe,C.P. and Brass,P.D., J.Polym.Sci., 1953, 24, 401.

5. Woods,M.E., Dodge,J.S., Krieger,I.M. and Pierce,P.E., J.Paint Tech., 1968, 40, 541.

6. Wachtel,R.E. and La Mer,V.K., J.Coll.Inter.Sci., 1962, 17, 531.

7. Bradford,E.B., Vanderhoff,J.W., J.Appl.Phys., 1955, 36, 864.

8. Davidson,J.A., Haller,H.S., J.Coll.Inter.Sci., 1974, 47, 459.

9. Napper,D.H., Ottewill,R.H., J.Coll.Inter.Sci., 1964, 19, 72.

10. Van der Hul,H.J. and Vanderhoff,J.W., Polymer Colloids, Ed.R.M.Fitch, Plenum, 1971.

11. Wims,A.M., Meyers,M.E., J.Coll.Inter.Sci., 1972, 39, 447.

12. Henke,B.L., du Mond,J.W.M., J.Appl.Phys., 1955, 26, 903.

13. Mason,S.H., Elder,M.E., Ulevitch,I.N., J.Coll.Inter.Sci., 1954, 9, 83.

14. Munro,D., Randle,K., Photo Correlation Spectroscopy and Velocimetry, NATO.Adv.Study Inst.,Series B, Plenum, N.Y., 1977, 23, 537.

15. Wilkinson,M.C., Ellis,R. and Callaway,R. The Microscope, 1974, 22(3), 229.

16. Goodall,A.R., Wilkinson,M.C., Hearn,J., J.Coll.Inter.Sci., 1975, 53, No.2.

17. Hearn,J., Ph.D. Thesis, Univ.Bristol, 1971.

18. Zhao,C.L., Holl,Y., Pith,T. and Lambla,M., Coll.Polym.Sci., 1987, 265, 823-829.

19. Bassett,D.R., Derderian,E.J., Johnston,J.E., MacRury,T.B., ACS Symp.Ser., "Emulsion Polymers and Emulsion Polym." No.165, (1980), 263-278.

20. Bartick,E.G., Messerschmidt,R.G., Amer.Lab.,Nov.,1984.

21. Tiefenthaler,A.M., Urban,M.W., Appl.Spectr., 1988, 42(1), 163.

22. Urban,M.W., Salazar-Rojas,E.M., Macromolecules, 1988, 21, 372.

23. Urban,M.W., Salazar-Rojas,E.M., J.Polym.Sci., Chem.Ed. in press, 1990.

24. Salazar-Rojas,E.M., Urban,M.W., Progr.Org.Coatings, 1989, 16, 4, 371-86.

25. Urban,M.W., J.Coat.Technol., 1987, 59(745), 29 and ref. therein.

26. Urban,M.W., Progr.Org.Coatings., 1989, 16, 4, 321-53.

27. Zhao,C.L., Holl,Y., Pith,T., Lambla,M., Br.Polym.J., 1989, 21, 155-160.

28. Tiefenthaler,A.M. and Urban,M.W., Appl.Spectrosc., 1988, 42(1), 163.

29. Lee,Y.J., Painter,P.C., Coleman,M.M., Macromolecules, 1988, 21, 346-354.

30. Socrates,G., Infrared Characteristic Group Frequencies, John Wiley & Sons, N.Y., 1980.

31. Vanderhoff,J.W., Br.Polym.J., 1970, 2, 161-173.

32. Evanson,K.W and Urban,M.W., work in progess.

33. Wake,W.C., Adhesion and the Formulation of Adhesives, 2nd ed., 1982, Applied Science Pub., Ltd., New York.

34. Adamson,A.W., Physical Chemistry of Surfaces, 4th ed., 1982, Wiley Interscience Pub., New York, NY.

HIGH RESOLUTION PARTICLE SIZE ANALYSIS OF COATING MATERIALS

John C. Thomas* and David Fairhurst

Brookhaven Instruments Corporation
Holtsville, NY 11742

The particle size and particle size distribution (*PSD*) of paints, inks etc. fundamentally affect coating performance. Small changes in the *PSD* can often dramatically alter product performance. Popular sizing methods based on laser light scattering and laser diffraction are fast, but have inherently low resolution. The Disc Centrifuge is a device which uses a rapidly spinning disc to determine the sedimentation time and, ultimately, the particle size of colloidal suspensions. This device requires no calibration since it relies on Stokes' law for centrifugal sedimentation. Recent advances in disc centrifuge design and data analysis mean that these devices can now yield high resolution results in the size range 0.01 to tens of micrometres. The basics of the disc centrifuge are outlined and data are presented which illustrate the utility of the technique to latex and carbon black size analysis.

INTRODUCTION

One of the oldest methods of determining particle size in colloidal dispersions is that of sedimentation. The particle size is determined from the rate at which a particle sediments through a viscous liquid under the influence of an external force. The rate at which even large colloidal particles sediment under weak forces such as gravity is so slow that, in practice, sedimentation measurements are usually made in centrifuges spinning at high angular velocity, and developing a field strength thousands of times that of the gravitational field. The disc centrifuge is a device which uses a rapidly spinning disc to determine sedimentation rates and, therefore, the particle size of colloidal suspensions. The centrifuge needs no calibration since it relies on Stokes' Law for centrifugal sedimentation.

Surface Phenomena and Fine Particles in Water-Based Coatings and Printing Technology
Edited by M.K. Sharma and F.J. Micale, Plenum Press, New York, 1991

Recent advances in disc centrifuge design and data analysis[1] mean that these devices can now yield high resolution particle size distributions in the micrometre size range. In the following, we outline the basics of the disc centrifuge technique by considering the operation of the Brookhaven Instruments *BI-DCP* Particle Size Analyzer, which is an advanced disc centrifuge system for high resolution particle size analysis. The *BI-DCP* uses disc centrifuge photosedimentometer (*DCP*) technology initially developed by and licensed from the Glidden Company, and a patented external gradient method of operation[2]. The *BI-DCP* yields high resolution results (peaks as close as 10% difference in size can be separated) in the size range 0.01 to 40μm. Typical analysis times range from 10 to 30 minutes. Disc rotation speed is continuously variable in the range 500 to 15,000rpm.

BASIC THEORY

Consider the arrangement of Fig. 1 which shows a spherical particle, *p*, sedimenting radially in a viscous liquid inside a spinning disc. The time, *t*, for the particle to travel from its initial position, r_i, to the detector at r_d, is readily calculated from first principles.

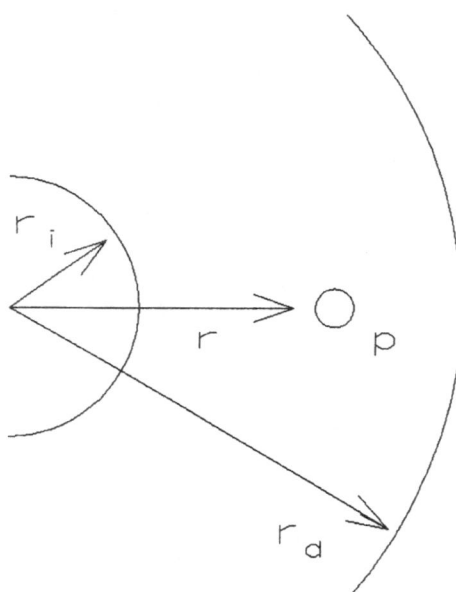

Figure 1 A spherical particle, *p*, sedimenting in a centrifugal force field.

The particle experiences a centrifugal force in a radial direction towards the periphery of the disc, a buoyant force due to the mass of the fluid displaced by the particle which acts in the opposite direction to the centrifugal force, and a viscous drag force arising from, and opposing, the motion of the particle through the liquid. The centrifugal force is $m_p w^2 r$ where m_p is the mass of the particle, *w* is the angular velocity of the particle (and disc) and r is the radial position of the particle. The buoyant force due to the displaced fluid is $m_f w^2 r$, where m_f is the mass of fluid displaced. The viscous drag force is $f(dr/dt)$, where f is the translational frictional coefficient of the particle.

The net force on the particle is

$$m_p d^2 r/dt^2 = m_p w^2 r - m_f w^2 r - f(dr/dt)$$

or

$$m_p d^2 r/dt^2 + f dr/dt - (m_p - m_f) w^2 r = 0 \qquad (1)$$

For spheres of diameter d and density ρ, the mass is

$$m_p = \pi \rho d^3/6$$

and the translational frictional coefficient is

$$f = 3\pi \eta d,$$

where η is the viscosity of the suspending fluid. Making these substitutions and rearranging equation (1) gives:

$$d^2 r/dt^2 + (18\eta/\rho d^2) dr/dt - \Delta\rho \, w^2 r = 0 \qquad (2)$$

Here $\Delta\rho$ is the difference in density between the particle and the solvent.

Equation (2) is a second order differential equation with constant coefficients. Its solution consists of two exponential terms, one of which decays on the order of milliseconds for typical conditions. The surviving term gives the time, t, for the particle to reach the detector as:

$$t = 18\eta \ln(r_d/r_i)/w^2 d^2 \Delta\rho \qquad (3)$$

This fundamental relationship, the Stokes equation for centrifugal sedimentation, reveals several important things.

1. Measurement of the appearance time, t, yields the diameter, d. For non-spherical particles this is an equivalent Stokes diameter.

2. Large particles appear first and small ones appear later in time. This inverse-square dependence on diameter means that, for example, a ratio of 10:1 in size in a sample means a ratio of 1:100 in appearance time. Thus there is excellent size resolution.

3. Large particles (greater than a few tens of micrometres), may be accommodated by using a more viscous spin fluid, by matching particle/spin fluid density more closely, by using a larger spin fluid volume, by using a lower spin speed or by using a combination of these.

4. The minimum size is determined by diffusion. When a particle diffuses an RMS distance comparable to that arising

from the centrifugation transport, the appearance time at
the detector is no longer a reliable guide to particle size.
For low density particles, such as polystyrene latex, the
minimum size is approximately $0.07\mu m$. For high density
particles, such as TiO_2, this size is approximately
$0.008\mu m$.

BASIC OPERATION

The instrument is shown schematically in Fig. 2 and consists
of two components; the motor/drive unit (which holds the
spinning disc) and a data system. Overall operation of the
instrument is controlled by the data system which is based on
a high speed IBM- AT™ compatible computer.

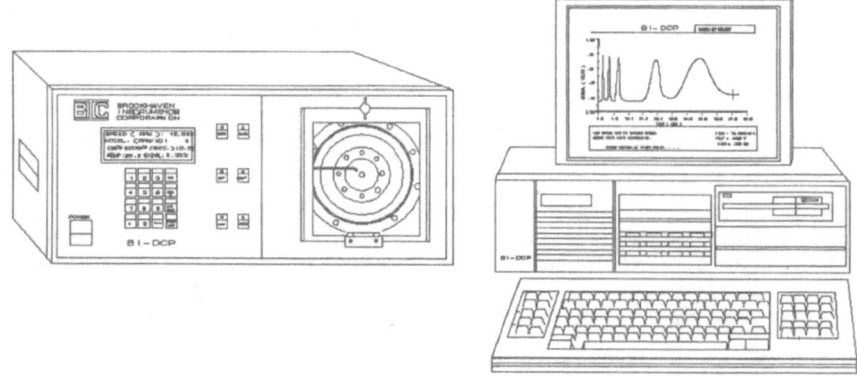

Figure 2 The *BI-DCP* Disc Centrifuge particle size analyzer.

With the disc spinning, a small volume (typically 0.25mL) of a
dilute suspension (typically less than 1% by weight
concentration) is injected onto the surface of the liquid (the
spin fluid), which was previously injected into the disc cavity.
The colloidal particles sediment through the fluid, radially
outward under the influence of the centrifugal force field.
Spin fluid volumes are of the order of 15mL.

White light from a tungsten/halogen lamp is passed through the
disc at a fixed radial position. As particles pass through the
light beam, the intensity decreases due to scattering and
absorption of the light. The extinction of the transmitted
light is measured with a photo-diode and recorded by the
computer system as a function of time. This is the basic
signal.

Hydrodynamically stable sedimentation is ensured by using either the external gradient method or the buffered line start method. Both methods introduce a density and viscosity gradient between the suspension and the spin fluid. This results in a smooth transition and subsequent laminar flow.

LINE START vs HOMOGENEOUS START TECHNIQUE

The method described above, where the material to be analyzed is injected on top of the spin fluid, is known as the line start (*LIST*) method. An alternative method is the homogeneous start (*HOST*) method in which the spin fluid initially contains a _uniform concentration_ of the material whose particle size is to be determined.

For the *LIST* method, all particles begin sedimenting from the same position, but, since different sized particles sediment at different rates (Eq. (3)), _fractionation occurs before the particles reach the detector_. In contrast, the *HOST* method never results in true fractionation. Rather, at any time, a mixture of particle sizes (large, small and in-between) will reach the detector. Consequently, the *LIST* method yields a true _differential_ detector response, whereas the *HOST* method yields a _cumulative_ response and suffers the attendant _loss in resolution_.

The merits of the *LIST* and *HOST* methods have been investigated in detail[3]. The *LIST* method invariably gives better results than the *HOST* method. While the modal diameters obtained from each of the methods are similar, the *HOST* method gives _significantly broader and sometimes distorted_ size distribution peaks compared with the *LIST* method. An additional problem with the *HOST* method is that it is mathematically much more difficult to apply extinction coefficient corrections to this type of data than to the differential data obtained by the *LIST* method. We note that extinction coefficients depend on both the size and nature (refractive index) of the material being measured. Furthermore, using incorrect or no extinction coefficient corrections, can greatly distort the *PSD*.

In order to obtain the highest resolution in size distribution determination, the *BI-DCP* uses the *LIST* method of operation. Many other devices available operate via the *HOST* method.

DATA ANALYSIS AND PRESENTATION

The *BI-DCP* Data System provides complete data accumulation, analysis, management and modeling capability for the *BI-DCP*. The graphic modeling utility allows users to optimize run time and disc speed from estimates of particle density, size and spin fluid properties.

The system is menu-driven and raw data are plotted in real-time on a high resolution display. Baseline selection is automatic or manual and can be done on the entire curve or peak-by-peak.

User-selectable extinction coefficient corrections are also provided.

Results that can be calculated from the data include:

Number, Area and Weight average diameters and distribution curves.

Turbidity and *PCS* average diameters and peak widths.

Polydispersity index, d_W/d_N.

The Data System will generate on-screen and hard-copy plots of raw data curve with baseline, and differential and cumulative forms of the Number, Area and Weight distributions. Experimental conditions and results are displayed and printed as full reports including user-selectable run numbers, titles and sample identifiers.

RESULTS AND DISCUSSION

Standard Latex Spheres

To demonstrate the performance of the *BI-DCP*, measurements were made on certified latex sphere standards from Duke Scientific. A spin fluid with a density and viscosity gradient was established by drawing 0.5mL of methanol into a syringe containing 10mL of water. With the disc spinning at 8,000rpm, the spin fluid was injected and 0.2mL of the sample was injected onto the meniscus. The sample was made by adding 3 drops of the Duke concentrate to 3mL of water containing 0.1% Triton-X, sonicating for 15 sec. and then adding 3mL of methanol.

Fig. 3 shows the weight distribution obtained for Duke Scientific catalog #3150 which has a certified mean diameter of 155 ± 4nm. The mean size obtained with the *BI-DCP* is 154nm which is in excellent agreement with the certified value. The polydispersity index is 1.046 and indicates a narrow size distribution. Clearly, the *BI-DCP* can accurately determine the particle size for narrow distributions.

Table I *BI-DCP* measurement results for a nine component latex sphere sample.

Duke Catalog No.	Certified Mean Diameter (nm)	Weight Average Diameter (nm)
3100	107 ± 3	111 ± 3
3200	220 ± 6	217 ± 6
3300	298 ± 3	294 ± 9
3400	398 ± 4	394 ± 10
3500	496 ± 4	498 ± 13
3600	597 ± 5	596 ± 16
3700	705 ± 6	703 ± 18
3895	895 ± 8	895 ± 22
4009	993 ± 21	992 ± 23

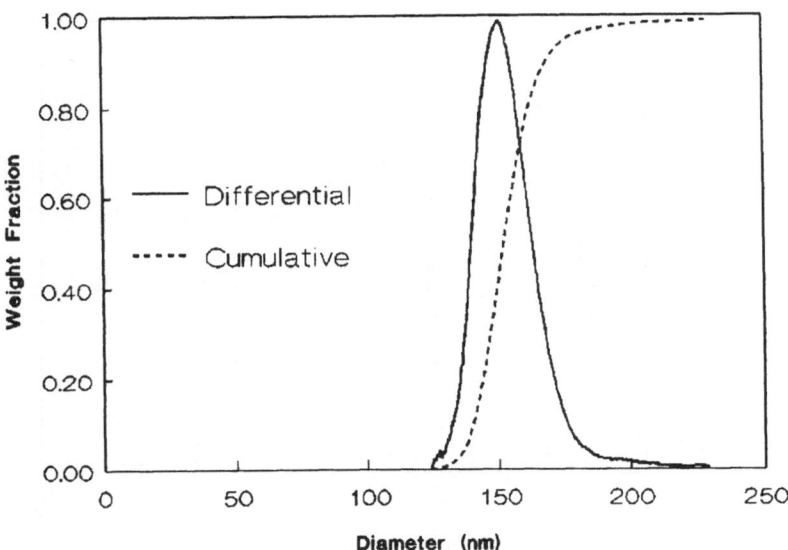

Figure 3 Weight distribution obtained with the *DI-DCP* for a 155 nm latex sphere standard.

To test the ability of the *BI-DCP* to resolve multi-component size distributions, a measurement was made on a sample containing similar weight fractions of nine latex standards. The sample was run at 7,000rpm and the spin fluid was 15mL of water. The components of the sample are given in Table I and Fig. 4 shows the resulting weight distribution. The nine fractions are unequivocally resolved. The weight-average size and full width at half height were determined for each peak and are also listed in Table I. Note the excellent agreement with the certified mean diameters. Note also that the two fractions at 895nm and 993nm only differ in size by 10% and that, even in the presence of the other components, these peaks are readily resolved. Furthermore, the cumulative undersize curve shows that the weight fractions of the components are basically the same, as expected.

Coating Materials

Latex is used extensively in the coatings industry and the coating properties (gloss, resilience etc.) depend fundamentally on the particle size and *PSD* of the latex. Both the size and *PSD* may be controlled with subtle changes in the conditions of the polymerization reaction.

Fig. 5 shows the weight distributions obtained with the *BI-DCP* for normal and modified samples of a commercial vinyl

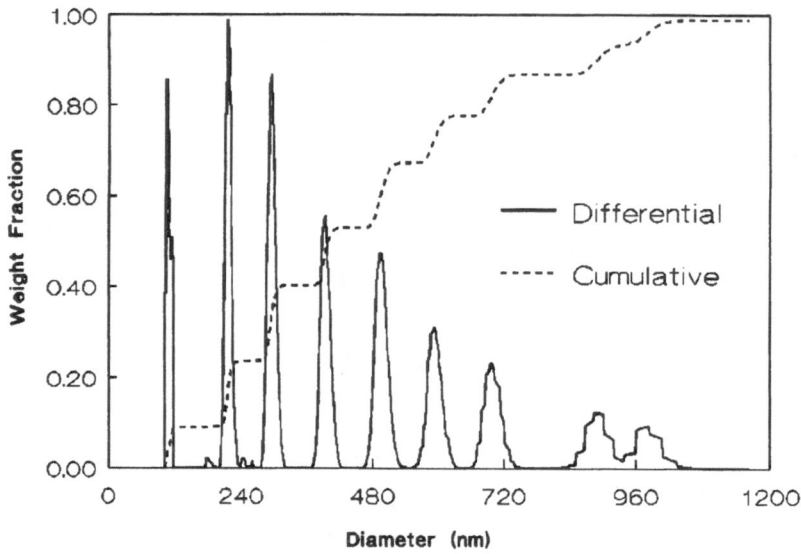

Figure 4 Weight distribution obtained with the *BI-DCP* for a nine component latex sphere sample.

acetate/butyl acrylate copolymer latex emulsion used in trade paint formulations. The modified product has an increased monomer feed time compared with the normal product. The feed time was increased in an attempt to improve the gloss and abrasion resistance of paints containing this material. The samples were run at 2,000rpm using 15mL of water as spin fluid.

According to the manufacturer, the normal product has a particle size of 300nm. The weight-average diameter and polydispersity obtained with the *BI-DCP* are 333nm and 1.399 respectively. It can also be seen from Fig. 5 that the particle size ranges from about 168nm to 800nm and that the *PSD* is bimodal with peaks at 214nm and 302nm. The modified product weight-average diameter and polydispersity were 335nm and 1.334 respectively. Evidently there is a small increase in the mean particle size and a small decrease in the polydispersity. These differences are indeed subtle. However, by examining the *PSD*, one can see significant differences in the modified product. Increasing the monomer feed time has produced a monomodal size distribution and removed the peak at 214nm. So there is a fundamental change in the <u>shape</u> of the *PSD*, yet the mean size and polydispersity are hardly affected. In this case it was essential to have the high resolution of the *BI-DCP* to see the effects of changing the manufacturing process.

The size of latex particles produced in a polymerization reaction varies with the amount and type of surfactant present in the reactor. We have examined three batches of polystyrene

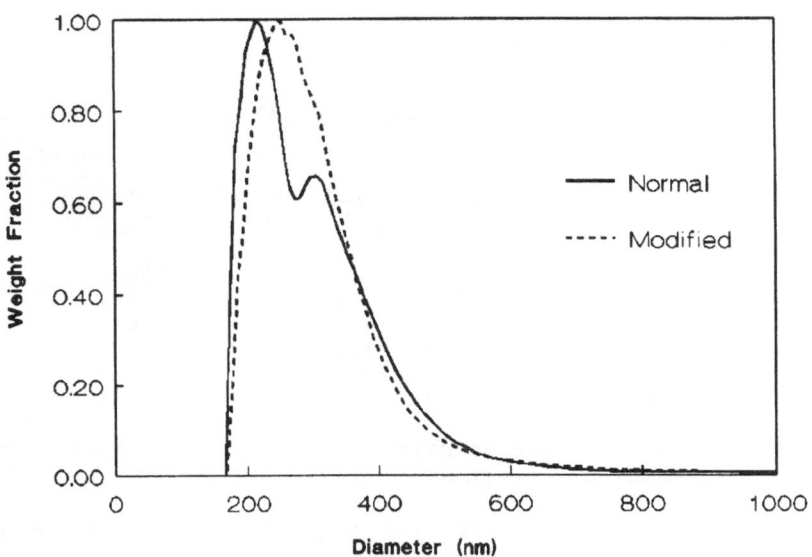

Figure 5 Weight distribution obtained with the *BI-DCP* for two vinyl-acrylic latexes. The modified product has an increased monomer feed time compared with the normal product.

latex used in commercial floor coating material, but prepared using different anionic surfactants. Samples A, B and C were prepared using a proprietary surfactant, Witconate AOS and SLS respectively. Particle size measurements were made with the *BI-DCP* and the *BI-90*, a particle sizing instrument based on the technique of photon correlation spectroscopy (*PCS*[4]). The *BI-DCP* measurements were made at 12,000rpm using 10mL of water as a spin fluid and took 25 mins. Table II shows the weight average diameter and polydispersity obtained with the *BI-DCP* and the *PCS* average diameter and polydispersity obtained with the *BI-90* for the three samples. The polydispersity values are shown in parentheses and, for the *BI-90*, the polydispersity is the normalized variance of the diffusion coefficient distribution[5].

It can be seen that, as the surfactant is varied from proprietary to Witconate AOS to SLS , the average particle size decreases whereas the polydispersity increases. Furthermore, both the *BI-DCP* and the *BI-90* track this behaviour. The *PSD's* obtained with the *BI-DCP* for these samples are shown in Fig. 6. where it is readily seen that the successive surfactants move the *PSD* to successively smaller sizes. Using the proprietary surfactant gives a *PSD* which is slightly bimodal (peaks at 145nm and 154nm) and skewed towards large sizes and with SLS the *PSD* is monomodal, somewhat broader and skewed towards smaller sizes. Thus, with the high resolution of the *BI-DCP*, one can follow the evolution of the *PSD* as the product formulation is varied.

Table II Effect of surfactant type on polystyrene latex *PSD*.

Sample	Surfactant	Particle Size (nm)	
		d_W (1)	d_{pcs} (2)
A	Proprietary	157	156
		(1.026)	(0.028)
B	Witconate AOS	141	140
		(1.037)	(0.047)
C	SLS	131	129
		(1.110)	(0.073)

1. d_W is the weight average size from the *BI-DCP*. The value in parentheses is the polydispersity, d_W/d_N.

2. d_{pcs} is the average size from the *BI-90*. The value in parentheses is the polydispersity, which is the normalized variance of the diffusion coefficient distribution[5].

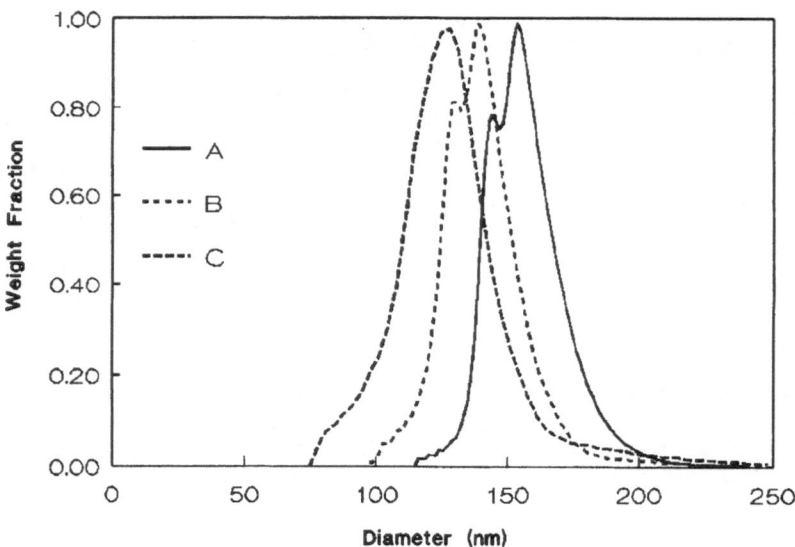

Figure 6 The effect of surfactant on the *PSD* of polystyrene latex. Samples A, B and C were made with a proprietary surfactant, Witconate AOS and SLS respectively.

Not all coating material has a spherical shape and a narrow
size distribution. Indeed, carbon black, which is a
constituent of SBR and many coatings products, has a plate-like
shape and its *PSD* is usually broad. The particle size is often
difficult to define since the dispersions are inevitably
comprised of aggregates and one is not measuring the primary
particles. Nevertheless, it is still important to be able to
determine the *PSD* accurately in order to predict the physical
characteristics of the blacks. Fig. 7 shows the *PSD's* obtained
from two different carbon blacks used in tire manufacture.
There are obvious similarities between the *PSD's*. *i.e.* they are
monomodal, broad and featureless and extend from a little under
30nm to around 150nm in size. The weight-average diameters are
62nm and 67nm for sample A and B respectively. The fundamental
difference between the samples is that sample B is about 25%
broader than sample A. The full width at half height of the
weight distribution curve is 38nm and 49nm respectively for
sample A and sample B. Again, this is a subtle but important
feature, since it will determine the end use of the black.

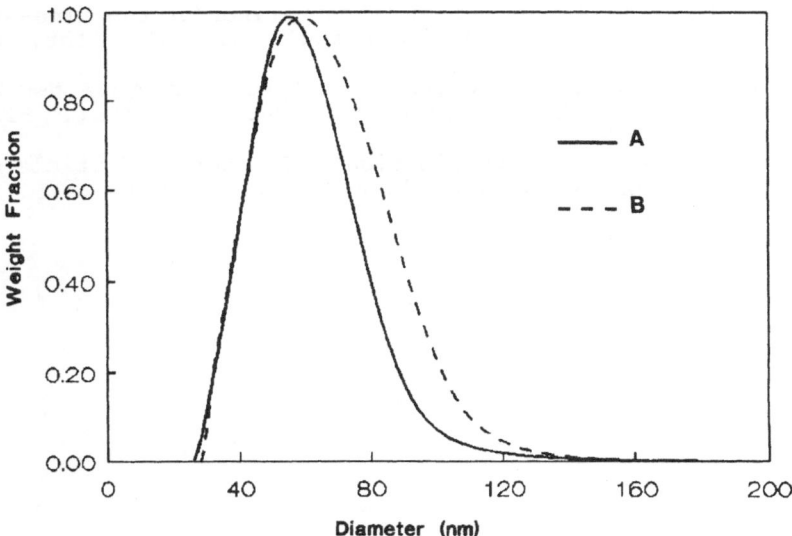

Figure 7 The **PSD** for 2 different carbon blacks. Sample B is
approximately 25% broader than sample A.

CONCLUSION

We have shown that the *BI-DCP* disc centrifuge is a high
resolution particle size analyzer which needs no calibration
and which is ideal for measurements on colloidal material.
Since coatings properties depend strongly on the size and *PSD*
of the constituent materials, it is essential that their *PSD's*
be accurately determined. The ability of the *BI-DCP* to yield
accurate, high resolution *PSD* data suggests that it will find
wide application in the coatings industry.

ACKNOWLEDGEMENT

We remain indebted to Prof. Karin Caldwell, Dept. Bioengineering, University of Utah and Larry Becker, Amoco Chemical Co., Naperville, IL. for providing the nine component latex sphere sample.

REFERENCES

1. Koehler, M.E., Zander, R.A., Gill T., Provder T. and Niemann, J.J., An Improved Disc Centrifuge Photosedimentometer and Data System for Particle Size Distribution Analysis, in: "Particle Size Distribution -- Assessment and Characterization," T. Provder, ed., ACS Symposium Series, (1987).
2. Holsworth, R.M., Provder, T. and Stransbrey, J.J., External-Gradient-Formation Method for Disc Centrifuge Photosedimentometric Particle Size Distribution Analysis, in: "Particle Size Distribution -- Assessment and Characterization," T. Provder, ed., ACS Symposium Series, (1987).
3. Coll, H. and Searle C.G., Particle Size Analysis with the Joyce-Loebl Disk Centrifuge: A Comparison of the Line-Start with the Homogeneous-Start Method, J. Coll. Int. Sci. 115:121 (1987).
4. Thomas, J. C., Photon Correlation Spectroscopy; A New Tool for Sub-Micron Particle Sizing, Chem. in Aust. 52:464 (1985).
5. Thomas, J. C., The Determination of Log Normal Particle Size Distributions by Dynamic Light Scattering , J. Coll. Int. Sci. 117:187 (1987).

CONCEPTS ON THE PREPARATION OF PARTICULATE FOR

MEASUREMENT BY LIGHT SCATTERING PARTICLE SIZE

ANALYZERS

Philip E. Plantz

Particle Technology Lab
3000 Old Roosevelt Blvd.
St.Petersburg, FL. 33716

ABSTRACT

Instrumental analysis of materials normally requires subjecting sample material to various types of treatment before introducing the sample to the instrument. Typical treatments include proper techniques of representative sampling and chemical preparation. These two treatment aspects are especially important to particle size measurements. Specifically, particulate in the sub-sieve size range is prone to agglomeration and thus require more attention to chemical preparation prior to analysis of size than coarser sizes. Coarser particulate, generally larger than 20 micrometers, is more subject to segregation and concomitant representative sampling variations. Decisions necessary to appropriate and proper sample chemical preparation by low-angle forward light scattering (Fraunhofer diffraction) include those of compatibility and effectiveness of surfactants, "mother liquors" and dispersive energy with regard to sample. This paper addresses practical considerations of chemical preparation concepts applicable to particle size measurements of water based coatings.

INTRODUCTION

Light scattering has been applied to particle size measurements for many years using the concept that small particles scatter light at wider angles than large particles. Mathematical treatment of the signals derived from the scattered light allow development of a distribution of particle sizes from as small as one-tenth micron to several hundred microns. While such instrumentation automatically performs its functions, other facets of particle size measurement are still dependent upon the operator. Obtaining samples that represent the entire process material is one very important aspect that is addressed by commercial equipment. Another important aspect is preparation of samples for measurement. Initially, a decision must be addressed as to the necessity or desirability of applying chemicals and energy to obtain a well dispersed particulate system. Knowledge of the presence of agglomeration may lend important process information while complete dispersion provides intra and interlaboratory databases.

Sample preparation normally includes two steps: surfactant selection and application of energy. Anionic, cationic and nonionic surfactants should be used with caution to prevent foam formation, surfactant interferences, micelle and flocculant formation, insolubility and competition for surfactant. An accepted practice is to develop surfactant activity curves for each material under study to ascertain optimum surfactant concentration. Consideration of these concepts permits selection of a surface active agent that will appropriately wet particulate. The second step in sample preparation involves application of energy to disrupt particulate agglomerates. Sources employed include ultrasonic baths and probes, blenders, stirrers and tissue homogenizers. Each device manifests characteristics that can cause particle attrition or insufficient deagglomeration if not used with care for particle integrity. Tissue homogenizers may damage particulate while blenders may cause contamination. Ultrasonic baths require stirred samples and consistently placed beakers. Consideration of all devices leads the author to a preference when considering flexibility of control and energy conditions.

EXPERIMENTAL

Various portions of the electromagnetic spectrum typically ranging from the near-infrared to visible regions are used in commercially available light scattering particle size analyzers. In some cases light obscured by a single particle [1] is monitored to determine size distribution but more common is the use of a mathematical approach to examine the light scatter pattern from a cloud or group of particles.[2-4]

The light scatter pattern is a result of the interaction of monochromatic light with particles larger than the wavelength of the incident radiation. Typically the source is a helium-neon gas or gallium-aluminum-arsenide solid state laser. Light striking a particle is scattered in many directions at various intensities while that scattered at low forward angles (up to 20 degrees) is collected by a lens and focused onto a means of separating the scattered light into distinct zones. Recent advancements somewhat redefine the forward angular scatter to wider angles to as much as 70 degrees requiring a second detector or a single large curvilinear detector to produce more zones thus increasing the effective measurement range. Each zone is a given distance from

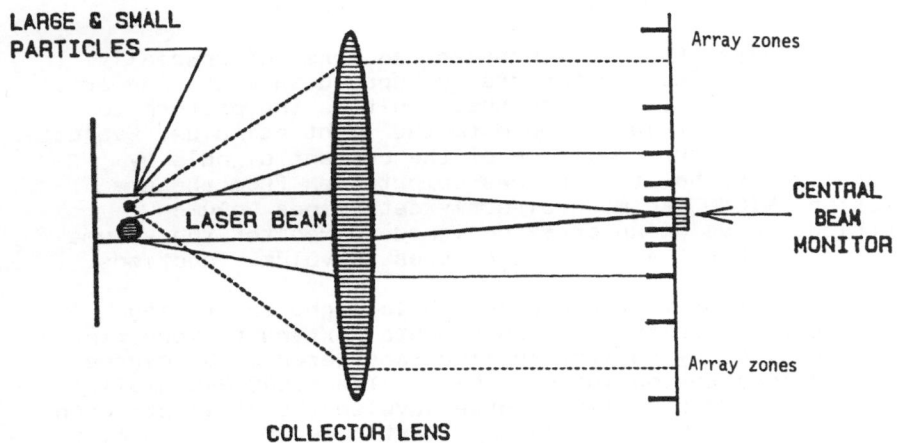

Figure 1. Smaller particles diffract light at larger angles
and lower intensity than larger particles.

TABLE   1

Surfactants used by MICROTRAC Particle Technology Laboratory

Surfactant	Potential Uses
Aerosol OT	Zirconium oxide, Graphite, Dry powder coatings
Triton X100	Metal powders, Glass frit
Sodium Metaphosphate	Alumina, Iron oxide
Sodium dodecyl sulfate	Carbon black
Gafac RE-610	Silver flake
Sodium pyrophosphate	Silicates, Metal oxides
Tetrasodium pyrophosphate	Calcium carbonate, Titanium dioxide
Aerosol OS	Cadmium
Twitchell base	Explosives, Propellants
Isopropyl alcohol	Tricalcium phosphate, Edible dyes
Methyl alcohol	Calcium sulfate Lead pigments
Lomar D	Alumina, Clays
Daxad 23	Inks
TAMOL SG-1	Low foaming surfactant for blending

the center of the light pattern. The distance from the center of the pattern to the zone is inversely related to the size of the particle while the intensity of the impinging light is directly proportional to the size (Figure 1).

To analyze the light pattern, an array of specially zone-shaped silicone detectors is spaced in either an area function or volume function that monitors the pattern to produce a current in response to the light stimulus. Evaluation of the pattern and amplitude of the current signals by specialized mathematics allows computation from the area function. MICROTRAC more directly determines volume distribution using an array designed to monitor the volume function rather than converting area to volume functions.

In addition to forward low angle light scattering (Fraunhofer diffraction) measurements extend to very fine particle sizes by monitoring light scattered at 90 degrees (wide angles) to the incident beam. The MICROTRAC Small Particle Analyzer utilizes three wavelengths of light, each polarized at orthogonal planes. Particles of a given size scatter light of a given wavelength at 90 degrees . The difference in polarized intensities for each wavelength relates to the concentration of 90 degree scattering particulate. By this means, particles as small as 0.12 micrometers can be measured as shown in Figure 2. In combination with Fraunhofer diffraction, particulate over the range .12 to 60 microns is measurable. Further explanation of the light scattering concepts employed may be found in the literature.[1,5,6]

RESULTS AND DISCUSSION

Prior to embarking upon procedure development to prepare samples for analysis, a decision should be made as to its necessity. Particulate can exist in a number of states in which the particles disperse as individual, discreet units or one of several situations where the particles attach to each other loosely or tightly. Preparation of particulate samples is directed to separating loosely bound units termed soft agglomerates. Often, agglomeration can affect color, hue or spreadability of paint or nail polish, or provide a prediction of ball point pen plugging. Since particle size and agglomeration are correlatable to pigment qualities, complete dispersal of sample prior to analysis may not reveal that product quality is in jeopardy. The presence of agglomerates, therefore can indicate process or manufacturing problems associated with reaction kinetics, concentration anomalies, milling characteristics or product stability. Measurement of these agglomerates, when employed, is generally useful for in-house quality control. By microscopic examination these are identifiable by attachment of particulate corners to the sides or edges of one or more particles. A more tightly bound group of particles termed an aggregate can be observed with large crystal faces adjoined. Aggregates such as these are often termed hard-agglomerates and may not succumb to de-agglomeration techniques. An example of MICROTRAC data showing a comparison of agglomerated and well dispersed particulate is shown in Figure 3.

Certain situations exist, however, that require complete dispersion. Agglomerates by nature of their loose structure and

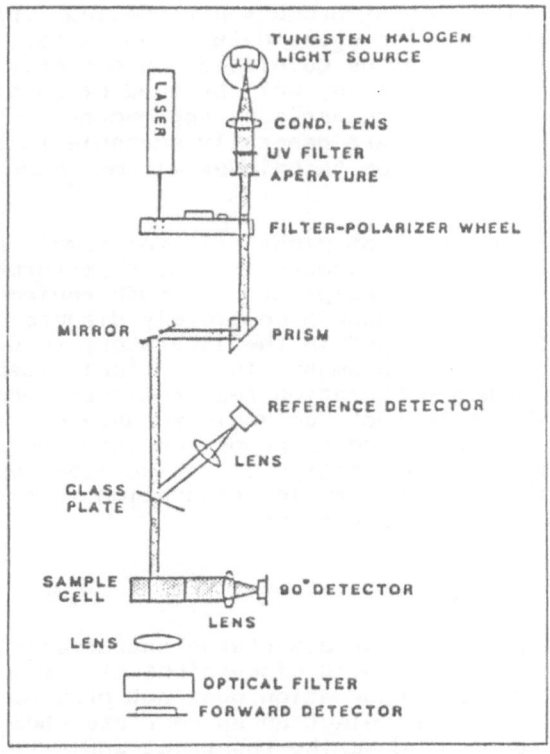

Small particle optical system.

90 DEGREE LIGHT SCATTERING

WAVELENGTH	.45	.60	.90
PARTICLE DIAMETER	.15	.2	.3

All units in micrometers

Figure 2. Orthogonal polarization of each wavelength provides concentration information for particles of a specific size.

low energy bonds can initially form and mildly disperse only to have individual particles interact with other particles to form smaller or larger assemblies. The randomness of this process can cause poorly reproducing particle size data. This becomes increasingly important when vendor and customer both analyze sample for particle size since agglomeration may cause an erroneous indication of product not meeting specifications. To avoid such a situation where data do not agree, particulate preparation methods must be developed and strictly followed. The methods provide a stable, well-behaved dispersion to provide a basis of data comparison. Agglomerates of particulate given above are generally amenable to sample preparation and dispersion techniques without causing damage or attrition.

Another situation that might call for sample preparation is in evaluation of final processing steps performed by in-house or customer equipment. If such equipment will, through its action, produce a completely dispersed product methods should be developed in the laboratory to duplicate the dispersion prior to measurement. In the final assessment, the necessity for sample preparation requires that knowledge of the ultimate use of the product, accumulated data or manufacturing process be acknowledged and incorporated into the sample preparation concept. In addition it is incumbent upon each analyst to decide whether sample preparation is required depending on the above situations.

SURFACTANT SELECTION

Concepts involving surface charge characteristics and free energy are often addressed in discussions of surfactant activity. Often such information does not provide the necessary guidelines to select an appropriate chemical treatment, nor does it describe the trial-and-error methods that must be undertaken. Some texts provide specific surfactants known to effectively wet specific particulate. Under such circumstances a broad selection of non-ionic, anionic and cationic surfactants should be made available. Some of those used by the MICROTRAC Particle Technology Laboratory and their application are shown in Table 1. The following discussion addresses the concepts of applying appropriate chemical and energy sources.

The terms wetting and dispersing should not be interchanged as each represents a specific step in sample preparation. Wetting in the context of this paper refers to the activity of preparing a solution or suspending medium, that will "coat" particles after they are separated often by a source of energy. Dispersion is the state in which the particles exist as separate, distinct entities that because of the modified surface properties are prevented from re-aggregation or re-agglomeration.

In concert with the above, consideration must be given to the fact that adsorption of surfactant or "wetting" chemicals to dry particles can sufficiently modify surface characteristics to affect product performance. This aspect of sample preparation must be considered as part of the decision

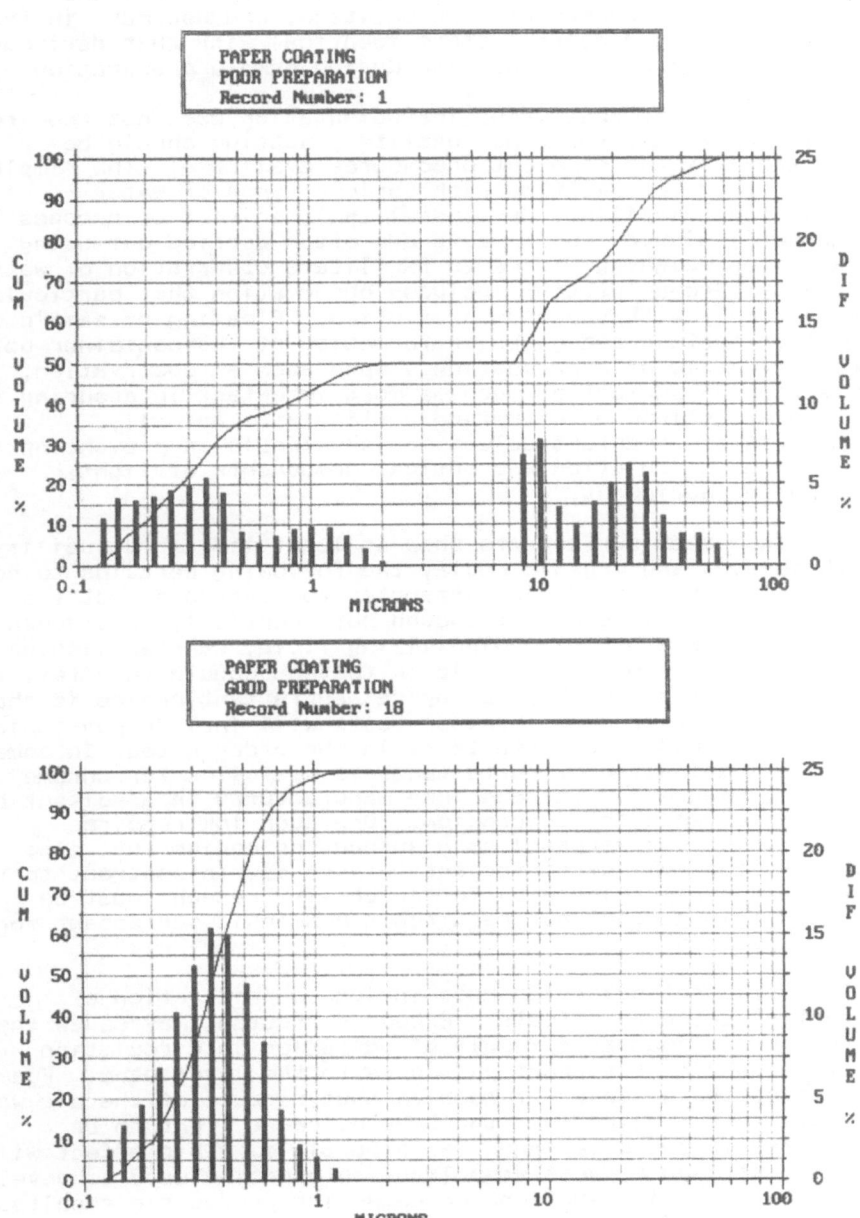

Figure 3

to address sample preparation. One should be fully aware that
this, as well as any other affective procedure, does not
necessarily provide a system that mimics particulate condition
of final product. Wetting and dispersion usually are the result
of modifying surface properties of product but these effects
notwithstanding, provide a means of quality control and
interlaboratory agreement. In addition, end use may, in fact,
produce a fully dispersed state identical with that derived by
surface modifications occurring during sample preparation.

        Since light scattering instrumentation does not require
special chemicals for functionality , wetting should be
approached by the simplest procedures available. The simplest
method is to test wetting with various types of water
(deionized, distilled, de-mineralized), solvent or process
("mother") liquid. Such tests are often carried out using
beakers or weighing dishes to facilitate observation of wetting
activity. Proper wetting includes observation that particulate
mixes freely with the chosen solution. Floating of sample on
solution surface and globular formation at the container bottom
are indicative of poor wetting. Upon such an observation,
selection of surfactant becomes most important in assuring that
data are obtained on dispersed particulate. As well,
appropriate surfactant selection is valuable in preventing
coating of circulation and optical components of light
scattering equipment.

        Since most surfactants show at least limited solubility in
both aqueous and organic media, the following pertains to both
environment types. Having attempted to suspend or wet the
sample in a fluid similar, though not identical, to process
conditions or other non-solubilizing fluid, carrier liquids
employed in the process should be tested. Should this fail to
wet particulate properly, a logical surfactant choice is that
is used in the process. Further tests will include non-ionic,
anionic and cationic surfactants in the order noted. In some
cases a combination of surfactants is useful as for chrome
dioxide (Figure 4). Various inks manufactured in a solvent base
require solvents for suspension. One such ink requires
non-solubilizing methanol as a suspending medium that also
portrays the role of surfactant (Figure 5). In contrast to the
inks, black dye (Figure 6) for solubility reasons must be
suspended in alcohol but requires a non-ionic surfactant for
proper wetting.

        Surfactant selection must include consideration of
potential untoward effects. Excess surfactant may cause the
formation of foam as a result of agitation in circulation
systems used to transport particles to the laser beam. Foam
indicates the presence of bubbles that because of the air-water
interface,that comprises the bubble, cause light to be
scattered in the same manner as a particle. This effect will
erroneously add to scattered light information used to develop
particle size distributions or cause irreproducible results.

        Excess surfactant also can cause a metastable dispersion
known as flocculant to form or may auto-interact to form
micelles. The former case is easily recognized by the
particulate mass "floating" on the <u>bottom</u> of a beaker with the
mass having the appearance of large aggregates easily and

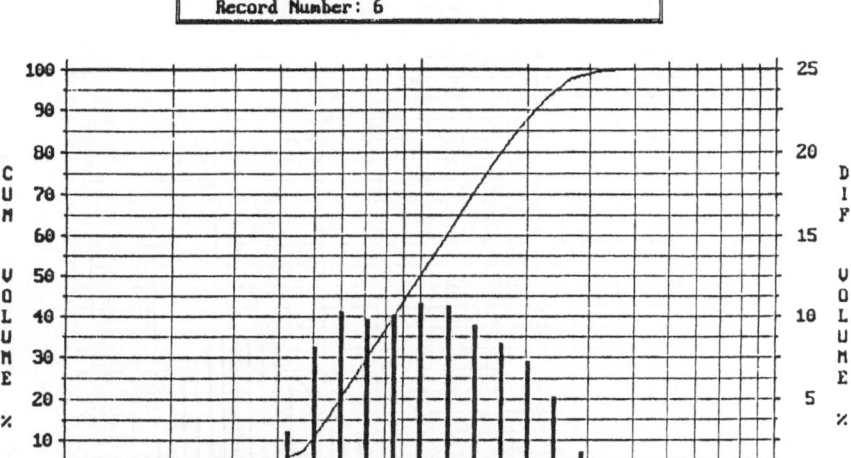

Figure 4.   Methanol acts to prevent dissolution
as well as wet red ink particulate.

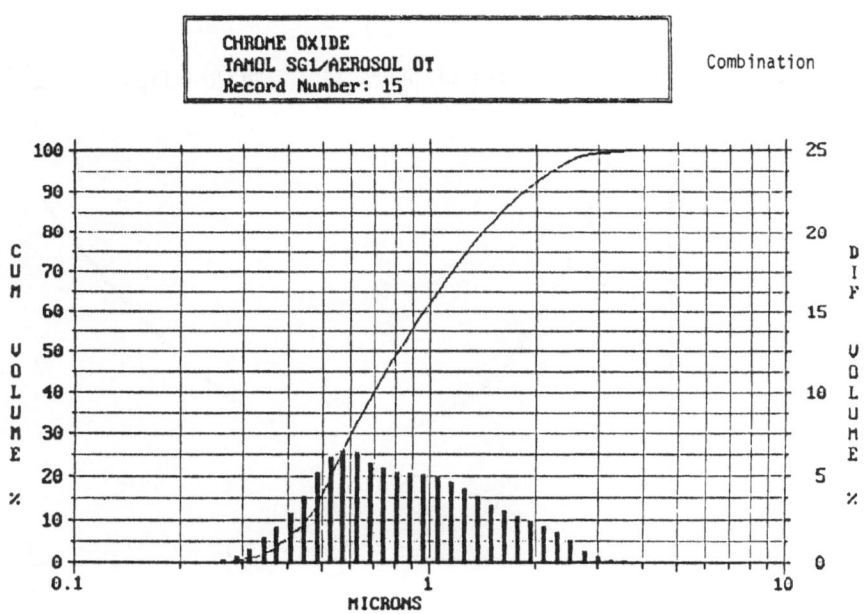

Figure 5.   Combination of surfactants to achieve
wetting of particulate.

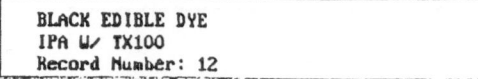

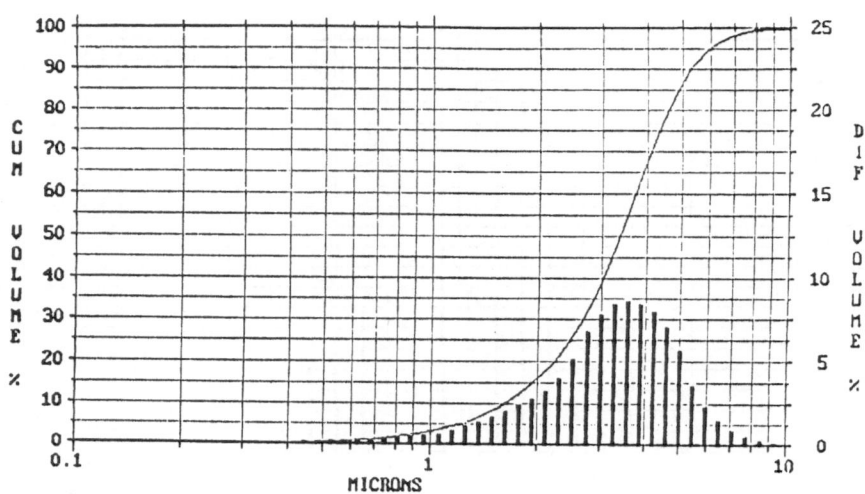

Figure 6. Dry requiring suspension in isopropyl
alcohol and wetting with Triton X-100.

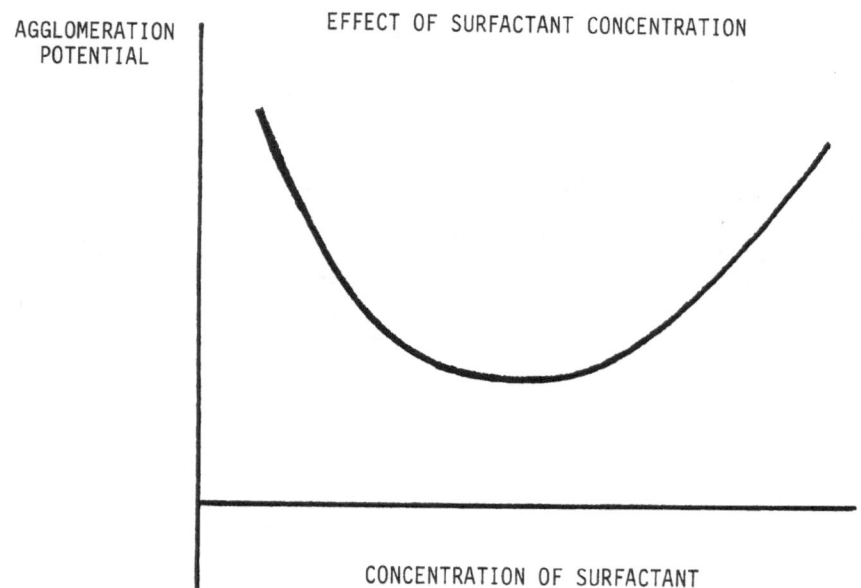

Figure 7. Particle size changes as a function
of surfactant concentration

promptly dispersed upon mild agitation. Micelles, on the other hand, are spontaneous organizations of surfactant molecules into spherical entities called vesicles that are measurable normally in the sub-micron region.

Each of the above situations can be avoided by using low surfactant concentration especially when employing water. Final circulation concentrations not exceeding 0.01% normally prevent such situations from occurring. Another easily avoidable situation involves careful selection of surfactant to prevent surfactant interactions. While non-ionic surfactants generally can be used with any solubilizing medium containing any other surfactant type, severe surfactant inactivating interactions can occur in solutions containing both, cationic and anionic species.[7] The consequent action is inter-nullification of the activity of both species with potential precipitate or gel formation and resultant data interference. Knowledge of surfactant type used in-process is necessary to avoid surfactant-surfactant interactions.

APPLICATION OF SURFACTANTS

As mentioned above, surfactants should be used with some caution to prevent misleading effects upon data. A means of avoiding such a situation is to develop surfactant activity curves by which optimum surfactant concentration can be determined. As shown in Figure 1, measurements of zeta potential, agglomeration potential, or particle size can be evaluated as a function of surfactant concentration. In this example particle size decrease (zeta potential increase) corresponds to an increased concentration of surfactant until minimum size is attained. Particle size then increases as a function of concentration. This can be explained by one of several mechanisms including flocculation and micelle formation as discussed above. A possible alternative explanation was given by Ottewill and Roastozi[9] who describe cationic surfactant adsorbed by the particle surface increasing zeta potential and promoting dispersive effects. At higher concentrations reported particle size then increases a result of decreased dispersion stability caused by compression of the electrical double layer.

From the surfactant curves, final concentration of chemical in the circulation system necessary for wetting may be determined. Appropriate calculations will reveal starting surfactant concentration. The initial concentration is expectably higher than that for circulation because of dilution effects encountered during transfer of prepared sample to the circulator. In general starting solutions in excess of 0.1% are not necessary but some materials require a paste to be prepared from a more concentrated mixture such as 10% or greater. The paste may be then gradually mixed with fluid selected as the carrier until a thin slurry is produced. Appropriate energy may then be applied to the sample slurry, as discussed below, to complete the dispersion process. Figure 8 shows an example of a pigment that required paste development with concentrated Triton X-100 and water as diluent and carrier fluid.

As noted above, surfactants are employed to modify surface

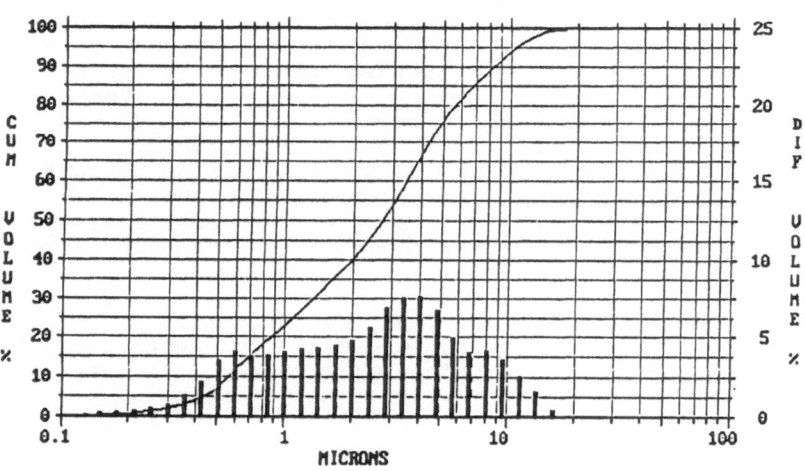

```
                              ch-top %pass %-ch        summary data
  up to 20 chars              59.69  100.0  0.0          dv:    0.0000
                              42.21  100.0  0.0          %10:   0.55
 ──────────────────────       29.85  100.0  0.0          %50:   2.71
  MICROTRAC DATA: DATA BASE UPDATE   21.10  100.0  0.6   %90:   8.53
 measurem't/present'n#:    10 -    1 14.92   99.4  4.3    mv:    3.59
 7997 0.12-59.6 SPA    res: NORMAL  10.55   95.1  7.8    cs:    4.429
 chan select bounds:    0.12- 59.69  7.46   87.3  9.1    mi:    0
 progression: geom/sqrt2  # chans:18 5.28   78.2 14.5    sd:    2.97
 sample date/time:  07/13/89 15:11   3.73   63.7 14.6    ma:    1.35
 sample ident: RED EDIBLE DYE        2.64   49.1 10.5
 sample   ID : TX100 PASTE & BOWL    1.87   38.6  8.8        parameters
 lot    code:                        1.32   29.8  8.4     name    value
 account #:4503        printer: OFF  0.93   21.4  7.6   smpl amt   0.000
 run time: 15 sec.  run #:   0/ 1    0.66   13.8  7.6   disprsnt
 sample preparation code:   0        0.47    6.2  3.6   disp amt   0.000
 notes: MAKE PASTE WITH              0.33    2.6  1.4   disp med
      TRITON X-100                   0.23    1.2  0.8   dmed amt   0.000
 TRITON X-100 IN BOWL                0.16    0.4  0.4   agitat'n
                                                        agit tim     0
                                                        circul'n
                                                        circ tim     0
                                                        param #1   0.000
    Esc-exit          F2-commands                       150 mesh   0.000
                                                        325 mesh   0.000
```

RED EDIBLE DYE
TX100 PASTE & BOWL
Record Number: 10

Figure 8. Mixing sample as a paste followed by
slow dilution with water.

characteristics. While particle shape or surface topography are not affected, the chemical microenvironment is necessarily changed in the attempt to achieve complete dispersion for measurement purposes.

Various processes may employ any of the charged or non-charged surfactant species. Caution should be exercised when charged species are known to be in use since only that particular charged species and/or non-ionic type may be used. Cationic and anionic species should never encounter each other.

SELECTION OF ENERGY

Surfactants provide an environment in which particles once separated will accept chemically induced surface changes that discourage attraction. Initial separation often is supplied by the charged dipole of water after which various chemicals assist wetting. Evaluation of the dispersion, by means to be discussed below, often will show that agglomerates of particulate are present. To alleviate this condition energy may be applied in various forms such as ultrasound or kinetic (shear).

Ultrasonic energy has largely been applied to biological systems[9-12] and even to welding of plastics. The ability to impart energy to suspensions and slurries stems from the production of microbubbles which radiate from a generator/vibrating plate combination. As the bubbles travel through a solution, growth and shrinkage of the bubble by continual flexing occur until a critical minimum size is attained upon which implosion releases energy. Particulate in the area of the energy release is vibrated and shaken apart allowing coating of particle surfaces by surfactant molecules.

There are several advantages and disadvantages to the use of ultrasound as a source of energy. As mentioned above, many surfactants foam considerably causing measurement interferences. Ultrasonic treatment provides a means of applying energy without severe agitation and mixing with air that causes foaming. Thus high, moderate and low foaming surfactants can be used without concerns of bubble-formation. Because particles are vibrated some shear may occur during treatment. Evaluation of particulate by microscopy prior to as well as after treatment is useful in determining any damaging effects wrought by excessive ultrasonic treatment. Typical effects are loss of sharp, pointy edges or general particle attrition. Both are easily observed by comparative microscopy. Should any of these observations be made, reduced intensity and treatment time must be considered.

Because of the design of ultrasonic devices, beakers are useful to contain sample, suspending fluid, and surfactant mixtures. As a result cross contamination of samples or contamination of the sample by equipment action does not occur. It is important, however, that any container employed be scrupulously clean. All residues of previous preparations, surfactants and fluids, must be removed. Containers should be thoroughly washed and rinsed with final rinsing in distilled or de-ionized water to avoid dry deposition of calcium and magnesium (hard water) salts.

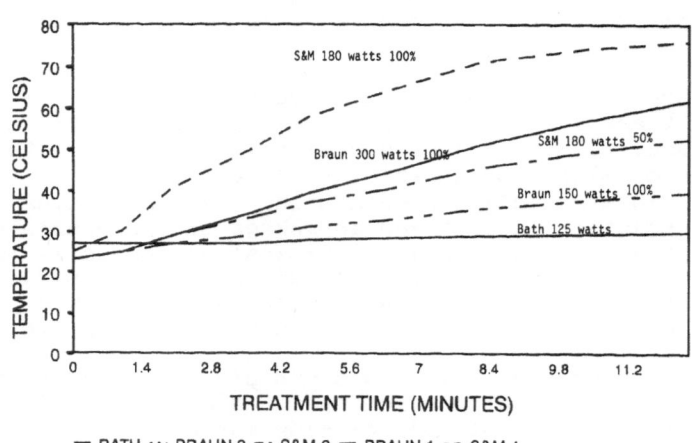

Figure 9. Energy imparted (measured as temperature change) as a function of time for two ultrasonic probes and an ultrasonic bath.

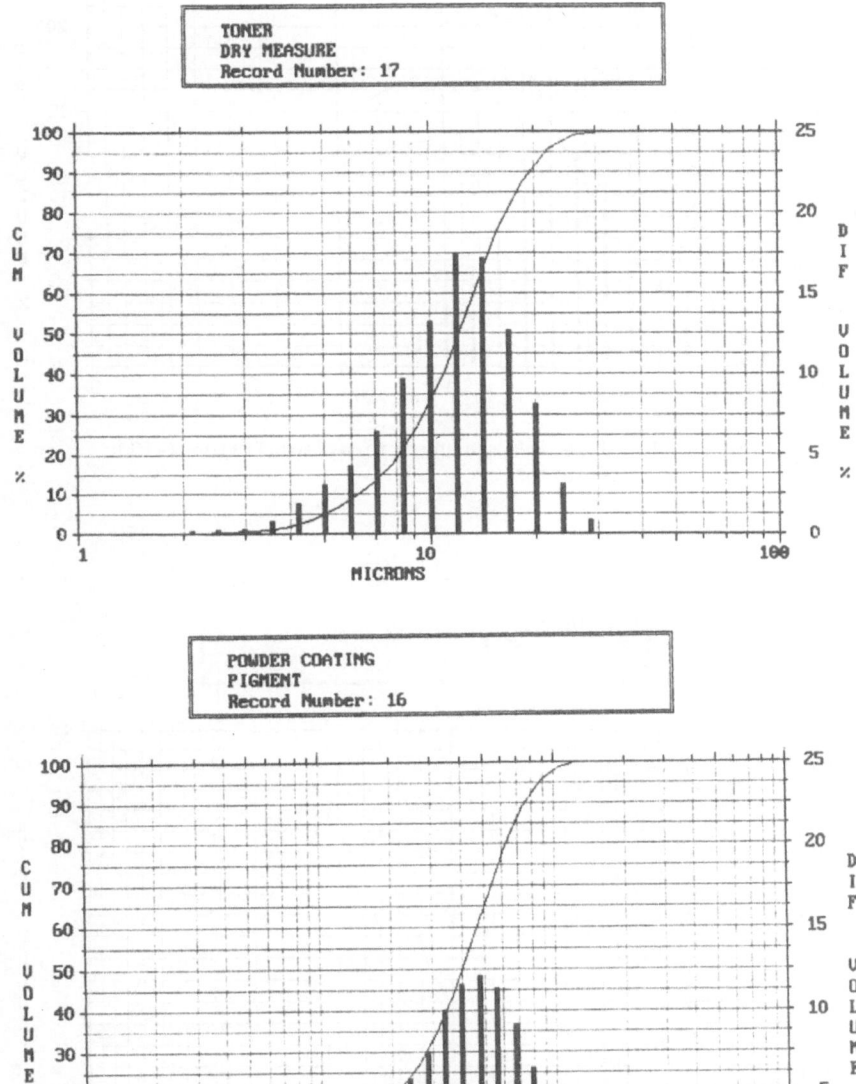

Figure 10

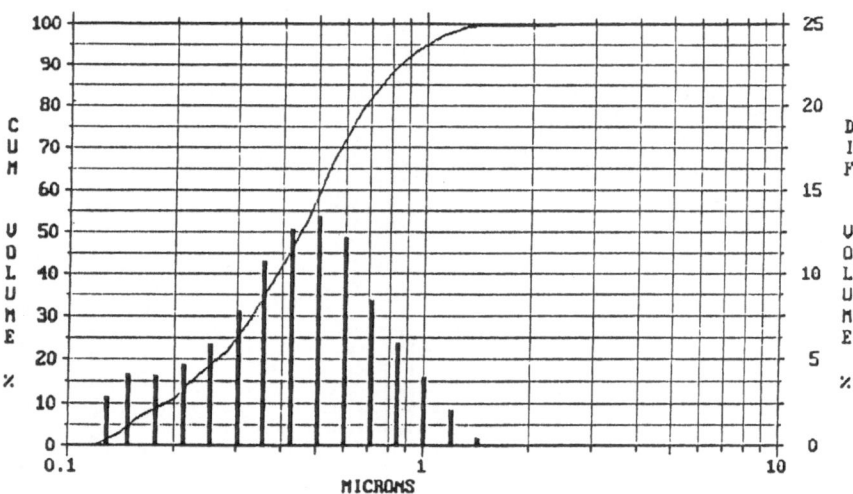

Figure 11. Blue ink measured in water/ Triton X-100
solution.

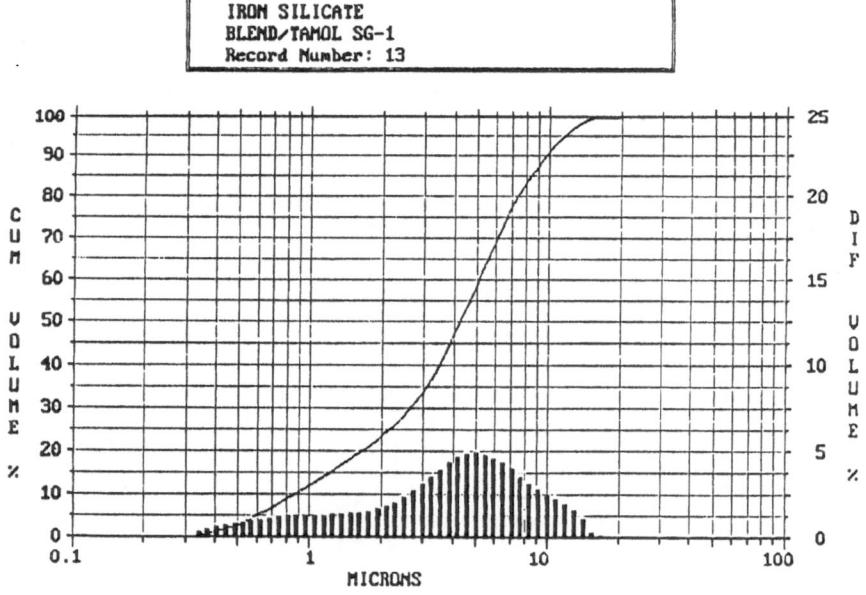

Figure 12. Iron silicate wetted with Tamol SG-1
and subjected to blending action to
complete dispersion.

Ultrasonic bath devices are available in a variety of sizes and power out-put. However, much of the energy is contained in the bath fluid and is not applied to the beaker contents. This loss of applied energy often results in poor dispersions and a requirement for extended treatment periods. Another facet of bath treatment includes a necessity to stir the sample during processing to prevent frictional attrition among particles resting on the beaker bottom. This is easily accommodated by use of an overhead stirrer. A particular problem in using an ultrasonic bath is that the energy profile across the bath surface is unequal. Beakers under treatment will migrate across the surface to a low energy region, thus effectively decreasing treatment. In other cases, the beaker is not placed in the same location of the bath for all preparations, causing non-repeatable dispersions.

An alternative to the bath is use of an ultrasonic probe that will provide considerable flexibility in treatment conditions. Consideration should be given to models featuring variable power out-put, timer, intensity feed-back circuitry and cycling. The latter two features are important to avoiding excessive treatment. Data obtained from sample preparations in this publication employed an ultrasonic probe manufactured by Sonics and Materials, Inc.

Cycling action where probe action is selectably discontinuous in a timed cycle allows particulate to escape null energy areas that form during treatment and effective reduction of treatment period and potential damage for all particles. Electronic feedback mechanisms provide constant intensity to provide repeatable treatment and are especially important when poorly dispersed particulate is attaining a dispersed state. Resulting production of finer particles could cause inconsistent intensity profile. Evaluation of conditions for use of the Sonics and Materials ultrasonic probe (3/8 inch probe tip) shows that treatment for 15 to 45 seconds at an output power of 100 watts does not attrite most materials. Generally, the shorter duration is sufficient for many materials, however, conditions should be evaluated for a specific material under investigation.

An alternative source of energy is that provided by movement of the particles. Blenders and tissue homogenizers represent such sources. Blenders provide energy as a result of the forces imparted to the particles by inducing high speed movement of particulate. As the agglomerated particulate travels through the suspending fluid, collisions with the container walls, blender blades, and each other result in particle separation into distinct units. Constant mixing action by the blades overcomes the possibly non-representative treatment caused by different particle speeds at the blade compared to those at the container wall.

Another shear device is the tissue homogenizer. This machine has an appearance similar to an ultrasonic probe and is used in a similar manner of inserting the probe directly into the slurry. The homogenizer tip is constructed to provide two concentric shafts the largest of which is approximately one-half inch in diameter. One portion of the shaft tip rotates at extremely high rpm while the other remains stationary. The concentric portions are very narrowly spaced and provide blades that makes the device extremely useful for shredding biological tissue samples. The same action of capturing sample between the rotating blades results in imparting extremely high shear to particulate with resulting particulate dispersal.

As with ultrasonic probes, both high speed mixing devices (blenders and homogenizers) have application limitations and may, in fact, be more damaging to particulate. The action of blenders and homogenizers can be considered to be similar to rod milling in that the particles may be forced against one another to impart dispersion. As such, particle attrition can occur. While dispersion of particulate is important, such milling action can result in erroneous or biased particle size reports. The high speed also causes extreme foaming problems when used with the presence of many surfactants in the suspension. As a result, restriction on the use of surfactant type must be accepted. Low-foaming surfactants or salts only may be used unless ample time, on the order of as much as two hours, is permitted for foam breakage. This would seem to be an undesirable effect when compared to instrument measuring periods as short as 15 seconds.

Where mechanical devices such as these high shear devices are implemented, movement of the parts causes wear on seals and the moving parts. Particles that arise as a result of this action can contaminate the sample. More abrasive particulate such as alumina or tungsten carbide will accelerate this action most easily observed as grey discoloration of white particulate under treatment.

An additional source of contamination is particulate sample trapped in the voids of seals. Future samples may be cross-contaminated when trapped sample washes into following slurries. Judicious cleaning and rinsing of equipment is an avoidance procedure for this common problem. However, the analyst should be apprised that rinsing equipment to achieve analytical transfer of sample to a beaker for sub-sampling can cause excessive dilution and destabilization of dispersion as a result of decreased surfactant concentrations.

An important aspect of sample preparation is reproducibility of technique and methodology. Often overlooked in studies of reproducibility in particle size measurements in this regard is the effect of differences in sources of energy discussed above. It is to be expected that use of a blender can produce results that are different from an ultrasonic bath but one should not assume that a particular ultrasonic probe will provide the same energy as that from a different manufacturer or that from the same manufacturer but with a different size probe tip. As shown in Figure 9, considerable difference can be expected in the total energy output of various devices. One must consider calibration of power produced by a particular type of device. Data for the experiments shown in the figure were performed using only water was subjected to various conditions of energy. Temperature measurements were taken as a function of time as a means to calculating the power resulting from each device by the equation:

$$H = M (T_f - T_i) S$$

Where H is the heat required to change a mass,
M having a specific heat, S, from an initial
temperature, Ti, to a final temperature Tf.

There are two aspects of the data that are of particular
interest. The Sonics and Materials ultrasonic probe shows a
different power profile than the Bransonic probe. This
suggests that calibration of the probes by the above or other
appropriate method is necessary. Ignoring this aspect can lead
to excessive or insufficient preparation with concomitant
particulate damage or continued agglomeration. In
addition, probe tips are subject to wear and degradation
producing less than expected power. Periodic evaluation of the
same probe, therefore, would seem necessary. This
consideration is especially important when a probe tip is
exposed to organic solvents that usually cause faster
degradation than aqueous environments. Similar arguments may
be forwarded for blenders and ultrasonic baths.

The second aspect is the comparison of power produced by
probe and blender. The probe produces more power and would be
expected to cause more damage to particulate than a blender.
However, we must consider that just as different milling
operations are necessary to produce desired results on various
materials, one must not tacitly assume that more power
correlates to a higher probability of particulate damage. Since
particulate is affected by shear differently than probe action
and the potential problems noted with a blender (above) may
cause a variety of effects including contamination, ultrasonic
energy may be a better choice of the two devices. This is
especially true for probes that provide flexibility of power,
cycling, time and intensity feed-back control.

It is the present opinion of the author that because of
the tremendous flexibility of ultrasonic probes, particle size
dispersions should be prepared using such a device. The other
devices described above may adversely affect reported particle
size and thus defeat the purpose of the measurement.

In some situations where sample preparation is
inappropriate, special devices can be used with laser particle
size analyzers to permit measurement of dry powders without
suspension in liquids.[14]   Dry powders are fed by mechanical
means to a position where particulate is gravity fed or
pneumatically accelerated across the path of the laser beam.
Depending upon the method of projection, particulate may be
dispersed very little thus leaving agglomerates intact. If
mechanical design provides sufficient energy, and agglomerates
are sufficiently loose, well dispersed particles may result.
The value of such dry analysis data should be weighed against
data by slurrying methods prior to measurement decision.   For
instance, dry powder coatings are applied in the dry state
while other pigment applications are slurried in solvent or
water. Dry measurement may be most practical in the former case
while slurry measurement, potentially employing sample
preparation, may be best in the latter.

The following (Figures 11-13) provide a variety of examples of particulate materials and associated sample preparations. From the foregoing discussion, appropriate surfactants, concentrations and particulars of energy devices must be evaluated for a given material. Additionally, it is not uncommon for these factors to vary for a particular material where process conditions including water source and plant equipment or plant location are different.

CONCLUSION

Information has been presented on various aspects and concerns of preparing sample materials for particle size measurement using low angle forward light scattering analyzers. Evaluation of appropriate surfactant includes wetting, foaming and concentration effects. The second step in preparing samples is selection of an energy device to separate agglomerated particles. Excessive use of any device can cause particle damage while insufficient application will not achieve complete sample dispersion. Selection of a device should include consideration of appropriate energy type, contamination effect and calibration of energy or power out-put. Similarly, careful evaluation of the necessity for sample preparation should be with due consideration of the objective of particle size measurement. Techniques such as these may be employed after deciding upon the necessity for preparation. All aspects of sample preparation as out lined must be scrupulously evaluated to achieve reproducible data.

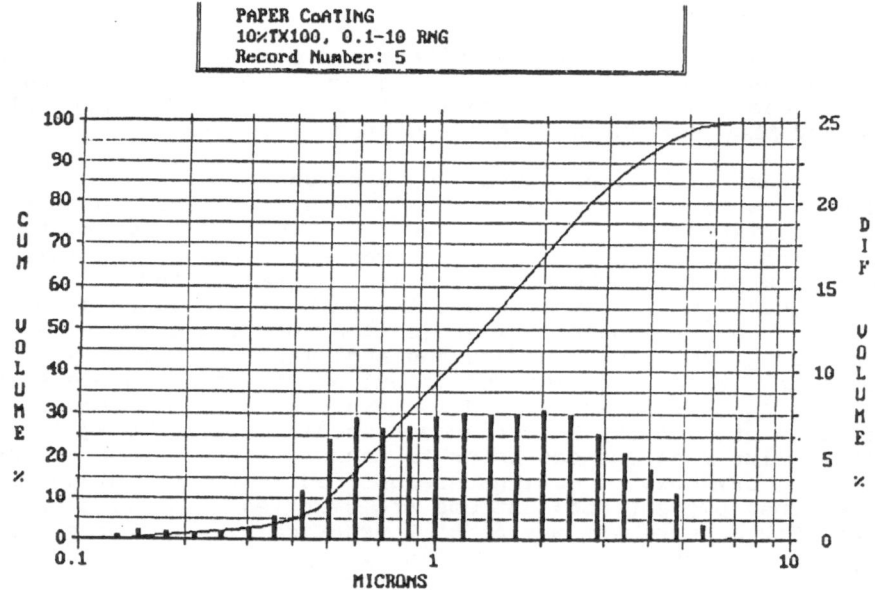

Figure 13. Paper coating treated with Triton X-100.

REFERENCES

1. Modern Methods of Particle Size Analysis
    Howard G. Barth, ed., 1984, John Wiley & Sons, Inc.

2. A.L. Wertheimer and W.L. Wilcock
    Appl. Opt. 15, 1616 (1976)

3. E.C. Muly, H.N. Frock, and E.L. Weiers, "The Application of
    Fourier Imaging Systems to Fine Particles", Seventh Annual
    Fine Particle Conference, Philadelphia, PA 1975

4. E.C. Muly and H.N. Frock, "Submicron Particle Size Analysis
    Using Light Scattering", Fine Particle Society Meeting,
    University of Maryland, College Park, MD., 1980.

5. G. Mie, Ann. Phys., 25 (3), 377 (1908)

6. H.C. Van de Hulst, "Light Scattering by Small Particles",
    Wiley, New York 1957, pp. 174-180.

7. Th.F. Tadros (ed.), Surfactants
    Academic Press, 1984.

8. R.H. Ottewill and M.C. Rastogi,
    Trans. Faraday Society 56,866,880 (1960).

9. A. Weissler, "Ultrasonics in Chemistry",
    J.Chem.Educ. 25 (28),(1948)

10. Cota-Robles, E.H., Marr,A.G. and Nelson, E.H.,
    "Submicroscopic particles in extracts of Azotobacter
    agilis", J. Bacterial. 75 (3), 243 (1958).

11. Apfel, R.E., "The tensile strength of liquids",
    Scientific American, 58-71 (Dec.1972).

12. Barenholz, Y. and Gibbs, D., "A simple method for the
    preparation of homogeneous phospholipid visicles",
    Brochem. 16, 2806 (1877).

13. W. H. Hart,"Description and Performance of a Dry Powder
    Pilot Scale Process Control System", Powder and Bulk
    Solids Conference, Chicago, IL., May 18, 1981.

FORMATION OF ULTRAFINE SILVER HALIDE PARTICLES IN MICROEMULSIONS USEFUL

FOR PHOTOGRAPHIC TECHNOLOGY

Mahendra K. Sharma

Research Laboratories
Eastman Chemical Company
Kingsport, TN 37662 (U.S.A.)

The formation of silver halide (AgX) particles in the
size range of 30-100 Å employing various microemulsions as
a reaction medium has been investigated. It has been found
that the size of AgX particles depends on the shape and
size of the dispersed phase of the reaction medium. As
shape and size of AgX particles play a prominent role in
the formation of latent image during photography, the
advantages of these ultrafine particles in the photographic
emulsions are also described. In order to understand the
process of ultrafine particle formation, several surfactant
association structures formed by surfactant, cosurfactant,
hydrocarbon and water are briefly discussed. An attempt
was made to correlate the formation of AgX particles with
dynamic structure of the reaction medium, e.g.
microemulsion.

INTRODUCTION

During the past two decades, much attention has been focused on the
formation of ultrafine metal or metal salt particles due to their
importance in several processes such as photography, reaction kinetics,
catalysis, performance of computer and electronic equipment, magnetic
tapes, automotive engines, etc.[1-7] A major problem of interest in the
formation of these particles is to achieve the monodispersed particles
in the size range smaller than 1000 Å in diameter. The discussion is
restricted in this article to the formation of ultrafine silver halide
particles and their importance in the photographic industry.

AgX Particles in Photography

The recording of an image by employing silver halide solutions has
been known since 1720. Among various chemicals incorporated to form a
photographic film, AgX particles play the most prominent role in the
development of an image during the photographic process. The scope and
importance of the AgX particles in color photography, as well as
complexity of the process have been discussed by Dahneke.[8] He has
demonstrated that a multi-layered color film before development shows
several gelatin layers coated onto various substrates such as film,

Surface Phenomena and Fine Particles in Water-Based Coatings and Printing Technology
Edited by M.K. Sharma and F.J. Micale, Plenum Press, New York, 1991

247

paper, etc. The color film contains silver halide grains coated with agents to sensitize them to light in selected spectral regions. Upon exposure and development, these AgX particles are dissolved and removed from the gelatin layers. These AgX particles convert to the silver atoms upon exposure to light by performing a photolytic reaction as follows:

$$Ag^+ X^- \xrightarrow{h\nu} Ag^0 + X^0 \qquad (1)$$

(Silver Atom)

The photolytic formation of very few silver atoms per grain results in an invisible change in the emulsion (e.g., latent image). This change produces an increased rate of image development in the locations in which latent image is formed. These silver atoms have sufficient mobility to form clusters at the grain surface in the partially developed photographic film.[8] During the process, the latent image acts as a microelectrode nucleus in the electrolytic reduction of the silver ions to elemental silver.

The product quality and process cost of the photography depends significantly on the size, shape and morphology of the AgX particles used in the photographic emulsions. For low cost and low graininess, the small AgX particle and low AgX concentration are required, whereas for sensitivity or speed, the high projected area to capture more photons and achieve a sufficient latent image at short exposure time is needed.

Several shapes and sizes such as cubic, octahedral, spherical, rod and 3D shaped AgX grains were formed in order to improve product quality and lower the cost of the photographic process.[8] As small AgX particles are important for photography, the ultrafine AgX particles were formed using microemulsions as a reaction medium. Before discussing the formation of these particles, various surfactant association structures are briefly described in order to understand the role of the composition and structure of reaction medium on formation of AgX particles.

Surfactant Association Structures

Surfactant molecules in solution beyond their critical micelle concentration (CMC) are widely known to form aggregates. Below CMC, surfactants are present in solution in the form of individual molecules. Figure 1 illustrates the formation of various association structures in solution with increasing surfactant concentration. It is likely that surfactant may form spherical, cylindrical, hexagonal, lamellar and reversed micelle (e.g. spherical) structures. If oil is present in the system, these association structures can solubilize oil, and can produce a clear, thermodynamically stable system. Depending on the nature of the oil phase and the oil/water ratio, the oil can be a continuous or disperse phase in the system.

The microemulsions are defined as the clear thermodynamically stable dispersions of two immiscible liquids containing appropriate amounts of surfactant or surfactant and cosurfactant. The dispersed phase stabilized by interfacial film of surface active molecules contains small droplets with diameter in the range of 100-1000 A.[9,10] At large oil/water ratios, an water-in-oil (w/o) microemulsion is formed, while at large water/oil ratios, an oil-in-water (o/w) microemulsion is formed. These two types of microemulsions are shown in

Figure 2. If the ratio of oil/water in the system is close to unity, it is likely that the structure of the system is quite different, and the system consists of microdomains of both liquids. Several studies on the structure of microemulsions[9-20] have been described by previous investigators.

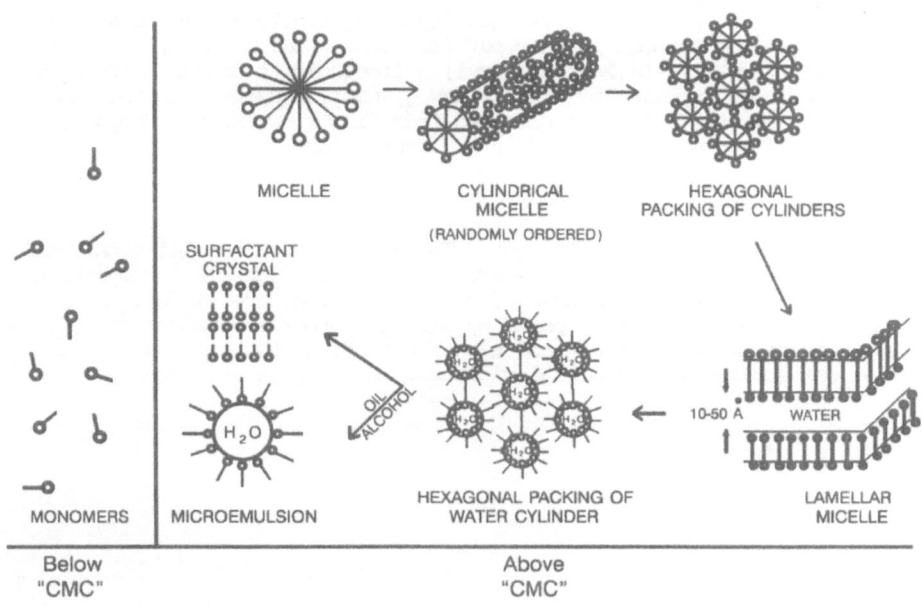

Fig 1. A Schematic Illustration of Surfactant Association Structures.

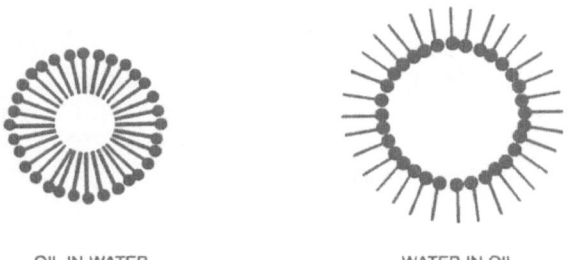

Fig 2. A Schematic Presentation of the Structure of Oil-in-Water (O/W) and Water-in-Oil (W/O) Microemulsions.

Figure 3 schematically illustrates the formation of silver chloride particles using microemulsions as a reaction medium. Two identical microemulsions were prepared except one contained aqueous AgNO₃ solution, whereas the second microemulsion contained aqueous NaCl solution as a dispersed phase. Upon mixing these two microemulsions, the AgCl particles are formed in the range of 30-100 A in diameter.

EXPERIMENTAL

Materials

The materials were of the purest quality commercially available and were used as received without further purification. Sodium dodecyl sulfate (SDS) was from BDH. Di(2-ethyl-hexyl) sulfosuccinate sodium salt (AOT, Aerosol OT) was supplied by Fluka. AgNO₃, NaCl, benzene, heptane and isopropyl alcohol (IPA) were obtained from Fisher Scientific Company. Double distilled water was used in all experiments.

Methods

Preparation of AgCl Particles. The aqueous solutions of 5 mM NaCl and of 5 mM AgNO₃ were first prepared separately. These solutions were stored for the preparation of microemulsions. The water-in-oil microemulsions were then prepared by mixing appropriate amount of SDS, IPA, benzene or AOT, heptane and either one of the two aqueous stored solutions so that two identical microemulsion samples at a desired

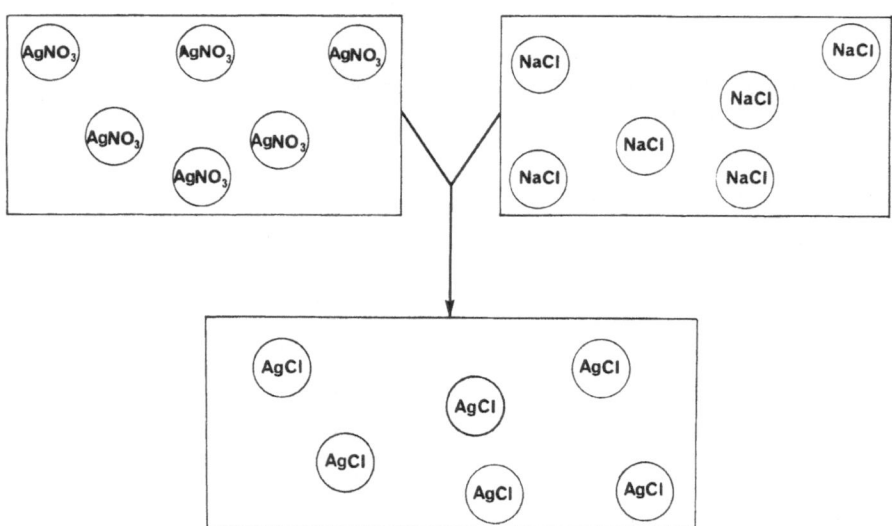

Fig. 3. The Schematic Illustration of the Formation of Ultrafine AgX Particles in Microemulsion Used as a Reaction Medium.

composition were prepared except one contained NaCl and the other contained AgNO$_3$. The formation of ultrafine AgCl particles is schematically illustrated in Figure 3. Upon mixing two microemulsions, the AgCl particles were formed in the range of 30-100 A in diameter. The coagulation rate of the AgCl sols can be determined using stopped-flow. The turbidity developed during mixing two microemulsions (e.g. coagulation of AgCl sols) was then monitored by transmittance measurements. The relative coagulation rate constant (K_{rel}) for AgCl[10] can be calculated by the following equation:

$$K_{rel} = \frac{\left[\dfrac{1}{Np} \dfrac{dln\ (I/Io)}{dt}\right]_{microemulsion}}{\left[\dfrac{1}{Np} \dfrac{dln\ (I/Io)}{dt}\right]_{water}} \qquad (2)$$

where Np is the number of particles per milliliter; I_o and I are the intensities of the incident and transmitted radiations respectively.

Electron Microphotographs. The microphotographs of AgCl particles were taken by transmission electron microscopy (TEM) with a Philips EM 300 G microscope. The microemulsion system containing AgCl particles was diluted with ethanol before deposition on the support because microemulsions used as a reaction medium contained high boiling solvents. Particles remaining in the suspension after dilution with ethanol could be deposited without extensive aggregation by evaporation of the solvent used as a continuous phase of the microemulsion. The microphotographs were taken by direct deposition of the particles on EM support or by preparing carbon replicas of particles deposited on carrier surface such as aluminium oxide and pumica. The microphotographs were analyzed for particle shape and size.

RESULTS AND DISCUSSION

Figure 4 shows the phase diagram of SDS/IPA/H$_2$O/C$_6$H$_6$ system. The w/o microemulsions selected for the formation of AgCl particles are shown by dotted line (PQ) in Figure 5. The volume of dispersed aqueous phase in the microemulsion system is higher close to Q as compared to P. The ratio of SDS/IPA is 0.5 by weight throughout the measurements to construct the phase diagram. It is observed that the turbidity occurs on mixing two microemulsions for the formation of AgCl particles. Based on earlier work,[11] the formation of the AgCl particles can be presented by a series of consecutive and simultaneous complex reactions (Figure 6). It is likely that AgCl particle formation includes nucleus formation (Step 1), crystal formation (Step II) and coagulation of small crystals (Steps III-IV) to form AgCl particles. As previous investigators[21] reported, the formation of AgCl nuclei involves about five ions, the variations in turbidity could not include the reaction Step I because small nuclei are most probably not perceivable.

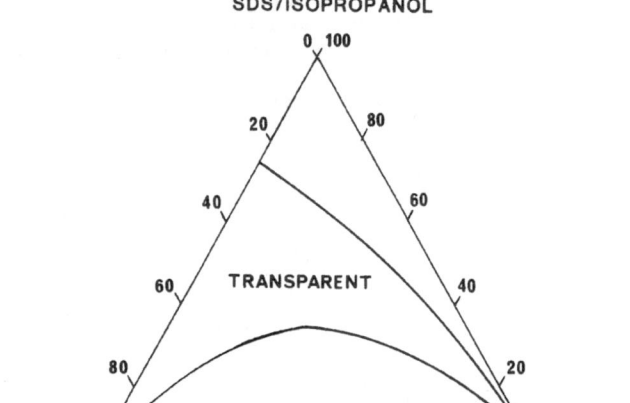

Fig. 4. Phase Diagram for SDS/IPA/Water/
Benzene System.

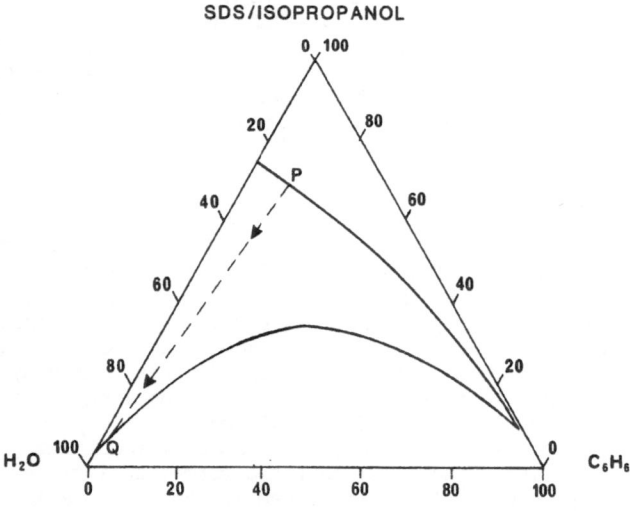

Fig. 5. The Selection of W/O Microemulsions
Containing SDS/IPA/Water/Benzene for the
Formation of AgCl Particles.

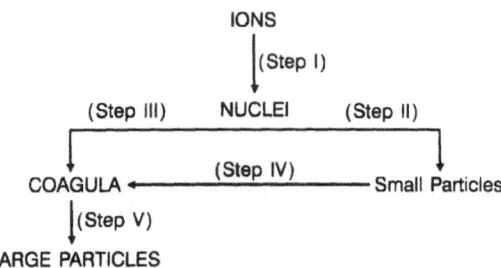

IONS

(Step I)

(Step III)   NUCLEI   (Step II)

COAGULA ←—— (Step IV) —— Small Particles

(Step V)

LARGE PARTICLES

Fig. 6.   Possible Reaction Steps for the
Formation of AgCl Particles.

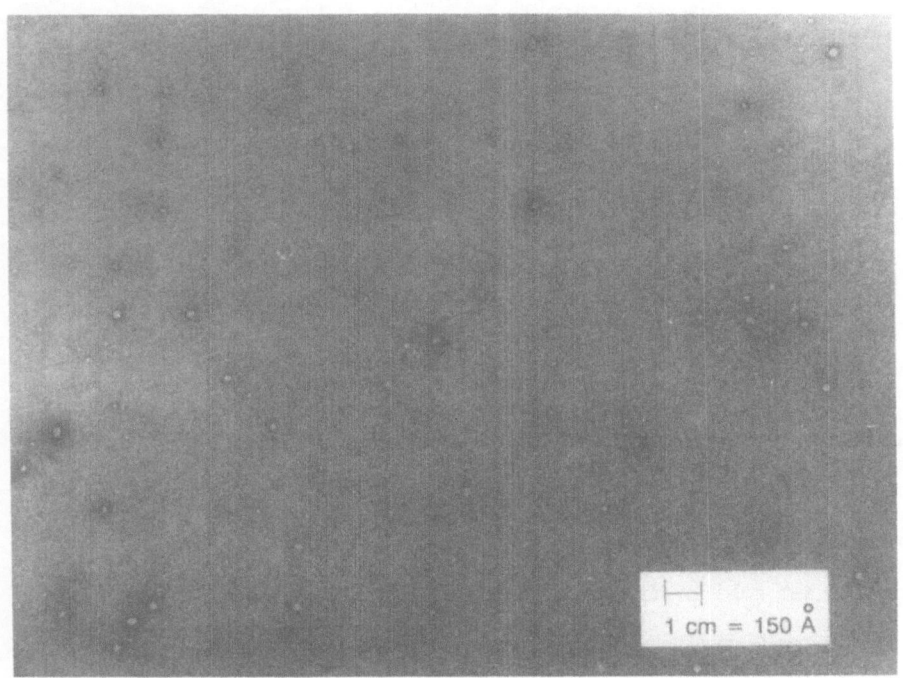

1 cm = 150 Å

Fig. 7.   Microphotographs AgCl Particles Formed in Microemulsions
Containing Water to Oil Ratio of 10 (W_o=10)

Therefore, the development in turbidity on mixing microemulsions for the formation of AgCl particles likely represents reaction Steps II-IV. The reaction kinetics of AgCl formation in microemulsion was studied in detail by Leung et al.[10]

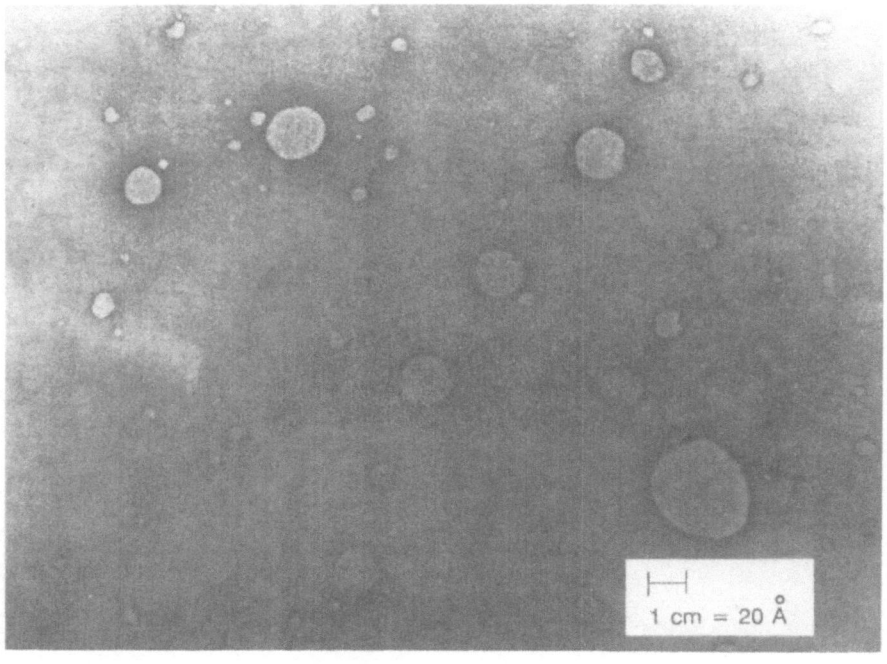

Fig. 8. Microphotograph of AgCl Particles Formed in Microemulsions Containing Water to Oil Ratio of 40 (W_o=40).

The electron microphotographs at different volumes of dispersed phase are shown in Figures 7 and 8. The size of the AgCl particles depends on the size of the dispersed phase of microemulsion. At low volume of dispersed phase, the size of the AgCl particles formed is small (Figure 7), whereas large AgCl particles are obtained with increasing volume of aqueous dispersed phase (Figure 8). In addition, it has been shown[10] that the dispersed phase kinetics play a significant role in the formation of AgCl particles in microemulsions.

Figure 9 shows the phase diagram of AOT/water/heptane system. The experimental results obtained on this system are listed in Table 1. It is observed that the size of the dispersed phase and AgCl particles increases with increase in the ratios of water/AOT and aqueous/oil. The size of the dispersed microemulsion droplets is larger in the presence of NaCl as compared to that in the absence of NaCl. The average diameter of AgCl particles is in the range of 30-80 A. A narrow size distribution (± 10 A) was observed for AgCl particles produced in microemulsion as a reaction medium. These results obtained are in good agreement with the literature values.[3-5,10]

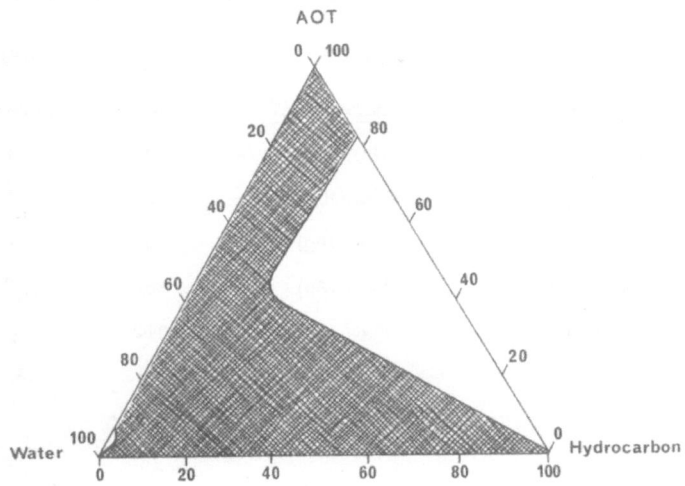

Fig. 9. Phase Diagram of AOT/Water/Heptane System

The size of the various metal or metal salt is recorded in Table 2. These metal or metal salt particles are produced in different surfactant systems used as a reaction medium by several investigators. These data demonstrate that the particles of metal/metal salts can be produced in the range of 20-100 A in diameter using different surfactant systems.

Table 1. Average Size of Dispersed Aqueous Phase and AgCl Particles Formed in AOT/Heptane/Water System.

Water/AOT Ratio (W_o)	Wt. Ratio of Aqueous/ Oil Phase	Diameter of Dispersed Phase[22] (Å)	Diameter of Dispersed Phase (5 mM NaCl in Water) (Å)	Average Diameter of AgCl Particles (Å)
10	0.018	90	108	34
20	0.036	114	170	42
40	0.072	206	201	70
60	0.108	294	299	76
75	0.135	-	404	-

AOT = 0.1 M, Heptane = 50 ml.

Table 2. Particle Size of Metal/Metal Salt in Different Microemulsion
          Systems.

Reaction Medium	Metal/Metal Salt	Particle Diameter Å	Reference
(1) PEGDE/Hexane/Water	Platinum (Pt), Rhodium (Rh)	30	(4)
(2) PEGDE/Hexane/Water	Palladium (Pd)	50	(4)
(3) AOT/Heptane/Water	Silver (Ag)	60	(5)
(4) SDS/IPA/Benzene/Water	Silver (Ag)	50	Present work & (10)
(5) AOT/Hexane/Water	Silver (Ag)	30-80	Present work
(6) AOT/Heptane/Water	Cadmium Sulfide (Cds)	<28	(3)

CONCLUSIONS

The microemulsions can be used as a reaction medium to produce
silver halide particles in the range of 30-80 A. The size distribution
of AgX particles produced in microemulsions is in the narrow range of
±10 A. It was observed that the size of the AgX particles depends on
the size of the dispersed phase of the microemulsion. These ultrafine
AgX particles may be useful for photographic technology.

REFERENCES

1. P. Lianos and J. K. Thomas, Chem. Phys. Letters; 125(3), 299
    (1986).
2. O. I. Micic, L. Zongglean, G. Mills, J. C. Sullivan, and D. Meisel,
    J. Phys. Chem. 91, 6221 (1987).
3. P. Lianos and J. K. Thomas, J. Colloid Inter. Sci., 117(2), 505
    (1987).
4. M. Boutonnet, J. Kizling, P. Stenius, and G. Maire, Colloid and
    Surfaces, 5, 209 (1982).
5. M. Dvolaitzky, R. Ober, C. Taupin, R. Anthore, X. Auvray,
    C. Petipas, and C. Williams, J. Disp. Sci. Tech., 4(1), 29
    (1983).
6. M. Dvolaitzky, M. Guyot, M. Laques, J. P. LePesant, R. Ober,
    C. Sauterey, and C. Taupin, J. Phys. Chem. 69, 3279 (1978).
7. R. Ober, and C. Taupin, J. Phys. Chem. 84, 2418 (1980).
8. B. Dahneke, Particulate Sci. Technol., 5(1), 1, (1987).
9. M. K. Sharma, and D. O. Shah, ACS Symposium Series No. 272, 1-18
    (1985).
10. R. Leung, M. J. Hou, C. Manohor, D. O. Shah, and P. W. Chun, ACS
    Symposium Series No. 272, 325-344 (1985).
11. R. H. Ottewill, and A. Watanabe, Kolloid-Z. 170, 38, 132 (1960);
    Kolloid-Z. 171, 33(1960); Kolloid-Z. 173, 7(1960).
12. R. Leung, D. O. Shah, and J. P. O'Connell, J. Colloid Interface
    Sei., 111, 286 (1986).
13. H. Soo, and C. J. Radke, J. Colloid Interface Sci., 102, 162
    (1986).

14. R. Leung, and D. O. Shah, J. Colloid Interface, Sci. 120(2), 320-329 (1987).
15. M. Kunedia, and S. E. Fnberg, Bull. Chem. Soc. Japan, 54, 1010 (1981).
16. M. K. Sharma, S. Y. Shiao, V. K. Bansal, and D. O. Shah, ACS Symposium Series 272, 87-103 (1985).
17. J. L. Salager, Ph D Dissertation, The University of Texas at Austin (1977).
18. F. Verzaro, M. Bourrel, and C. Chambu, Solubilization in Microemulsion. In: Surfactants in Solution, Vol. 6, K. L. Mittal and Bothorel, P. Editors, Plenum Press, New York, pp 1137-1157 (1986).
19. P. Becher, Ed. Encyclopedia of Emulsion Technology, Vol. 1, Marcel Dekker, Inc., New York (1983).
20. W. Grerbacia, and N. L. Rosano, J. Colloid Interface Sci., 44, 262 (1973).
21. D. H. Klein, L. Gordon, and T. H. Walnut, Talanta, 3, 107(1959).
22. H. F. Eicke, Pure and Applied Chem. 52, 1349 (1980).

ELECTROKINETIC AND OPTICAL STUDY OF ORDERED POLYMER COLLOIDS

Y. P. Lee and F. J. Micale

Zettlemoyer Center for Surface Studies
Sinclair Lab. #7, Lehigh University
Bethlehem, PA 18015, U. S. A.

The electrokinetic measurements demonstrate the necessity of pH control and the potential influence of salt concentration on the ordering phenomena. According to the modified concepts of effective hard sphere model and Kirkwood-Alder transition theory for ordered dispersions, the predicted phase diagram with the input of electrokinetic potential information agrees qualitatively with the literature data. This agreement implies that the order-disorder transition of a monodisperse latex dispersion results from electrostatic repulsive forces and has the same origin as the hard-sphere transition. Reflection spectrophotometry was employed to study the particle concentration and temperature dependence on the ordered structure of deionized monodisperse latices. Based on Bragg's law, the experimental nearest interparticle distance was determined from the wavelength at the reflection peaks. The results show that the peaks become broader and are moved toward smaller wavelength with rising particle concentration. The experimental nearest interparticle distance (De) agrees reasonably well with the theoretical distance (D_t) calculated on the assumption of a perfect crystal lattice. The order-disorder transition temperature estimated from the attenuation of reflection intensity increased with increasing latex concentration. The shift of peak position with temperature was interpreted as lattice structure transition, i.e. f.c.c. to b.c.c. with increasing temperature. The similarity of the structure suggests that the ordered latex behaves as a crystal.

INTRODUCTION

A polymer latex colloid is a surfactant-stabilized colloidal dispersion of spherical polymer particles in a continuous medium. When formed by emulsion polymerization, a typical synthetic latex is polydisperse (has a broad particle size distribution) and appears milky white . The first monodisperse latices were of polystyrene and were produced by the Dow Chemical Co. in the early 1950's. Since then, uniform polymer latices have attained considerable interest as a model system for the investigation of various colloidal phenomena.

Surface Phenomena and Fine Particles in Water-Based Coatings and Printing Technology
Edited by M.K. Sharma and F.J. Micale, Plenum Press, New York, 1991

259

One of the remarkable features of monodisperse latices is the iridescence due to Bragg diffraction of visible light from ordered arrays of particles[3]. This ordering is destroyed with the addition of excess salt [22], the reduction of particle concentration [23], or an increase in temperature [24]. It is believed that each of the above factors suppresses the electrostatic repulsions localizing the particles within the lattice, allowing Brownian motion to collapse the structure. Hence, Hachisu and coworkers[14] attempted to map out the phase diagram, showing order-disorder transition, by simple visual observation of iridescence.

It is of theoretical interest to relate the microstructure and macroscopic rheological properties of dispersion to interparticle forces and thence to the effect of variables, such as particle concentration, temperature and salt concentration. Although ordered polymer colloids have limited practical importance, they do provide an excellent model for the study of multiparticle electrostatic and hydrodynamic interactions. This paper represents the first effort in a series of studies to elucidate the interplay among colloidal forces, dispersion microstructure and resulting rheological properties of ordered polymer colloids and to investigate the order-disorder phase transition behaviors of monodisperse latex.

EXPERIMENTAL

Materials

Dow polystyrene latices were used in this study and designated A. Serum replacement technique[1] was employed to remove electrolytes, oligomers and any surfactants in the latex. It has been shown[1] that this cleaning method gives characterization results for latexes in good agreement with the ion-exchange technique. The water used for the latex cleaning and for dispersion preparation was distilled and deionized by Milli-Q water system. The filtration cell used in serum replacement can be applied to concentrate the latex. The volume fractions were determined gravimetrically.

Transmission Electron Microscopy

Transmission electron micrographs of latex A particles were obtained with the Philips EM-300. The magnification of the instrument was calibrated by a grating replica.
Approximately 400 particles were measured on a Zeiss MOP3 image analyzer in order to obtain the particle size distribution.

Conductometric Titration

After serum replacement, the cleaned latex A was weighed into the titration beaker, diluted by distilled deionized water, and then sparged with nitrogen for 20 minutes. The latex was then titrated conductometrically under a nitrogen blanket with 0.02 N standard sodium hydroxide solution added continuously at a rate of 1 cc/min under continuous stirring. Other details of this surface characterization technique have been reported earlier [1]

Fourier Transform Infrared Spectroscopy

Samples of the purified latices A were examined with a Sirius 100 FTIR Spectrometer (Mattson Instrument, U.S.A.). Spectra were obtained on the solid polymer in the form of a thin film. The film was made by casting a solution of freeze-dried latex in carbon tetrachloride onto the rock salt window and removing the solvent *in vacuo*.

A pure polymeric film (Analect Instrument-IR standard) designated as standard polystyrene was used in order to establish a standard in spectra band assignments. The spectrum of this standard sample did not display any bands which could be attributed to surface functional groups.

Electrophoresis

Electrophoretic mobilities of latex A were measured directly by microelectrophoresis apparatus, Pen Kem System 3000 (Pen Kem, Inc., U.S.A.), at infinite dilution for a range of pH and NaCl concentrations. pH was adjusted by the addition of acid or alkali.

Reflection Spectrophotometry

Reflection spectra of latices A in visible and near-infrared ranges were recorded on a Cary 2300 spectrophotometer. Quartz optical cells were used, as will be described later. Regarding the study of temperature dependence on the reflection spectra, the spectra measurements were executed ten minutes after sample cells immersed in a thermostatic waterbath had reached the desired temperature.

Viscometry

Bohlin VOR Rheometer (Bohlin Reologi, Inc., U.S.A.) was used to measure the steady state viscosity of the latex samples. A couette measuring geometry was chosen. Calibration of the instrument was carried out using Newtonian mineral oil (Cannon Instrument Co., U.S.A.). In order to avoid errors due to evaporative drying, high humidity was maintained around the sample by an aqueous aerosol spray .

THEORY

Bragg Diffraction

(a) Experimental Nearest Interparticle Distance (De)

If the distance between the particle centers is of the order of the wavelength of visible light, the monodisperse latex will show a brilliant iridescence due to Bragg diffraction. In order to analyze light diffraction data, Krieger[3] introduced equation [1] by analogy of the Bragg's law in X-ray diffraction.

$$2d_{(hkl)} \sin \theta = m \left(\frac{\lambda_0}{n_s} \right)$$
[1]

where θ is the diffraction angle in the latex, $d_{(h k l)}$ is the interplanar spacing in a crystal structure, h,k,l is the Miller indices, m is the diffraction order, λ_0 is the wavelength of the light in air, n_s is the mean refractive index of latex suspension.

The optical cell used for reflection spectra measurements is shown in Fig.1. The refractive indices of air, quartz glass, water and polystyrene particle are designated by n_a, n_g, n_w and n_p respectively. n_s can be estimated from the empirical equation [2][3] where ϕ denotes the volume fraction of the latex.

$$n_s = (1-\phi) n_w + \phi n_p$$
[2]

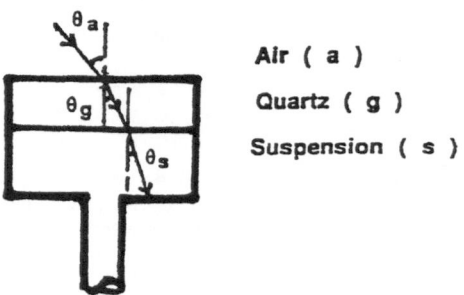

Fig. 1   Optical cell for reflection spectra measurements

Applying Snell's law at the air-quartz-latex interface, we have

$$\frac{\sin \theta_a}{\sin \theta_g} = \frac{n_g}{n_a} \qquad [3]$$

$$\frac{\sin \theta_g}{\sin \theta_s} = \frac{n_s}{n_g} \qquad [4]$$

[3] x[4]
$$\frac{\sin \theta_a}{\sin \theta_s} = \frac{n_s}{n_a} \qquad [5]$$

So
$$\sin \theta = \sin(\frac{\pi}{2} - \theta_i) = \cos \theta_s = [1 - (\frac{n_a}{n_s})^2 \sin^2 \theta_a]^{\frac{1}{2}} \qquad [6]$$

Because the type of crystal lattice formed in the ordered latex is usually face-centered cubic (f.c.c.) or body-centered cubic (b.c.c.)[4], following analysis will be based on these structures. Furthermore, best fit between experimental and theoretical nearest inter- particle distance (De and D_t, respectively) was found if (111) plane of the f.c.c. lattice or (110) plane of b.c.c. lattice was parallel to the interface of optical cell.

Therefore
$$d_{(111)} \text{ for fcc} = (\frac{1}{3})^{\frac{1}{2}} b = (\frac{2}{3})^{\frac{1}{2}} De \qquad [7]$$

$$d_{(110)} \text{ for bcc} = (\frac{1}{2})^{\frac{1}{2}} b = (\frac{2}{3})^{\frac{1}{2}} De$$

where b is the unit cell dimension.   Substitute equation [6], [7] into equation [1], we have

$$2(\frac{2}{3})^{\frac{1}{2}} De [1 - (\frac{n_a}{n_s} \sin \theta_a)^2]^{\frac{1}{2}} = \frac{m\lambda_0}{n_s} \qquad [8]$$

In the present study, θ_a is equal to 3.33° and n_a, n_w and n_p were taken as 1.00, 1.33 and 1.60, respectively.   De was then determined from the wavelength at the reflection peak.

(b) Theoretical Nearest Interparticle Distance (D_t)

The experimental nearest interparticle distance (De) will be compared with the theoretical one (D_t) computed on the assumption of perfect crystalline lattice. Consider the volume occupied by the spheres and unit cell, we have

$$\text{for fcc}: \quad \frac{4\frac{4}{3}\pi\left(\frac{D_0}{2}\right)^3}{b^3} = \phi \qquad [9]$$

$$\text{for bcc}: \quad \frac{2\frac{4}{3}\pi\left(\frac{D_0}{2}\right)^3}{b^3} = \phi \qquad [10]$$

where Do is the particle diameter. The relationship between the theoretical nearest interparticle distance and the unit cell dimension is already known as

$$\text{for fcc}: \quad D_t = \frac{\sqrt{2}}{2}b \qquad [11]$$

$$\text{for bcc}: \quad D_t = \frac{\sqrt{3}}{2}b \qquad [12]$$

Substitute equation [11], [12] into equation [9], [10], respectively, we have

$$\text{for fcc}: \quad \left(\frac{D_0}{D_t}\right)^3 = \frac{\phi}{0.74} \qquad [13]$$

$$\text{for bcc}: \quad \left(\frac{D_0}{D_t}\right)^3 = \frac{\phi}{0.68} \qquad [14]$$

Phase Diagram

To date, several attempts have been made to account theoretically for the phase transition of monodisperse latices. One approach is to assume that colloidal particles interact via a DLVO pair potential[5] and then the system properties are examined by the Monte Carlo simulations[6]. Unfortunately, such an approach cannot account for the order-disorder coexistence regime. Another approach is concerned with the Kirkwood-Alder Transitions[7] by specifying effective hard sphere diameters in terms of the double layer thickness ($1/\kappa_0$, see equation [15])[8]. Barnes and coworkers[8] provide no dependable criterion for selecting magnitude of Debye Length.

Fig. 2   Schematic representation of effective hard sphere concept

$$(a\kappa_0)^2 = \frac{a^2 e^2 (\sum n_{i0} Z_i^2)}{\varepsilon_0 DkT} \qquad [15]$$

where a is particle radius, D dielectric constant of continuous medium (78.54 for water at 25° C), ε_0 permittivity of free space (8.854×10^{-12} coul/v.m), k Boltzmann constant (1.38×10^{-16} erg/molecule °K), T absolute temperature (°K), n_{i0} ion concentration in the bulk of the solution (mole/l) and Z_i charge per ion.

In the present analysis, the second approach is used but with corrected double layer thickness (l/κ, equation [16]). Russel[9] has made a correction to the Debye length by considering the counterion density and the reduction in the fluid volume by the presence of particles. Consequently l/κ is shorter than that at infinite dilution, l/κ_0.

$$(a\kappa)^2 = (\frac{1}{1 - \phi}) [(a\kappa_0)^2 + 3\phi (1+a\kappa_0) (\frac{e\phi_0}{kT})] \qquad [16]$$

where ϕ_0 is the surface potential of particle in mv.

Fig.2 is the schematic representation of effective hard sphere concept. The Kirkwood-Alder transition theory[7] has located the melting point of the hard sphere assembly. There-fore, prediction of the order-disorder transition requires recognizing that there exists a solid state when

$$\phi_{eff} = (\frac{a + \frac{1}{\kappa}}{a})^3 \phi_s = (1 + \frac{1}{\kappa a})^3 \phi_s \geq 0.55 \qquad [17]$$

and a fluid state when

$$\phi_{eff} = (\frac{a + \frac{1}{\kappa}}{a})^3 \phi_l = (1 + \frac{1}{\kappa a})^3 \phi_l \leq 0.50 \qquad [18]$$

Combining equations [15]-[18], a phase diagram can be constructed under various conditions with the assumption that the zeta potential obtained from electrophoretic measurement approximates surface potential of the particle.

Table 1. Characterization of latex A

Diameter	233	nm
Standard Deviation	±14	nm
Surface Charge Density		
Strong Acidic Group	2.46	$\mu coul/cm^2$
Weak Acidic Group	9.70	$\mu coul/cm^2$

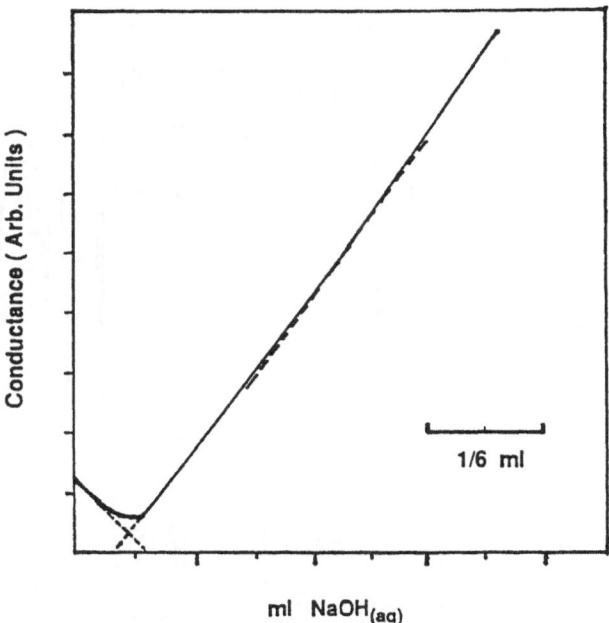

Fig. 3   Conductometric titration of cleaned latex A (0.2474 g solids in 60 g dispersion) with 0.02 N $NaOH_{(aq)}$

EXPERIMENTAL RESULTS AND DISCUSSIONS

Characterizations of Model Colloids

Model colloids are defined as stable dispersions of rigid, monodisperse spherical particles of constant surface charge. The surface of model colloids must also be characterized regarding the nature, concentration, and distribution of functional groups. Table 1   summarizes the particle size and surface charge density based on the TEM and conductometric titration results. In view of the small coefficient of variation on the mean particle diameter, the particle size distribution was considered to be monodisperse. The two inflection points observed in the titration curve of cleaned latex A (Fig. 3) indicate the coexistence of strong and weak acidic group on the particle surface. Additional evidence for this point will be offered in the IR spectra study.

Comparing the IR spectra between (a) standard and (b) latex A (Fig. 4), very weak but significant absorptions at 1200, 1230, 1300 cm^{-1} in (b) were found and attributed to the vibrations of surface sulfate group (10). This is expected since latex A was synthesized by the $K_2S_2O_8$ initiated emulsion polymerization process. The absorption frequency shift from 1675 in (a) to 1657 cm^{-1} in (b) ascribed to the carbonyl group vibrations provides complementary support to the observation of extra absorption band at 1770 cm^{-1} in (b). The latter absorption has been assigned to the C=O stretching of phenylacetic acid[11]. The absorption at 3590-3650 cm^{-1} in (b) is considered to be free OH stretching. Based on the above discussions, it is strongly suggested that latex particle carries both strong (sulfate) and weak (carboxyl) acidic group and possibly hydroxyl group.

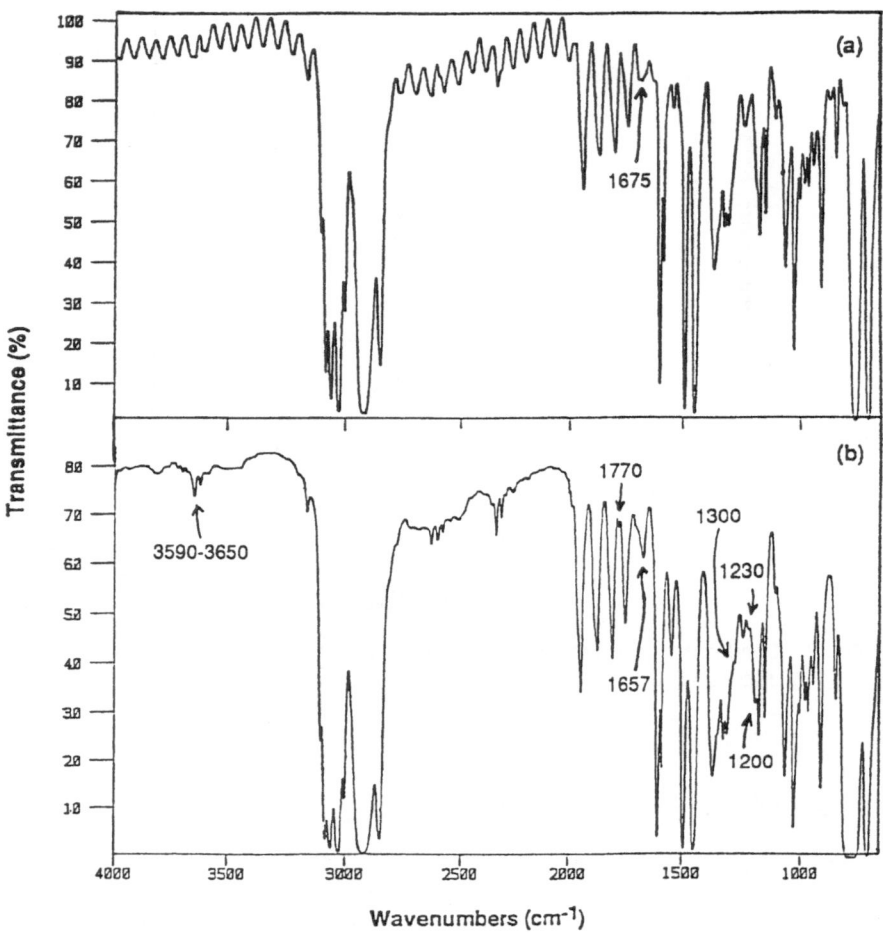

Fig. 4 Infrared spectra of (a) standard polystyrene sample (b) latex A

Investigations on the Particle Interactions

Based on the Derjaguin-Landau-Verwey-Overbeek theory (DLVO theory) of colloidal stability[5], the interparticle interaction consists of electrostatic repulsion and van der Waals attraction. In the ordered latex, the short range attractive van der waals force can be neglected when compared with the electrostatic repulsion due to overlay of electrical double layers[12]. The important properties of the electrical double layer that are experimentally available are the zeta potential and the thickness of double layer. Unfortunately, the electrokinetic potential or other relevant measure of the particle charge is rarely reported on the ordered dispersion system. In the present study, preliminary electrophoretic mobility measurements of latex A were carried out to examine the potential impact made by the pH and sodium chloride concentration on the ordered latex system.

Fig. 5 shows the effect of pH on the electrophoretic mobility of latex A at various NaCl concentration. The general patterns were essentially identical for latices at three different

NaCl concentrations. At low pH, the latex particle adsorbed enough protons to undergo charge reversal, the trend being stronger for samples with higher NaCl concentration, since the isoelectric point increases with increasing solution NaCl concentration. With further increasing pH , particle mobility increases due to the dissociation of surface acidic groups. At intermediate pH ranges ca. 5-9, the mobility reaches a phateau where complete ionization occurs. Also, the number of added ions was too small to cause any appreciable changes in mobility within this pH range. At higher pH, the mobility climbs up to a maximum followed by a rapid decrease. The increase was attributed to the preferential adsorption of OH⁻ on latex particle. The final decrease of mobility was ascribed to the compression of the electrical double layer at higher ionic concentrations-similar to that at extremely low pH.

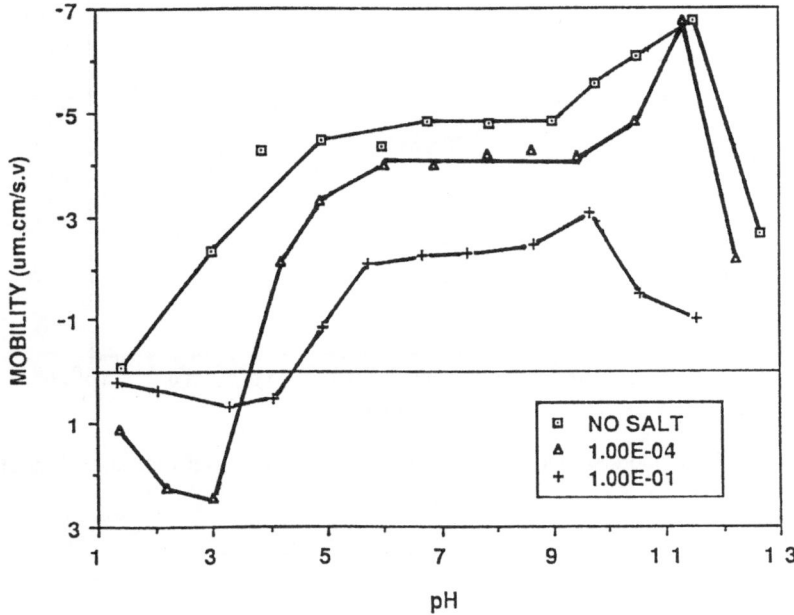

Fig. 5   Electrophoretic mobility vs. pH for latex A at various NaCl Concentration (mole/l)

The effect of NaCl concentration on the electrophoretic mobility of latex A is illustrated in Fig. 6. It is shown that there is a gradual decrease in mobility up to a NaCl concentration of 10^{-4} M, after which the mobility increases to a maximum at about 10^{-2} M followed by a dramatic decrease. The initial decrease was attributed to the decreasing ionization potential of surface groups with increasing concentration of Na^+ ions. The maximum was believed to be resulted from the adsorption of Cl^- ions on the hydrophobic parts of the particle surface. The final drop was associated with the compression of diffuse double layer.

The corresponding zeta potentials for the measured electrophoretic mobilities are given in Fig. 7. Zeta potentials for the latex particle at different NaCl concentration were obtained from Wiersema's numerical tables[2] which take into acount the electrophoretic retardation due to backward flow of the solvent and the relaxation effect due to distortion of the ionic atmosphere surrounding the particle. The zeta potentials decrease monotonically with increasing ionic strength or decreasing double layer thickness $1/\kappa_0$ except at very low concentration of NaCl.

The results in Fig. 5 and 6 support the concept of a dual charging mechanism on polymer latex surfaces which was proposed by Micale and co-workers[13]. One mechanism is controlled by the ionization of surface groups. The other mechanism is controlled by the nature of the bare polymer surface which preferentially adsorbs anions.

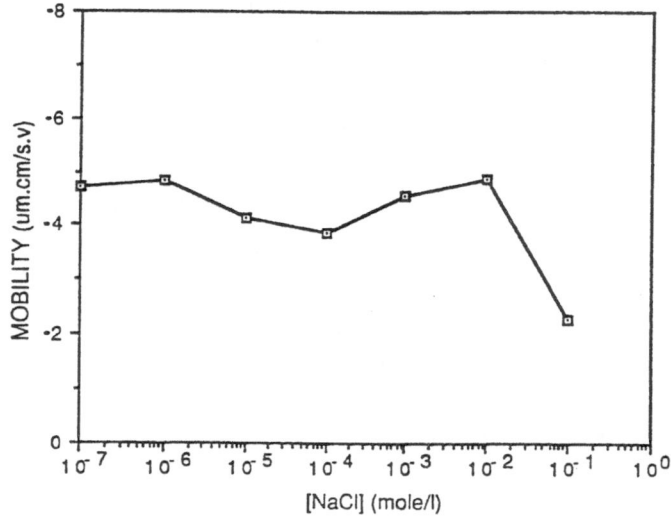

Fig. 6   Electrophoretic mobility vs. molar concentration of NaCl for latex A at 25 °C

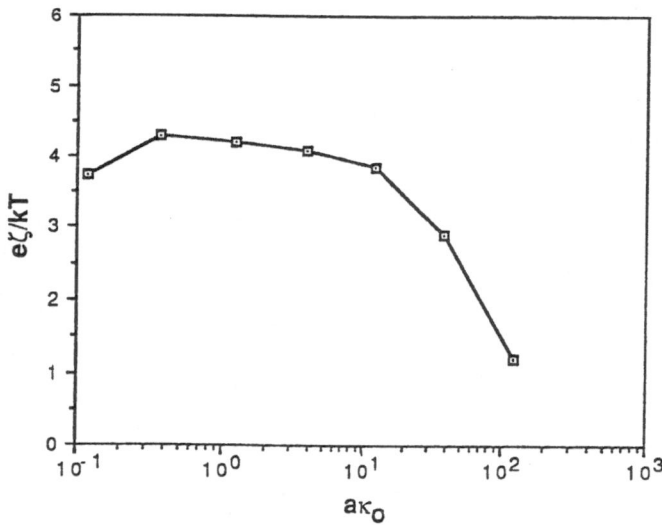

Fig. 7   Dimensionless zeta potential vs. $a\kappa_0$ for latex A in NaCl aqueous solution at 25 °C

With the information of zeta potential for latex A, prediction of the phase diagram from the theory discussed previously can be made, as shown in Fig. 8. Although no detailed experimental data are available to test the validity of these predictions, these phase boundaries conform qualitatively with the data of Hachisu[14] in terms of the sigmoidal trend and the width of the coexistence region. It is also revealed in Fig. 8 that the ordered phase will occur at lower latex particle concentrations with lower salt concentration. Moreover, phase transition lines are almost independent of the ionic concentration at very high concentration. This is anticipated since the double layer thickness around a particle under this condition is small in comparison with the particle radius. It is believed that further optical and rheological studies on the monodisperse latices will cast new light on the understanding of order-disorder phase transition.

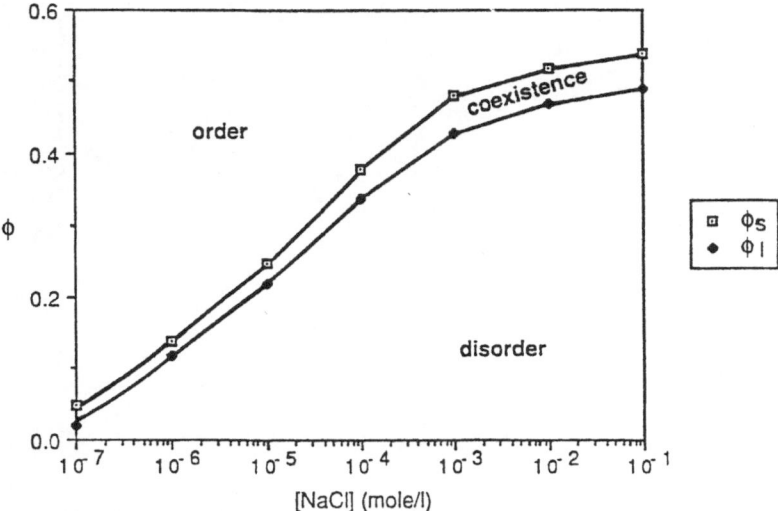

Fig. 8 Phase diagram showing order-disorder transition for latex A at 25 °C, as predicted by effective hard sphere model

Explorations of Dispersion Microstructure

The crystal-like structure of ordered latices has been investigated by several workers. Hachisu[15] demonstrated unequivocally the presence of a regular latticelike distribution of the particles in dilute dispersion by using a metallurgical microscope. The photograph taken by Hachisu[15] showed clearly the ordered region coexisting with the disordered region. Williams and Crandall[16] found, by the means of Bragg diffraction of laser light, that at low particle concentraions only a body-centered cubic structure exists whereas it coexisted with a face-centered cubic structure at high concentrations. The choice of experimental technique for observing ordering depends on the dimensions of the supra-molecular species under observation and may include one of the many radiation scattering techniques such as X-ray scattering[17].

The effect of particle concentration and temperature on the ordered structure of deionized monodisperse latexes were studied in this work by reflection spectrophotometry. Fig. 9 exhibits the reflection spectra of deionized latices A at various particle concentrations. It is shown that the reflection peaks become progressively broader and are shifted in the direction of smaller interparticle spacings with increasing particle concentration. The increase in peak sharpness with dilution is attributed to the increase of crystal thickness[18]. Because irradiations are strongly scattered by the particles in the dispersion, it is anticipated that intensity of the reflection peak becomes weaker with higher diffraction order. Indeed, secondary peaks (m=2), although weak, appear at half the wavelength of primary peaks (m=1) and tertiary peaks (m=3) are too weak to be observed. It is also interesting to note that the reflection peak was found at a volume fraction as low as 4.37 % where the particles are 2.5 diameters apart. This implies that long-range electrostatic repulsion between particles is related to the existence of order in dilute dispersion.

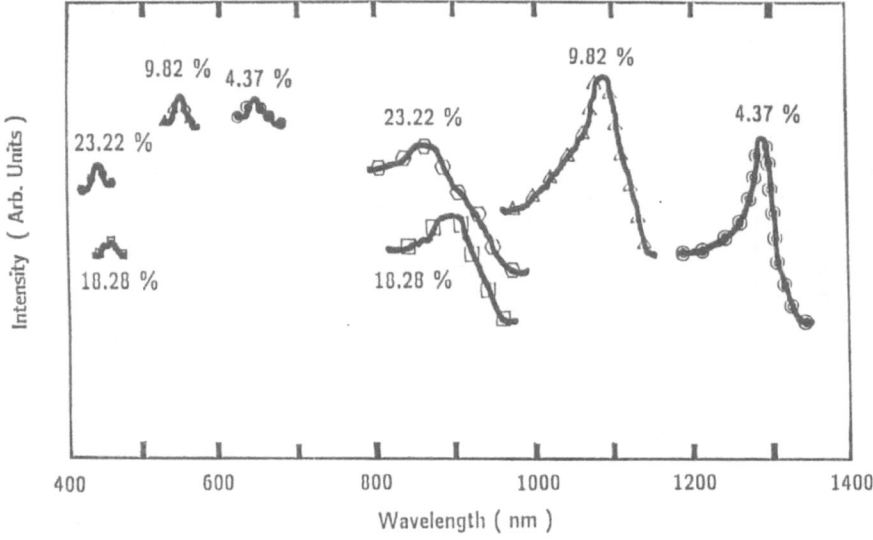

Fig. 9   Reflection spectra of deionized latexes A at various volume fractions at 22 °C (intensity x 4 for secondary peaks)

Fig. 10 provided the comparisons between De and D_t calculated from equations [8], [13] and [14]. The relative difference $(De-D_t)/D_t$ increases with increasing the particle concentraion. More crystal defects and coexistence of ordered and disordered regions might explain the larger deviation at higher particle concentration. Closer match between De and D_t of face-centered cubic agrees with the experimental findings[16] that f.c.c. structure is much more stable than b.c.c. at high particle concentration.

Temperature dependence on the reflection spectra of deionized latices A at various particle concentrations is illustrated in Fig. 11. Generally speaking, the reflection intensity   decreased with elevating temperature. The displacement of peak position toward smaller wavelength with higher temperature was ascribed to lattice structure transition, i.e. f.c.c. to b.c.c. with rising temperature. This is in agreement with the thermodynamic predictions[19]. The order-disorder transition temperature, which can be estimated by observing the decrease in reflection intensity, increased with increasing particle concentration. A higher melting temperature is not unexpected with closer packing of particles since the similarity of structure suggests that the ordered latex behaves as a crystal. With regard to the temperature-dependence of electrostatic repulsion between particles, it is more complicated than what can be illustrated from the optical experiments. Not only is the dielectric constant of dispersion medium temperature dependent, but it also affects the dissociation of particle surface groups. Further electrokinetic studies as a function of temperature are required to elucidate the complex situations.

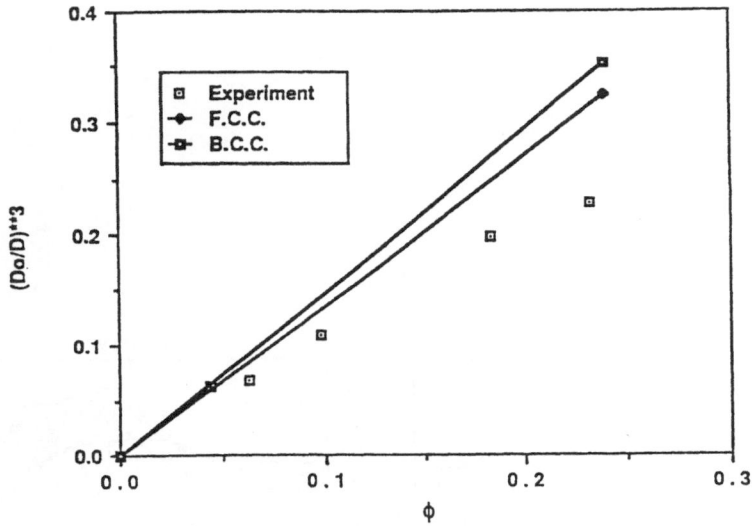

Fig. 10 Dimensionless interparticle distance vs. volume fraction for deionized
latexes A at 22 °C, □ : experimental data from reflection spectra, solid
line : theoretical calculation for perfect crystal lattice

Rheological Studies

The rheological behavior of ordered dispersions are of theoretical interest because of
the viscoelasticity caused by the screened electrostatic forces between particles.
Crandall and Williams [20] obtained the Young's Modulus by measuring the vertical
compression of the   ordered structure due to gravity. The high frequency limit of the s
torage modulus (G_∞), and the low-shear limit of viscosity ($\eta(0)$) have been measured by
shear wave propagation and by creep compliance measurements [21] respectively.

Fig. 12 shows the viscosity curve of deionized latex A at various particle
concentrations at 25 °C. Due to the sensitivity limitations of the instrument, the anticipated
infinite viscosity at low shear is not observed. Instead, typical pseudoplastic behavior and
Newtonian high-shear limit were seen for samples with high and low particle concentration,
respectively. The rising viscosity with higher particle concentration can be explained
qualitatively by the increase in energy dissipation resulting from alterations of particle
trajectories due to the presence of electrical charges, which cause the particles to avoid one
another. These results highlight the sensitivity of the dispersion viscosity to the nature and
strength of the interparticle forces. The picture presented here remains incomplete and
requires further investigation on the viscoelasticity of ordered latex dispersions.

CONCLUSIONS

We have demonstrated that ordered polymer colloids present a unique system for
studying the interrelationships among particle interaction, dispersion microstructure and
macroscopic viscoelasticity. Convenient probes, such as salt concentration, temperature
and particle concentration, controlled the strength and range of interparticle force.
Furthermore, electrokinetic, optical and rheological analyses provide an integrated
approach to achieve a better understanding of colloidal crystals.

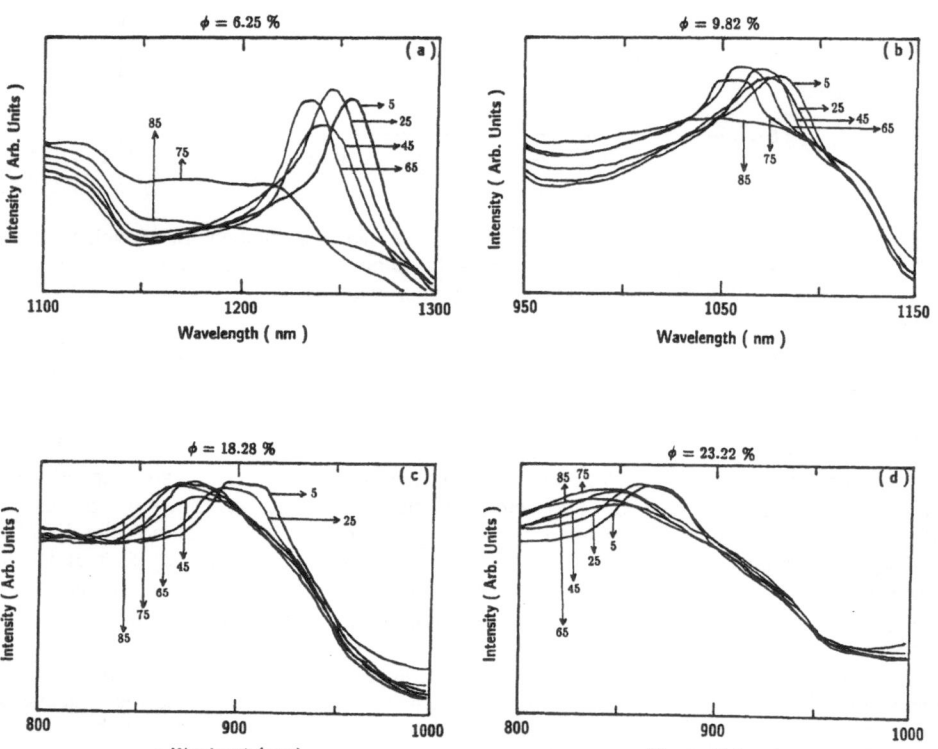

Fig. 11 Effect of temperature (in °C) on reflection spectra of deionized latexes A
at various volume fractions φ (a) 6.25 % (b) 9.82 % (c) 18.28% (d)
23.22% (numbers on curves indicate temperature)

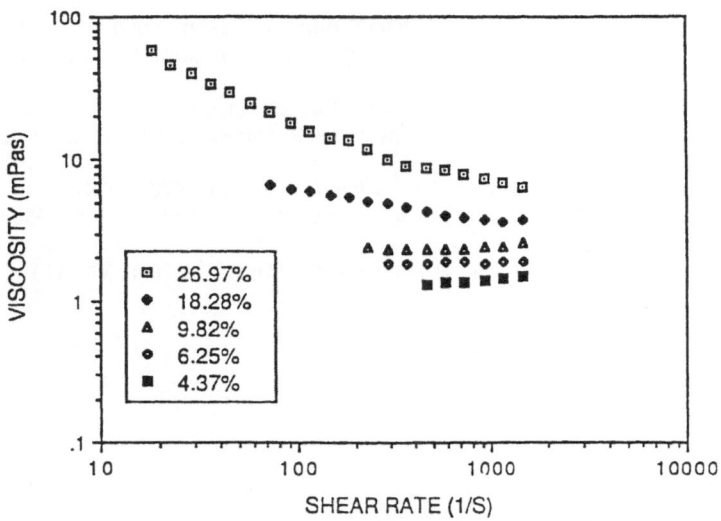

Fig. 12 The viscosity profile of deionized latex A at various volume fractions at 25 °C

REFERENCES

1. S.M. Ahmed, M.S. El-Aasser, G.H. Pauli, G.W. Poehlein and J.W. Vanderhoff, *J. Colloid Interface Sci.* **73**, 388 (1980)
2. P.H. Wiersema, A.L. Loeb, and J. Th. G. Overbeen, *J. Colloid Interface Sci.* **22**, 78 (1966)
3. P.A. Hiltner and I.M. Krieger, *J. Phys. Chem.* **73**, 2386 (1969)
4. R. Williams and R.S. Crandall, *Phys. Lett.* A, **48**, 225 (1974)
5. E.J.W. Verwey and J. Th. G. Overbeek, *Theory of the Stability of Lyophobic Colloids*, Elsevier, Amsterdam, 1948
6. w. Van Megen and I. Snook, *Adv. Colloid Interface Sci.* **21**, 119 (1984)
7. J.M. Ziman, *Models of Disorder*, N.Y., Cambridge University Press, 1979, P.232
8. C.J. Baraes, D.Y.C. Chan, D.H. Everett, and D.E. Yates, *J. Chem. Soc. Faraday Trans. II*, **74**,136(1978)
9. W.B. Russel, *The Dynamics of Colloidal Systems*, Madison, The University of Wisconsin Press, 1987
10. L.J. Bellamy, *Infrared Spectra of Complex Molecules*, Methuen, London, Wiley, N.Y., 1958
11. J.N. Shaw and M.C. Marshall, *J. Polym. Sci.*, **6A1**, 449 (1968)
12. M. Wadati and M. Toda, *J. Phys. Soc. Japan*, **32**, 1147 (1972)
13. C.M. Ma, F.J. Micale, M.S. El-Aasser and J.W. Vanderhoff, in "*Emulsion Polymers and Emulsion Polymerization*" (D.R. Bassett and A.E. Hamielec, ed.), P.251, ACS Symposium Series No. 165, Washington D.C., 1981
14. S. Hachisu, Y. Kobayashi, and A. Kose, *J. Colloid Interface Sci.* **42**, 342 (1973)
15. A. Kose, M. Ozaki, K. Takano, Y. Kobayashi, and S. Hachisu *J. Colloid Interface Sci.* **44**, 330(1973)
16. R. Williams and R.S. Crandall, *Phys. lett.* A, **48**, 225 (1974)
17. E.B. Sirota, H.D. Ou-Yang, S.K. Sinha, P.M. Chaikin, J.D. Axe, and Y. Fujii, *Phys. Rev. Lett.* **27**,1524(1989)

18. B.D. Cullity, *Elements of X-ray Diffraction*, Addison-Wesley, London, 1978
19. P.M. Chaikin, P. Pincus, S. Alexander and D. Hone, *J. Colloid Interface Sci.* **89**, 555 (1982)
20. R.S. Crandall, and R. Williams, *Science*, **198**, 293 (1977)
21. R. Buscall, J.W. Goodwin, M.W. Hawkins and R.H. Ottewill, *J. Chem. Soc. Faraday I*, **78**, 2873(1982)
22. S. Hachisu and Y. Kobayashi, *J. Colloid Interface Sci.* **46**, 470 (1974)
23. I.M. Krieger and P.A. Hiltner in "*Polymer Colloids*" (R. Fitch, Ed.) P.63, Plenum, N.Y., 1971
24. R. Williams, R.S. Crandall, and P.J. Wojtowicz, *Phys. Rev. Lett.* **37**, 348 (1976)

MICROSTRUCTURE OF A WATER DISPERSIBLE POLYESTER

Peter W. Raynolds

Research Laboratories,
Eastman Chemical Company,
Kingsport, TN, 37662 (USA)

Eastman's water dispersible AQ® polyesters are linear, relatively high molecular weight ionomers that are based on isophthalic acid. These polyesters form colloidal dispersions in water that consist of charge stabilized spheres with a diameter of ca. 22 nm. In the presence of 5% n-propanol the colloidal particle partially deaggregates and individual molecules can be observed along with aggregates of approximately 100 molecules. A microgel - like model for the colloidal particle is suggested.

INTRODUCTION

Eastman AQ® polyesters, which are produced by the Eastman Chemicals Division of the Eastman Kodak Company, have been used for many years in as adhesives and temporary coatings. The three polyesters, Eastman AQ 29, AQ 38 and AQ 55, are composed of varying amounts of isophthalic acid, sodium sulfoisophthalic acid, diethylene glycol, and 1,4-cyclohexanedimethanol. The three commercial products listed in Table I were developed in response to the need for products with different water resistance and blocking properties.

TABLE I. Eastman AQ® Water Dispersible Polyesters
Composition in Mole Percent and Properties.

	AQ 29	AQ 38	AQ 55
Isophthalic Acid	89	89	82
Sodium sulfo-isophthalic acid	11	11	18
Diethylene glycol	100	78	54
1,4-Cyclohexane-dimethanol	-	22	46
Molecular weight (Mw)	16,000	14,000	14,000
Acid Number	<2	<2	<2
Hydroxyl Number	<10	<10	<10
Tg, °C	29	38	55

Surface Phenomena and Fine Particles in Water-Based Coatings and Printing Technology
Edited by M.K. Sharma and F.J. Micale, Plenum Press, New York, 1991

275

These thermoplastic, amorphous polyesters disperse readily in water at 70 °C with minimal agitation to give translucent, low viscosity dispersions containing no added surfactants or solvents. Although the ionic content is high, the bulk of the polymer is hydrophobic. The dispersions consist of rigid spheres, each containing approximately 100 polymer molecules. The typical dispersion properties for various polyesters are presented in Table II.

TABLE II. Dispersion Properties of Various Polyesters.

	AQ 29	AQ 38	AQ 55
Percent Solids	30	35	28
Viscosity, cps, Brookfield	45	30	42
Density, lbs./gal.	8.83	8.84	8.97
pH	5-6	5-6	5-6
Particle dia., nm, by PCS	34	27	20

A detailed understanding of the nature of the dispersion became necessary when these polyesters were used in other applications, particularly in dispersions containing added organic solvents. This paper reports the colloidal properties of these linear, high molecular weight, water-dispersible polyesters, and a model for the microstructure of the colloidal particles.

EXPERIMENTAL

Dispersion of AQ 55 in Water: A 1000 ml 3 neck flask equipped with a mechanical stirrer, thermometer, reflux condenser and heating mantle was charged with 408 g of distilled water. The water was heated to 70-80° and 192 g of AQ 55 polyester was added in four equal portions, 15 minutes apart. Heating was continued for an additional hour. The translucent dispersion was cooled to room temperature.

Low-volume-fraction relative viscosity measurements were made with a Cannon-Fenske capillary viscometer at 25.00 °C, using ASTM method D 445-83. Particle sizes were determined in 0.01 M sodium chloride by photon correlation spectroscopy, using a Brookhaven instrument. Electrophoretic mobilities were determined in 0.01 M sodium chloride using a PenKem 3000 electrokinetic analyzer. Mobilities were converted to zeta potentials using the O'Brien and White algorithm[3]. Dialysis of the dispersions was done with Spectra/Por® semi-permeable membrane with a 6,000-8,000 molecular weight cutoff.

RESULTS AND DISCUSSION

The viscosity of a dilute, colloidal dispersion of rigid, uncharged spheres depends on the volume fraction, ϕ, as given by the Einstein equation:

$$\text{Relative Viscosity} = 1 + 2.5\phi$$

When the particles are charged, as is the case here, electrostatic interactions between the particles become important, and a second order term is needed to explain the viscosity behavior at low volume fraction:

$$\text{Relative Viscosity} = 1 + a\phi + b\phi^2$$

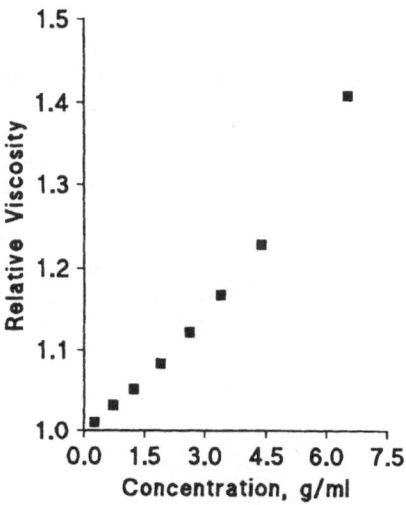

Figure 1. Relative Viscosity as a Function of AQ 55
Dispersion Concentration in the Presence of 0.01 M NaCl.

The coefficients, a and b, may be calculated from the particle size, the zeta potential, and system properties[1]. The "a" term corrects for the primary electroviscous effect, which is the viscosity increase due to the energy dissipated as the ion cloud around the particle is distorted by shear. The "b" term corrects for the secondary electroviscous effect, which is the viscosity increase due to the interaction between the charged particles. Parameters a and b can be calculated from system properties such as the zeta potential, ionic strength, and ion properties. The low-shear viscosity of AQ 55 in 0.01 M sodium chloride is shown in Figure 1.

The line is slightly curved, due to the secondary electroviscous effect. Since the primary objective here is to monitor the change in swelling induced by added n-propanol, we calculated, instead of the volume fraction, ϕ, the dispersed density of the polymer. This value, which is a characteristic of the polymer dispersed in a particular solvent, is the weight of polymer per unit volume of dispersed polymer. It is related to ϕ in the following way:

$$\phi = \frac{\text{ml polymer}}{\text{ml dispersion}} = \frac{\dfrac{\text{g polymer}}{\text{ml dispersion}}}{\dfrac{\text{g polymer}}{\text{ml polymer}}} = \frac{\text{Concentration}}{\text{Dispersed Density}}$$

A non-linear least squares fit of the nine data points to the quadratic equation, with dispersed density as the adjustable parameter, yielded a density in dispersion of 0.70 g/ml. Since the density of the dry polymer is 1.326 g/ml, the polymer, on a volume basis, occupies (0.70/1.326)*100% = 53%, or about half of the total volume of the solid phase. The remainder is the water that swells the polymer.

From the density of the polymer in dispersion and the molecular weight, straightforward geometrical calculations showed that a 22 nm colloidal particle of Eastman AQ® 55 polymer consisted of roughly 100 individual polymer chains. Three possible structures for the arrangement of the individual chains in the particles are shown in Figure 2. A micellar model would segregate most of the charged groups on the outside of the particle, leaving the interior relatively hydrophobic. We felt that this arrangement was unlikely, since the coulombic charge on a typical particle, calculated from the zeta potential (-65 mv), only accounted for about 10% of the ionic groups that must be present from the calculated mass; a large proportion of the charge must be hidden in the interior[2]. The ionic content of AQ 55 is so high that not all of the ionized groups can place themselves on the surface of a 22 nm particle. We considered two other models in more detail: a collection of tangled, uncoiled chains and a microgel-like aggregate of individually coiled chains. The distinction between the tangled chain and microgel models is that in the former, the hydrophobic interactions of the polymer are mostly intermolecular, while in the latter, the hydrophobic interactions are mostly intramolecular.

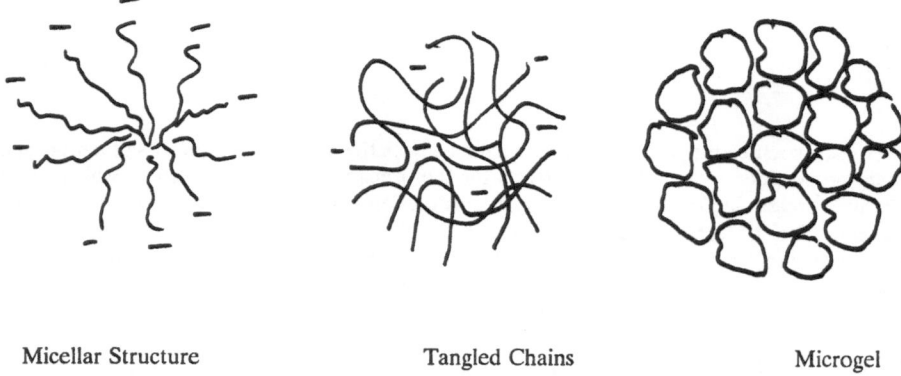

Micellar Structure   Tangled Chains   Microgel

Figure 2. Three Possible Models for Colloidal AQ 55 Polyester.

When a dispersion of AQ 55 polyester in water was added to 0.01 M sodium chloride in water containing 5% by weight of n-propanol, a change occurred in the particle size distribution over a period of 40 minutes at 25 °C (Figure 3). The initial distribution of diameters centered at ca. 22 nm was replaced with a bimodal distribution of particles with diameters of ca 16 nm and 6 nm. The 6 nm particles were approximately the size that would be expected for a single polymer chain. Based on the intensity of the scattered light, the 6 nm distribution represented about 1/3 of the total weight. The same sort of change occurred more rapidly with 10% n-propanol, and almost instantly with 0.01 M sodium chloride containing 50% n-propanol. No 6 nm particles were observed in the absence of n-propanol. These results point to an equilibrium between individual polymer chains and aggregates of roughly 100 molecules. The equilibrium favors individual molecules as the n-propanol content is increased.

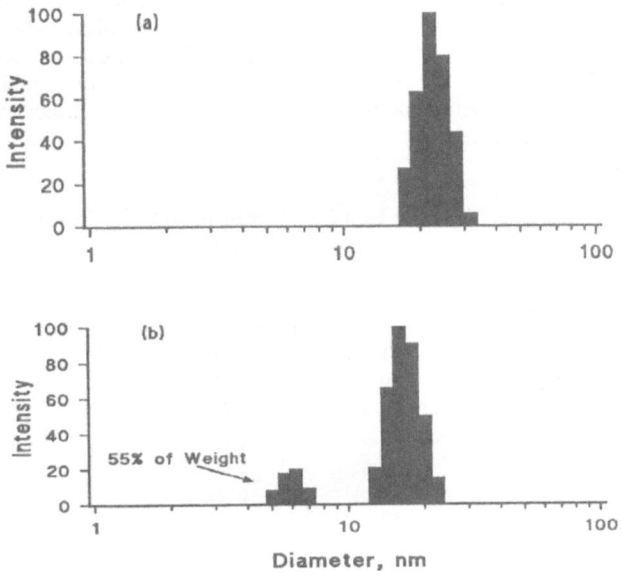

Figure 3. Particle Size Distribution of AQ 55 Polyester Dispersion
in Water - n-Propanol (90/10) mixture (a) Immediately after
mixing the AQ 55 with the Solvent and (b) After 40 Minutes.

TABLE III. Distribution of Particle Sizes
In AQ 55 Polyester Dispersions Containing n-Propanol.

Percent n-Propanol	Approx. Weight % 12-30 nm	Approx. Weight % 4-7 nm
0	100	0
5	65	35
10	45	55
50	10	90

Attention was next turned to the nature of the individual polymer chains once they were freed from the aggregate; in particular, whether they were solvated or not. Pure water is not a good solvent for AQ 55: no individual polymer chains are observed by light scattering measurements. Although it seemed unlikely to us, it was possible that the 5% n-propanol - water mixture was a much better solvent for the polymer than water itself, and that the light scattering results simply reflected a transition from a clear, colloidal system to a solution of relatively uncoiled polymer chains. Although too much weight should not be placed on the exact percents of 5 nm and 20 nm particles shown in Table III, it seems probable that a substantial proportion, perhaps half, of the weight of the polymer, at the 1% solids level, exists as individual molecules. If the chains were starting to become solvated and uncoiled in 5% or 10% n-propanol, the viscosity of the dispersions would be expected to increase, since the viscosity of polymer solutions depends, in part, on the end-to-end distance, or the mean free radius of gyration, of the individual polymer molecule. The low-volume-fraction, low-shear viscosity

279

behavior of AQ 55 dispersions containing various amounts of n-propanol was analyzed to determine if the individual polymer molecules in dispersion were coiled. The viscosities were determined in the same concentration range - 0.5 to 5% - that was used in the light scattering experiments. Curves similar to the one shown above for AQ 55 were generated, and were analyzed in the same way to determine the density of the polymer in dispersion. The density results are summarized in Table IV.

TABLE IV. Dispersed Density of Eastman AQ® 55 Polyester
In n-Propanol - Water.

Weight Percent n-Propanol	Density of Polymer in Dispersion, g/ml
0	0.70
10	0.68
14	0.69
20	0.54
30	0.47
40	0.40
50	0.35

At n-propanol levels of 20% and above, the liquid phase began to solvate the non-polar portions of the polymer chain, the polymer swelled, the density in dispersion decreased, the volume fraction increased and the relative viscosity increased dramatically. Propanol levels below 20% had virtually no effect on the volume fraction or relative viscosity of the dispersion: at low levels of n-propanol, the polymer chains are in essentially the same state - coiled - as in pure water.

The foregoing indicates that at low n-propanol levels (1) individual AQ 55 molecules exist in rapid equilibrium with the colloidal particle; (2) the individual molecules are coiled and (3) the density of the individual molecules is about the same as the colloidal particles. These observations do not, rule out the "tangled chain" model for the colloidal particle that was described above but do, in our view, make it less likely than an aggregate of individually coiled molecules. It is difficult for us to envision a thermodynamically reasonable sequence of events that, within a few minutes at 25°C, would induce an uncoiled polymer chain in a colloidal particle to break many inter-chain hydrophobic bonds, eject itself into the continuous medium and then, finding the continuous medium unable to solvate it, coil. The enthalpy of the transition appears unfavorable, since more of the hydrophobic molecule is exposed to the water, and the entropy seems to be unfavorable, since many degrees of freedom are lost in the coiled form. Needless to say, what seems reasonable is not always true, and we merely present the "raisin cluster" model as a simple but unproven way of explaining the facile equilibrium between individual polyester molecules and aggregates under mild conditions.

CONCLUSION

Aqueous dispersions of Eastman AQ® 55 polyester are translucent, consisting of particles with an average diameter of 22 nm in 0.01 M NaCl. In the presence of small (5%) quantities of n-propanol, an equilibrium occurs within minutes at 25°C consisting of 6 nm particles (individual polymer molecules) and 16 nm aggregates. Viscosity measurements showed that the density of the polymer in dispersion does not change with this small an amount of alcohol; thus, the individual particles are not solvated, but are as coiled as they are in pure water. Although proof is lacking, the simplest explanation for the facile deaggregation of the AQ 55 polyester particle in the presence of small amounts of n-propanol is that the aggregates consist of individually coiled polymer molecules loosely held together by a relatively few hydrophobic interactions with nearest neighbors.

REFERENCES

1. R. J. Hunter, "Zeta Potential in Colloid Science", Academic Press, New York, 1981, pp. 192-196.

2. Reference 1, pp. 49.

3. R. W. O'Brien and L. R. White, J. Chem. Soc. Faraday II, **74,** 1607 (1978).

THE DYNAMIC SURFACE PROPERTIES OF SURFACTANTS

Paul D. Berger and Christie Berger

Witco Corporation
Organics Division
Houston, Texas

The Dynamic Surface Tension and Surface
Dilatational Viscosity od surfactant
solutions can be easily and accurately
measured using the Maximum Bubble
Pressure Technique (MBPT). Modifications
to the classical technique have been
made to enable one to determine dynamic
surface properties for opaque as well
as transparent solutions and dispersions.
Surface properties can be determined down
10 milliseconds and easily extropolated
to smaller time intervals. The average size
and distribution of bubbles and droplets
can readily be determined. The results of
Dyanamic Surface Property measurements can
be applied to the analysis of wetting,
penetration, spreading and other surface
phenomena which are time dependent.

Introduction

Surfactants play a major role in many industrial applications. Among
their many applications one finds them used as emulsifiers, wetting
agents, penetrants, dispersants, defoamers, antistatic agents, corrosion
inhibitors, flow modifiers, and leveling agents. The performance and
value of surfactants is largely due to their ability to significantly

alter surface and interfacial properties at low concentrations. Many
studies have been performed on the surface tension lowering properties
of surfactants and on their structural/ performance relationships. Only
recently, however, has it been possible to include the dimension of time
to these studies. We have used the maximum bubble pressure technique,
MBP[1] to study the dynamic surface tension and surface dilational
viscosity of various surfactants and have correlated these properties to
several time related applications such as penetration and wetting [2]. The
dynamic surface tension is the non-equilibrium surface tension of a
system containing a surface active agent. The surface dilational
viscosity is the component of surface viscosity which resists changes in
surface area. We have used dynamic surface measurements to better
understand and to develop products for use as agricultural spray tank
adjuvants, emulsion polymerization surfactants, wetting agents and
dispersants for paints and inks, crude oil demulsifiers, textile dye
carriers and soil penetrants. We believe the study and understanding of
the dynamic surface properties of surfactants will play a significant
part in the development of new and improved surface active agents. It
can also provide a better understanding of the benefits and problems
surfactants create.

Apparatus for the Maximum Bubble Pressure Technique

Figure 1 shows the apparatus we used to determine the dynamic surface
properties of various surfactants using the maximum bubble pressure
technique. This procedure was recently reviewed by Mysels [3]. We have
used a design based on that of Professor Wasan's group at IIT. The
solution of material containing the surfactant under study is contained
in the specially designed tensiometer vessel (Reliance Glass Works Inc,
Bensenville Il). This is fitted with a cooling jacket to keep
temperature constant and a precision bore capillary tube through which
air or a second liquid is pumped. When measuring dynamic surface
properties, air or another gas is accurately metered by a dual stage
regulator and a precision flow meter (Gilmont Instruments Inc, Great
Neck, N.Y.). The gas passes through the capillary creating a bubble at
the tip. The pressure within the bubble is measured using a precision
digital monometer (Meriam Instrument Co., Cleveland OH). The bubble
formed at the tip of the capillary passes between a photodiode and a
light source after detachment from the tip of the capillary. The bubble
disrupts the light source and the frequency of bubble detachment is

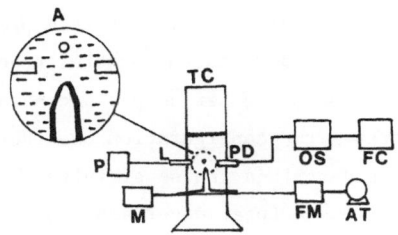

Figure 1.   Schematic Diagram of Maximum Bubble Pressure Apparatus:
A) Capillary Tube, FM) Flow Meter, AT) Air Tank,
L) Light Diode, PD) Photo Detector, OS) Oscilloscope,
FC) Frequency Counter, M) Manometer, P) Power Supply,
TC) Tensiometer Container.

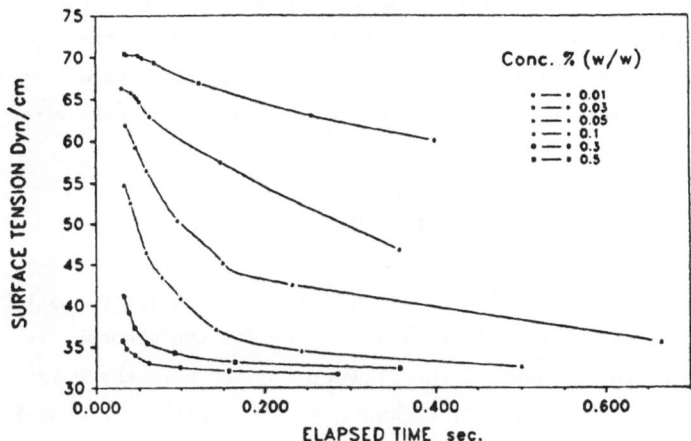

Figure 2.   Effect of Conc. on DST of NP-100.

measured using a frequency counter (John Flude Mfg Co, Everett WA) which
is channelled through an oscilloscope (Leader Instrument Corp,
Hauppauge, N.Y.).

For opaque liquids a differential pressure transducer (Baratron Model
225AD, MKS Instruments, Richardson Tx.) can be substituted for the
photodiode detector. Here the actual maximum bubble pressure can be
recorded on a strip chart recorder and the bubble frequency can be
determined from the number of peaks per cm per second on the strip
chart. Figure 1a illustrates the configuration used for opaque liquids.

Using the apparatus as described the sensitivity of the method was
found to be ±0.2 dyne/cm with a total uncertainty of ±0.30 dyne/cm.

Theory of Maximum Bubble Pressure Technique

The pressure within a bubble is maximum when the bubble is spherical.
A bubble emerging from the tip of a clean, precision bore capillary
reaches its maximum pressure when the diameter of the bubble is equal to
the internal diameter of the capillary. In the MBP technique the
diameter of the capillary is determined by optical measurement and by
calibration with a known liquid such as pure water. We found the radius
of our capillary to be 87.30 μm using distilled deaerated water as a
standard. The pressure within the bubble is measured using the monometer
and corrected for the hydrostatic head of solution above the capillary.
Knowing the radius of the capillary and the maximum pressure within the
bubble, the surface tension can be calculated using the following
relationship:

$$\sigma = (P_m - dh) \, gr_c \, / \, 2, \qquad\qquad (1)$$

where σ is the dynamic surface tension in dyne/cm, P_m is the maximum
pressure within the bubble in g/cm^2 corrected for the atmospheric
pressure, d is the density of the liquid in g/cm^3, h is the height of
the liquid above the capillary in cm, r_c is the capillary radius in cm
and g is the gravitational constant in cm/sec^2.

By measuring the maximum bubble pressure at various flow rates, a
series of surface tensions can be obtained each for a particular rate of
bubble formation. The frequency counter is used to determine the rate of
bubble detachment in kHz or reciprocal milliseconds. Knowing the
frequency one can plot the surface tension on the Y-axes against the
inverse frequency on the X-axes to obtain a plot of the surface tension

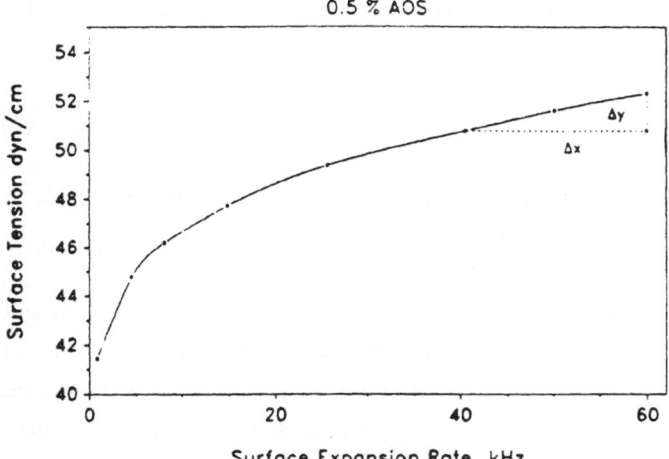

Figure 3. Det. of Surface Dilational VISC.

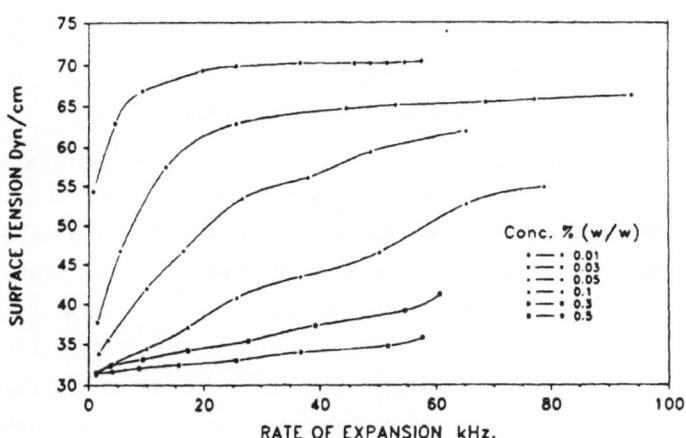

Figure 4. Effect of CONC. on DST of NP-100.

as a function of elapsed time in milliseconds. This is shown in Figure 2
for a series of solutions of 10 mole ethoxylated nonylphenol of
differing concentrations.

Knowing the flow rate one can calculate the expansion rate of each
bubble using the relationship:

$$\epsilon = Q/\pi r_c^3 ,$$ (2)

where ϵ is the surface expansion rate in Hz, Q is the flow rate in
cc/sec, and r_c is the capillary radius in cm as before.

The surface dilational viscosity can be obtained by examining the
dynamic surface tension values at various rates of surface expansion.
The gradient of the dynamic surface tension with respect to expansion
rate can be obtained in the linear region of large expansion rates as
shown in Figure 3. In this region the rates of surface expansion are
much larger than the rate of surfactant adsorption[4] . The gradient of
the curve calculated in this area is the surface dilational viscosity in
surface poise, SP x 10^{-3}.

A much more detailed theoretical treatment of dynamic surface tension
can be found in the work of Joos and Rillaerts[5] and the thesis of Kao [4].

TABLE I

SURFACTANTS AND THEIR APPLICATIONS

NAME	DESCRIPTION	APPLICATIONS
WITCONOL NP-60	6 Mole Ethoxylated Nonylphenol	Oil Wet.,Grind
WITCONOL NP-80	8 Mole Ethoxylated Nonylphenol	Wetting,Grind.
WITCONOL NP-100	10 Mole Ethoxylated Nonylphenol	Wet,Grind,Level
WITCONATE AOS	C14-16 α-Olefin Sulfonate, Na	Foam.,Detergent
EMCOL 4161L	Monooleamido Sulfosuccinate	Pigment Disp.
EMCOL 4300	Monolaureth Sulfosuccinate	Oil Wet.,Grind
EMCOL 4500	Di 2-ethylhexyl Sulfosuccinate	Wet., Grinding
EMCOL 4910	Polypropoxy,C12-15 Alkyl "	Pigment Disp.
EMCOL K-8300	Mono alkanolamide Sulfosuccinate	Pigment Susp.

Experimental Results & Discussion

Dynamic surface tensions and surface dilational viscosities were obtained for several surfactants described in Table I.

These properties were compared in some cases to observed performance characteristics to determine the importance of dynamic surface properties to surfactant performance.

Effect of Concentration on Dynamic Surface Tension

Referring back to Figure 2 one can see the result of measuring the surface tension at various frequencies for different concentrations of a 10 mole ethoxylated nonylphenol solution (WITCONOL NP-100). The results show that the surface tension values decrease as the concentration increases and that the surface tension reaches its minimum value faster as the concentration is increased. From the curves it can be seen that the minimum concentration necessary to give maximum surface tension lowering and probably maximum performance changes with bubble formation frequency. As the speed of bubble formation is increased, more surfactant is necessary in order to reach a minimum surface tension. The CMC of WITCONOL NP-100 was determine by plotting the static surface tensions against concentration and determining the concentration above which the surface tension does not decrease. This value was found to be 0.02%. If all the curves containing concentrations at or above the static CMC (critical micelle concentration) are extended to infinite time, they will all reach a surface tension value equal to the static surface tension found to be 31.6 dyne/cm using a Du Nuoy ring[6]. Notice that in Figure 2 the dynamic surface tension curves continue to change far above the CMC. This is because the CMC is determined for a static system and we are looking at a dynamic system during our measurements. The curve for 0.01% concentration which is below the CMC will never reach the minimum surface tension. The CMC can be looked at as the concentration whose dynamic surface tension curve, when approaching time = ∞ ,gives the minimum surface tension found for the product.

Much of the literature describes the CMC as the concentration range where many surface related phenomena are maximized or minimized [7]. It is claimed that interfacial tension and surface tension is minimum at the CMC and that foaming, detergency and wetting are maximized at or above the CMC.

If one wishes to choose a concentration for a surfactant which gives

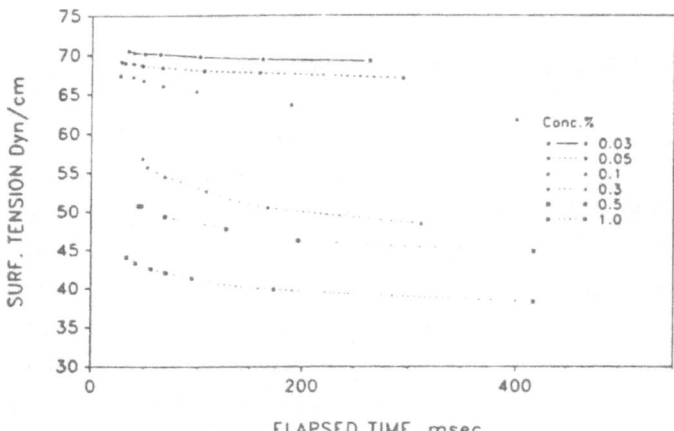

Figure 5.   DST of Witconate AOS.

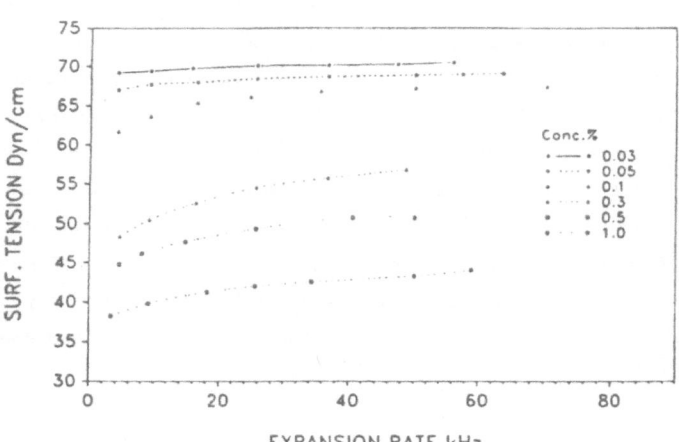

Figure 6.   DST of Witconate AOS.

optimum surface tension lowering in a dynamic process such as high speed
jet printing, spray coating or wetting, they can examine a series of
curves such as Figure 2 and determine the concentration necessary to
give maximum lowering in the period of time required by the process.

Figure 4 shows the expansion rates against dynamic surface tension,
for various concentrations of WITCONOL NP-100. From these curves the
surface dilational viscosities for various concentrations of WITCONOL
NP-100 can be determined. These values will be used later to determine
the effect of surface dilational viscosity on foam stability.

Figures 5 and 6 presents similar data as in Figures 2 and 4 but using
values determined for WITCONATE AOS ,a C14-16 α- olefin sulfonate, as
the surfactant. From the linear region of the slope of Figures 4 and 6
at high expansion rates, (> 40 kHz), the surface dilational viscosities
can be calculated. If the curves are extended to expansion rates = 0,
they will intersect the Y-axes at the lowest surface tension found by
static measurements for concentrations at or above the CMC. The CMC for
AOS was found to be around 0.1% on an "as is" basis and its minimum
surface tension was found to be 35.1 dyne/cm. Those curves representing
concentrations at or above the CMC will intersect the Y-axes in Figure 6
at this surface tension value where expansion rate is 0. Likewise they
will all reach this value in Figure 4 if the curves are extended to
infinite time allowing the surfactant adequate time to come to its
static surface tension value.

TABLE II
EFFECT OF CONCENTRATION ON WETTING
ADSEE 801 SPRAY TANK ADJUVANT

Concentration (% by wt)	Static (dyne/cm)	Dynamic (dyne/cm)	Wetting (seconds)
1.0	29.6	27.4	0
0.1	29.0	44.0	8
0.01	29.4	66.3	267

Table II shows the effect of increasing the concentration of
surfactant above the CMC on wetting. The surfactant used was an alcohol
ether based agricultural wetting agent ADSEE 801. Static surface tension
measurements show the CMC to be at or below 0.01 %. Note that the
dynamic surface tension measured after 30 milliseconds continues to

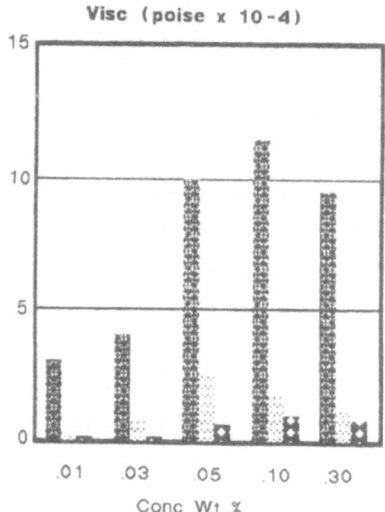

Figure 7

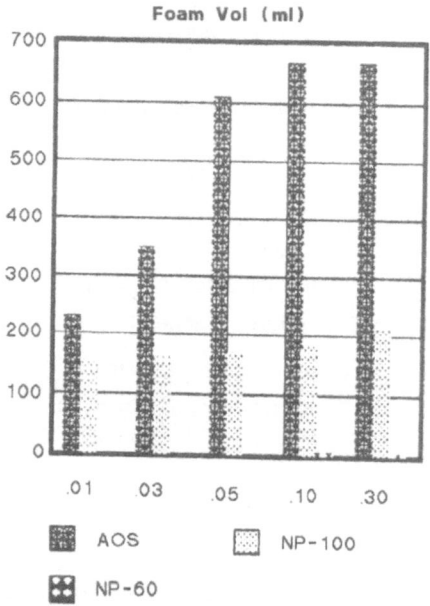

Figure 8

decrease as the surfactant concentration is further increased 10 and 100 times above the CMC. Also the wetting properties continue to improve as the concentration is increased. At or slightly above the CMC the product does not appear to be a good wetting agent. In fact, this product used at concentrations of 0.5-1.0% is found to be an excellent wetting agent in the field. The dynamic surface tension values predict that this material should be used at these higher concentrations.

Surface Dilatational Viscosity and Foaming

Figure 7 shows the surface dilational viscosities for WITCONOL NP-60 (6 MOLE EO), WITCONOL NP-100 (10 MOLE EO) and WITCONATE AOS (C14-16 α-olefin sulfonate). The AOS gives higher surface dilational viscosities across the entire concentration range. All products have been adjusted to a percent active basis since the original activity of WITCONATE AOS is 40% compared to 100% for the two WITCONOLS.

Figure 8 compares the results of foaming tests on various concentrations of WITCONATE AOS at 25°C with those for WITCONOL NP-60 and NP-100. The tests were run by mixing 100 ml of each solution in a Waring blender jar at high speed for 30 seconds and pouring the resulting foam into a 1000 ml graduated cylinder. The initial foam volume and the time elapsed when 50 ml of liquid appear (half-life) is recorded. There is a good correlation between the concentration giving the maximum surface dilational viscosity and the concentration giving the best foam performance. This indicates that an optimum surfactant concentration exist for a dynamic process like foam generation and that performance falls off when concentrations are below that concentration. WITCONOL NP-100 is a low foaming nonionic surfactant giving a maximum foam height of 150ml compared to the 670ml maximum produced by AOS. WITCONOL NP-60 is a non-foaming surfactant and gives the lowest surface dilational viscosities.

Table III shows the effect of addition of defoamer on the dynamic surface tension and surface dilatational viscosity of a foaming agent. The surface tension of the foaming agent, AOS, is 48.3 dyne/cm and 47.4 dyne/cm after addition of defoamer. These are both the values after 30 msec. There is no appreciable change in the dynamic surface tensions with or without defoamer. The surface dilatational viscosity however; changes drastically. Before addition of defoamer in is 12×10^{-4} and it is reduced to less than 1×10^{-4} cps after addition of defoamer. The defoamer reduces the surface dilational viscosity, hence the ability of the film to withstand stretching and therefore destabilizes any bubbles that form.

TABLE III

EFFECT OF DEFOAMER ON DYNAMIC SURFACE PROPERTIES

TEST SOLUTION	SURFACE TENSION	SURF DIL VISC	FOAM VOL
in DI H_2O	30 msec, dyne/cm	SP x 10^4	ml
0.1% AOS	48.3	12	650
0.1% AOS w/0.01% Defoamer	47.6	<1	75

Effect of Static and Dynamic Surface Properties on Wetting

Table IV shows the results of various test to determine several properties of a series of sulfosuccinates. These have been found to be either good wetting or dispersing agents in coating systems. The structures of these anionic surfactants were previously described in Table I. Properties measured for each sulfosuccinate were as follows: wetting by the Draves Method[8], surface tension after 33 milliseconds by the Maximum Bubble Pressure Procedure, static surface tension by the De Nuoy ring method and surface dilational viscosity also by the Maximum Bubble Pressure Procedure.

TABLE IV

PROPERTIES OF EMCOL SULFOSUCCINATES

0.10% @ 25°C

PRODUCT	FUNCTION	WETTING	SURF. TENS dyne/cm		SURF VISC
		sec	33 msec	static	surf poise
EMCOL 4161L	Pigment Disp	90	70.2	34.0	10 x 10^{-4}
EMCOL 4300	Oil Wet.,Grind.	18	69.8	28.1	20 x 10^{-4}
EMCOL 4500	Wetting,Grind.	<1	49.5	34.5	15 x 10^{-4}
EMCOL 4910	Pigment Disp.	10	63.6	26.8	90 x 10^{-4}
EMCOL K-8300	Pigment Disp.	90	70.3	42.1	20 x 10^{-4}

The data from Table IV shows that a relationship exists between the dynamic surface tension at 33 millisecond for each of the sulfosuccinates and their wetting times in seconds. From this data it is evident that surfactants which lower surface tension rapidly will wet

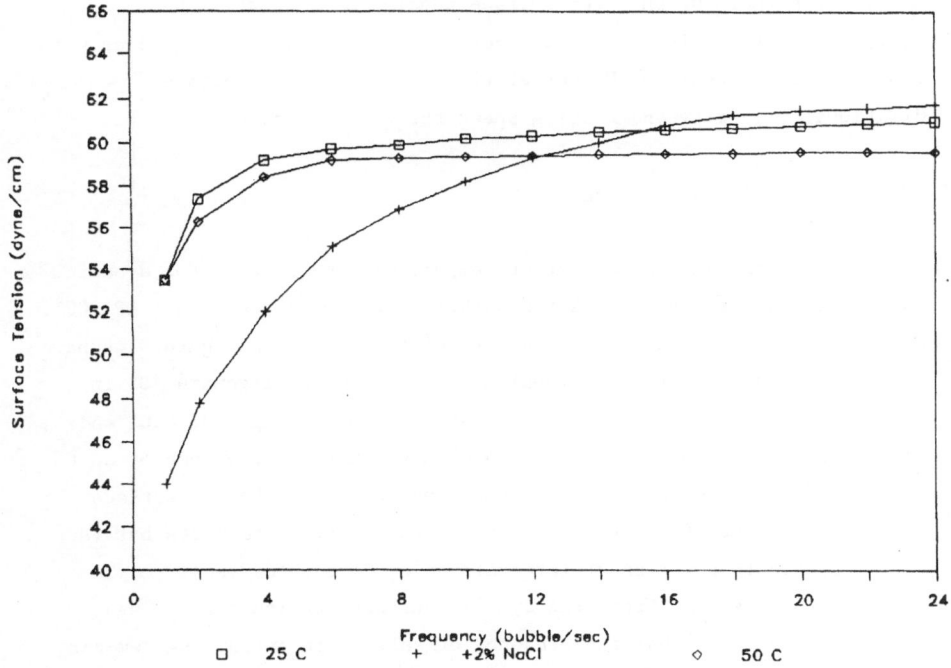

Figure 9. DST of Witconate AOS. Effect of Temp. and Salt.

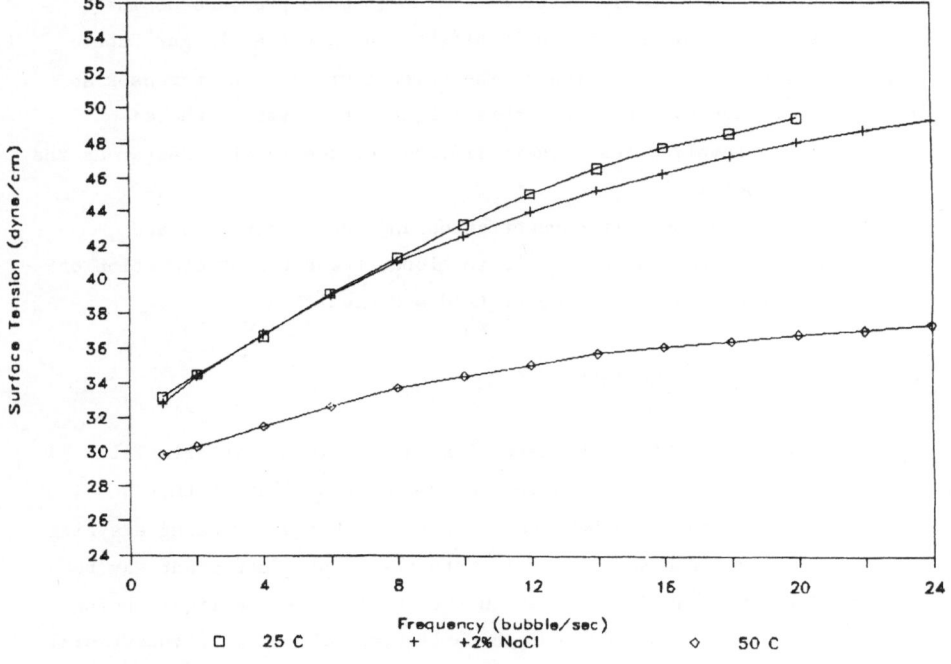

Figure 10. DST of Witconol NP-100. Effect of Temp. and Salt.

faster than those which lower surface tension more slowly. The
correlation of wetting and static surface tension is not as good as with
dynamic surface tension. No correlation exists between surface
dilational viscosity and wetting speed for these products.

Effect of Temperature and Salt

We have determine the effect of temperature and salt on the dynamic
surface tension for two surfactants (Witconate AOS and Witconol NP-100).
Figures 9 and 10 summarize the results of this work. In Figure 9 we have
compared the dynamic surface tension curves of 0.1% Witconate AOS in
deionized water at 25°C with results at 25°C containing 2.0% NaCl and
results at 50°C with no salt. The data shows that the presence of salt
reduces the surface tension by driving the surfactant to the surface
more rapidly. The effect becomes greater as the time increases but is
very appreciable after short intervals (i.e. 24 bubbles/sec).
Temperature does not effect the dynamic surface tension curve. The
implication here is that the presence of salt in an anionic system can
speed up the rate at which the surface tension attains its equilibrium
value thus giving lower surface tension values in shorter time periods.
Figure 10 shows that for a nonionic surfactant, the effect of
temperature on the dynamic surface tension is more pronounced. The
effect of salt is minimal and only slightly apparent at larger bubble
frequency (shorter bubble times). The temperature effect increases as
the bubble frequency increases. This implies that systems where
nonionics are involved reach their minimum surface tension faster as the
temperature increases.
By controlling the salt concentration and the temperature and
selecting the proper nonionic/anionic blend, the rates of migration of
each to the interface can be controlled and predicted.

Dynamic Interfacial Tensions

Recently we have begun the determination of dynamic interfacial
tensions between two liquids using an apparatus similar to that
described in Figure 1. We have modified the technique by using a syringe
pump to introduce the second liquid into the first. Surfactant may be
dissolved in the liquid contained in the chamber, in the liquid being
introduced into the chamber or in both liquids. The dynamic interfacial
tension is determined using a program involving successive iterations.
The theory is based on work by Kumar[9] who presented a unified approach

to bubble and drop formation in his paper. The dynamic surface tension is calculated using the program by entering the densities of the liquids, the flow rate, the bubble frequency, the viscosity of the continuous phase, the capillary diameter and the hydrostatic head above the capillary. This technique has been applied to the study of the mechanism of demulsification and to the spontaneity of emulsifiers. Other applications where liquid/liquid systems are involved are apparent.

Conclusions

The Maximum Bubble Pressure technique is a rapid and accurate means of measuring dynamic surface tension and surface dilational viscosities of surfactants. Many properties of surfactants, including wetting and foaming are dependent on dynamic surface properties. Dynamic surface property measurements show that the conventional concept of CMC is not a good predictor of surfactant effectiveness especially where short periods of time are involved. The technique can be used to screen surfactants and determine their optimum concentrations for systems where speed as well as effectiveness is of importance. It can also be used to determine dynamic interfacial tensions between two liquids. Also surfactant mixtures can be studied to determine the rates of surface migration for individual components.

References

1- Huang, D., Nikolov, A. and Wasan, D.T., Langmuir J., 1986, 2, 672.

2- Berger, P.D., Hsu, C., Jimenez, A.J., Wasan, D.T. and Chung, S., ACS Symposium Series #137 1988, ACS Washington D.C. Chapter 13.

3- Mysels, K.J., Langmuir J. 1986, 2, 428.,Langmuir J. 1989 5, 442.

4- Kao, R.L., Ph.D. Thesis, Illinois Institute of Technology, Chicago, 1987.

5- Joos, P., Rillaerts, E., J. Colloid & Int. Sci., 1981, 79, 96.

6 -Harkins, W.D., Jordan. H.F., JACS, 1930, 52, 1751.

7- Preston, W.C., J. Phys. Colloid Chem, 1948, 52, 84.

8 -Draves, C.Z., *Amer Dyest. Rep*, 1939, <u>28</u>, 421.

9 -Kumar, R., *Chem Eng Sci.*, 1971, <u>26</u>, 177.

WETTABILITY AND SURFACE ENERGIES OF POLYMER

SUBSTRATES

Dhirendra Kumar[*] and S.N. Srivastava

Department of chemistry, Agra College
Agra-282002, India

The coating and printing of polymer films with
water-based formulations are relatively difficult as compared
to solvent-based formulations. The surface tension of water
is higher than that of the solvents. In addition, the surface
energy of polymer surfaces is in the range of 25-40 ergs/cm^2.
In order to understand the wetting and spreading behavior of
coating materials, the polar and non-polar surface energies
were evaluated by measuring contact angle of water and
methylene iodide on various non-porous substrates. These
results were utilized to explain the spreading, wettability
and adhesion phenomena in order to understand the interactions
between water-based coating/printing materials and non-porous
substrates.

INTRODUCTION

The term wetting applied to the displacement of air from a liquid or
solid surface by water or an aqueous solution. Wetting is a process
involving surfaces and interfaces. When the surface to be wet is small,
then equilibrium conditions can be attained during the wetting process and
the free energy changes involved in the process determine the degree of
wetting attained. On the other hand, when the surface to be wet is large,
then equilibrium conditions are often not reached during the time allowed
for wetting and the degree of wetting is determined by the kinetics rather
than the thermodynamics of the wetting process.

It is known that the wetting phenomena play a significant role in the
coating and printing processes as described by previous workers[1-15]. For
adhesion to occur, an excellent wetting of the substrate by the coating or
ink formulation is required. In addition, the excellent wetting can lead
to good or poor adhesion depending on various interactions between film
forming material and substrate. Wetting is only one of the most important
parameters of coating and printing processes, however; any variation in
the wetting properties and surface characteristics of the coatings/inks
can affect the interactions as well as rheology of the water-based
coating/ink systems. The role of wetting in different typers of printing
processes has been reported recently by Micale et al.[11,12].

[*]Lunglei Government College,
Lunglei-796701, Mizoram (India)

Surface Phenomena and Fine Particles in Water-Based Coatings and Printing Technology   299
Edited by M.K. Sharma and F.J. Micale, Plenum Press, New York, 1991

The wetting of a substrate by a given liquid phase can be quantitively studied by measuring contact angle. The drop formation of a liquid on the substrate depends on the interactions occur at the surfaces such as air/liquid, liquid/solid and air/solid. The contact angle and surface tension of various liquids are studied in order to understand the wetting behavior of the coating/printing substrates by the liquids. These data are used to evaluate dispersive, polar and total energies of several non-porous substrates.

THEORETICAL

The theory of wetting is reflected in various relations applicable to experimental data. These relations were employed to draw general conclusions regarding wetting behavior and mechanisms involve to provide necessary interactions among various surfaces for achieving proper adhesion.

Contact Angle: When a liquid drop is placed on the clean surface of a solid, there is a solid-liquid interface between these two phases, whereas the bare surface of the solid adsorbs the vapor of the liquid until the fugacity of the adsorbed material is equal to that of the vapor and the liquid. One of the methods to measure contact angle by tilting plate was used by Adam and Jessop[16] and modified by Fowkes and Harkins[17]. The plate is tilted until a position is found at which the water surface remains undistorted upto the line of contact with the solid. When the tilting plate is lowered to expose a fresh portion of the plate to the liquid, the angle measured is referred to as the advancing contact angle. When the plate is raised partially out of the liquid, the angle measured is referred to as the receding contact angle.

Another method for measuring the contact angle of a liquid on a solid surface is the sessile drop method of Poynting and Thomson[18]. If a small quantity of a liquid is placed on a flat solid surface, provided that the liquid does not spread spontaneously, a sessile drop will be formed. On addition of more liquid to the drop, the height increases until it reaches a maximum value. Further addition of the liquid increases the diameter of the drop, but not its height. A relationship between the maximum height and contact angle is expressed by the following equation.

$$1 - \cos \theta = h^2 \, d \, g/2 \, \gamma_L \qquad (1)$$

where,
$$d = \text{the density of the liquid.}$$
$$g = \text{acceleration of gravity.}$$
$$\gamma_L = \text{surface tension of the liquid.}$$
$$h = \text{height of the sessile drop.}$$
$$\theta = \text{contact angle.}$$

The contact angle can be determined by measuring the height of the drop, and liquid density. The contact angle can also be measured experimently using goniometer[19]. In order to better understand the wetting process, the wetting phenomena can be categorized as follows:

Wetting Phenomena

Three type of wetting have been known[20] as: (i) spreading wetting, (ii) adhesional wetting (iii) immersional wetting. Among these types of wetting, only spreading and adhesional wettings are involved in the coating/printing processes. Therefore, the discussion is restricted to spreading and adhesional wettings.

(i) <u>Spreading wetting</u>: In spreading wetting, a liquid in contact with a substrate spreads over the substrate and displaces another fluid, such as air, from the surface. For the spreading to occur, the surface free energy of the system must decrease during the spreading process. When a liquid (L) spreads covering an area (a), then the decrease in the surface free energy of the system due to decrease in area of the substrate/air interface is a $\times$ γ_{SA} (where γ_{SA} is the interfacial free energy per unit area of the substrate). At the same time the free energy of the system has been increased because of the increase in liquid/substrate and liquid/air interface. The increase in surface free energy of the system due to the increase in the liquid/substrate interface is a $\times$ γ_{SL} (where γ_{SL} is the interfacial free energy per unit area at the liquid/substrate interface). Since the liquid/air interface has also been increased by area (a), the increase in surface free energy due to increase in the interface is a $\times$ γ_{LA} (where γ_{LA} is the surface tension of liquid). The total decrease in surface free energy per unit area of the system due to the spreading wetting is

$$- \Delta G/a = \gamma_{SA} - (\gamma_{SL} + \gamma_{LA}) \qquad (2)$$

if the R.H.S. of the above equation is positive, the system decreases in surface free energy during the spreading process.

This quantity is a measure of the driving force behind the spreading process, and is usually called the spreading coefficient (S).

$$S = \gamma_{SA} - (\gamma_{SL} + \gamma_{LA}) \qquad (3)$$

Since liquid (L) will not spread on substrate (S) unless $(\gamma_{LA} + \gamma_{SL})$ is less than γ_{SA}. Therefore, if S is positive, spreading can occur spontaneously. The situation in which S is negative, the liquid will not spread spontaneously over the substrate. When the contact angle is larger than zero degree, then the spreading coefficient can not be positive or zero. Equation (3) can be written as follows:

$$S = \gamma_{LA}(\cos \theta - 1) \qquad (4)$$

or $$\cos \theta = [(\gamma_{SA} - \gamma_{SL})/\gamma_{LA}]$$

Hence for complete spreading wetting to occur contact angle should be zero.

(ii) <u>Adhesional wetting</u>: In adhesional wetting, a liquid not originally in contact with a substrate makes contact with that substrate, and adheres to it. In this case the change in surface free energy is as follows:

$$- \Delta G/a = \gamma_{SA} + \gamma_{LA} - \gamma_{SL} \qquad (5)$$

where (a) is the surface area of the substrate in contact with (an equal) surface area of the liquid after adhesion. The driving force of adhesional wetting phenomenon is known as work of adhesion (W_a), which is expressed by following equation.

$$W_a = \gamma_{SA} + \gamma_{LA} - \gamma_{SL} \qquad (6)$$

In this process the reduction of the interfacial tension between substrate and the wetting liquid results in an increased tendency for adhesion to occur, but reduction of either the surface tension of liquid or the surface tension of the substrate decreases the tendency of adhesion to occur.

301

When the contact angle between liquid, substrate and air, after adhesion, is larger than $0°$, then the work of adhesion can not be positive or zero. Equation (6) can be written as follows

$$W_a = \gamma_{LA}(\cos\theta + 1) \tag{7}$$

or $\quad \cos\theta = [(\gamma_{SA} - \gamma_{SL})/\gamma_{LA}]$

It is clear from equation (7) that an increase in the surface tension of the wetting liquid always causes increased adhesional wetting, whereas, an increase in the contact angle obtained after wetting may indicate decreased tendency for adhesion to occur. For good wetting, adhesion may be good or poor depending on suface interactions, but for poor wetting always adhesion will be poor.

Evaluation of Surface Energies: The energy associated to a liquid drop adhering to the solid surface is expressed as a sum of the existing interfacial tensions minus that of the newly created interface:

$$W_a = \gamma_{SA} + \gamma_{LA} - \gamma_{SL} \tag{8}$$

From combining equations described above, and rearranging, one can obtain a relation as follows:[21]

$$1 + \cos\theta = 2\{(\gamma_i^d \gamma_s^d)^{1/2} + (\gamma_i^p \gamma_s^p)^{1/2}\}/\gamma_{LA} \tag{9}$$

By measuring contact angles of two different liquids (e.g. water and methylene iodide) of known dispersive and polar components, these components can be evaluated for polymer films or dried coating/ink films. The water and methylene iodide are selected because of their high total surface energies resulting the formation of the droplet, while differ greatly in their functional polar components of the surface energies.

The total surface energy (γ^t) of the coating/printing substrates can be evaluated from the summation of their polar (γ^p) and dispersive (γ^d) components of the surface energies as follows:

$$\gamma^t = \gamma^d + \gamma^p \tag{10}$$

EXPERIMENTAL

Materials: Several polymer films were used to measure contact angle of liquids. These films were washed with water and completely dried before use. Methylene iodide and other liquids were BDH (Analar grade), and used as received without further purification. Deionized, distilled water was used in all experiments.

Methods: Surface tension of various liquids was measured by drop volume method using Agla micrometer syringe[22]. The micrometer syringe was cleaned using acetone, and completely dried before measuring volume of the liquid drop. An average of three readings was taken to evaluate surface tension of the liquid. Contact angle measurements were performed by measuring height of a drop on a solid clean surface or by using Goniometer[19]. Results are reported as an average of three readings. From the contact angle data of water and methylene iodide on a given substrate, surface energies such as dispersive, polar and total energies of the substrates were evaluated by using relations described in theoretical section of this article.

RESULTS AND DISCUSSION

Figure 1 schematically shows the surface forces acting on a liquid drop. If the surface tension of solid is greater than that of the liquid, the contact angle of liquid drop on that surface will be less than 90 degree, and liquid will spread (figure 1A). If surface tension of solid is less than that of the liquid, the contact angle will be greater than 90 degree, and liquid will never spread (figure 1B). Therefore, surface tension of the formulated coating/ ink, as mentioned earlier, must be less than the surface tension of the substrate. If the coating/printing substrate is polymer film (e.g. low energy substrate), the coating/printing with water-based system is relatively difficult as compared to the solvent-based system due to high surface tension of water (e. g. 72.8 dynes/cm). In general, surface energy of the polymer substrates lies in the range of 25-40 dynes/cm, which can be raised to 35-50 dynes/cm range by corona discharge or coating the polymer substrates.

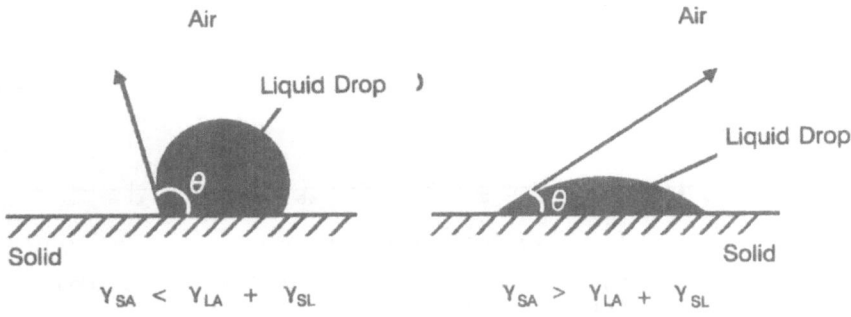

Figure 1. A schematic illustration of the forces acting on a liquid drop.

In order to reduce the surface tension of water-based coating/ink formulations, alcohol/water mixture is preferred instead of water. The surface tension can also be reduced to the desired level by incorporating surfactants. As surfactant molecules are larger than that of alcohol molecules, diffusion of surfactant molecules in the liquid medium is slower than alcohol molecules. Therefore, the mixture of water/alcohol containing surfactant can provide optimized coating/ink formulation with excellent properties such as improved gloss, fast drying rate, improved film forming ability, minimize coating and/or printing defects and good adhesion.

The surface tension of different alcohol/water mixtures and their contact angle on polymer film is presented in figure 2. A linear relationship is observed between surface tension and contact angle. It is evident from the figure that liquid drop with low surface tension can spread faster as compared to the liquid drop with high surface tension on the same substrate. For coating/printing polymer substrates, the surface tension of the coating/ink needs to be in the range of 25-40 dynes/cm range, which can be achieved by incorporating alcohol and/or surfactant in the water for water-based formulations.

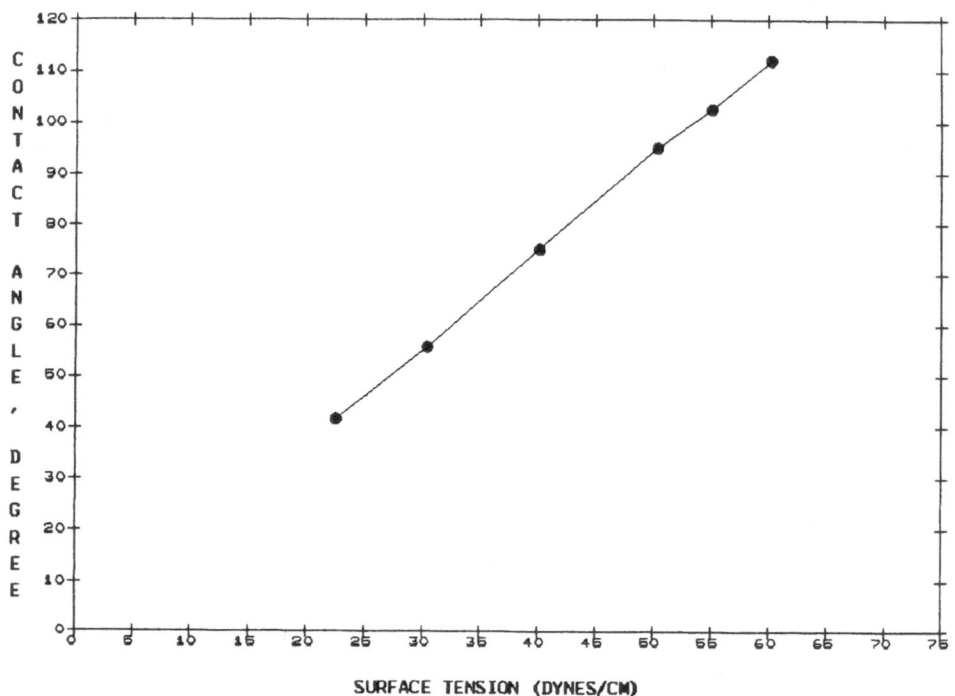

Figure 2. Surface tension of liquids as a function of their contact angle on polymer films.

For evaluating surface energies of the polymer substrates, contact angles of water and methylene iodide were measured, and data obtained are recorded in table 1. Contact angle of water is higher as compared to that of the methylene iodide on various polymer films, indicating films are hydrophobic in nature, and can not wet with water.

The contact angle of water on several polymer films is in the range of 95-115 dynes/cm, whereas the contact angle of methylene iodide is in the range of 45-70 dynes/cm. The surface energies of the polymer films were evaluated by using data of table 1. For evaluating surface energies, polar and dispersive energies for water used were 51.0 and 21.8 ergs/sq cm, respectively, while polar and dispersive surface energies for methylene iodide used were 1.3 and 49.5 ergs/sq cm. The dispersive, polar and total surface energies for various polymer films are recorded in table 2.

Table 1. Contact Angle of Water and Methylene Iodide on various
Polymer Substrates

Polymer Film	Contact Angle	
	Water	Methylene Iodide
A	96.0	59.0
B	102.0	51.0
C	98.0	45.o
D	115.0	66.0
E	109.0	57.0

Table 2. Surface Energies of Various Polymer films

Polymer Films	Surface Energies			Polar/dispersive Ratio
	Total	Polar	Dispersive	
A	29.2	1.2	28.0	0.044
B	34.2	0.0	34.2	0.001
C	37.4	0.2	37.2	0.004
D	26.7	0.2	26.5	0.007
E	31.6	0.1	31.5	0.002

It is evident from table 2 that the polar surface energy is very low
for the polymer films examined. The dispersive surface energies are in
the range of 25-40 ergs/sq cm, while polar surface energies are in the
range of 0-2 ergs/sq cm. In order to adhere coating/ink to these polymer
substrates, the polar and dispersive surface energies of the formulated
coating/ink should be the same as that of the polymer substrate.

CONCLUSIONS

The wettability of the polymer films was studied by measuring contact
angle of alcohol/water mixtures, water and methylene iodide. A linear
relationship was observed between surface tension of liquids and their
contact angle on the given polymer film. The polar, dispersive and total
surface energies were evaluated using contact angle data of water and
methylene iodide. Results demonstrate that the dispersive surface energies
mainly contribute to the total surface energies of the polymer films.

REFERENCES

1. Vash, R., Wetting and dispersing, in "Handbook of Coatings Additives" Calbo, L. J. (Editor), Marcel Dekker, Inc., p.511-539 (1987).

2. Terkowitz, A., Vehicles and ink performance, American Ink Maker, 19-35, December (1989).

3. Podhajny, R. M., TRPPI, Polymers, Laminations, and Coatings Conference, Proceedings Book-2, p.333 (1987).

4. Smith, N. C., TAPPI Proceedings on Polymers, Laminations, and Coatings Conference, Book-2, p.583-585 (1987).

5. Markgraf, D. A., Treatment required for printing with water-based inks, Proceedings on Polymers, Laminations, and Coatings Conference, Book-1, p.333-336 (1987).

6. Seefried, Jr., C. G. and Mier, M. A., Surface characterization of corona treated polyethylene film; ANTEC p.269-272 (1985).

7. Blythe, A. R., Briggs, D., Kendall, C. R., Rance, D. G. and Zichy, V. J. I., Surface modification of polyethylene by electrical discharge treatment and the mechanism of autoadhesion; Polymer, 19, 1273-1278 (1978).

8. Oss, C. J. van, Good, R. J. and Busscher, H. J., Estimate of the polar surface tension parameters of glycol and formamide, for use in contact angle measurements on polar solids; J. Disp. Sci. Technology, 11(1), 75-81 (1990).

9. Kaeble, D. H., Physical Chemistry of Adhesion; Wiley- Interscience, (1971).

10. Fowkes, F. M., Treatise on Adhesion and Cohesion; Patrick, R. L. (Editor), Marcel Dekker, New York, NY (1967).

11. Micale, F. J., Iwasa, S., Lavelle, J., Sunday, S. and Fetsko, J. M., The role of wetting: Part-1; American Ink Maker, p.44-54, September (1998).

12. Micale, F. J., Iwasa, S., Lavelle, J., Sunday, S. and Fetsko, J. M., The role of wetting: Part-2; American Ink Maker, p.25-35, October (1998).

13. Owens, D. K. and Wendt, R. C., Estimation of surface free energy of polymers; J. Applied Polymer Science, 13, 1741-1747 (1969).

14. Gould, R.; Editor- Contact Angle: Wettability and Adhesion, Library of Congress #63-14481 (1964).

15. Van der Linden, R., Adhesion and absorption of polymers; Polymer Science Technology: Part-B. p.563 (1980).

16. Adam, N. K., The Physics and Chemistry of Surfaces; London, Oxford University Press, London (1941).

17. Fowkes, F. M. and Harkins, W. D., J. Am. Chem. Soc., 62, 3377 (1940).

18. Poynting and Thomson, "Properties of matters"- Text Book of Physics, London, p.156 (1905).

19. Sharma, M. K., Shiao, S. Y., Bansal, V. K. and Shah, D. O., Effect of chain length compatibility on monolayers, foams, and macro- and microemulsions; in Macro- and Microemulsions: Theory and Practice; ACS Symosium Series, 272, 87-103 (1985).

20. Rosen, M. J., Surfactants and Interfacial Phenomena; John Wiley and Sons, Inc., New York, NY, p.240-274 (1989).

21. Watson, W. M., Adhering to polypropylene with water-based inks; American Ink Maker, 38, October (1984).

22. Sharma, M. K., Sharma, M., Jain, S. P. and Srivastava, S. N.; Adsorption studies of sodium deoxycholate at various liquid/liquid interfaces in the presence of added salt and their role in emulsion stability, J. Colloid and Interface Sci., 64, 179-184 (1978).

PARTICLE-SURFACE DEPOSITION IN THE PRESENCE OF ADSORBED POLYMER LAYERS

Th F Tadros[1], P Warzynski[2] and M Zembala[2]

[1]ICI Agrochemicals, Jealotts Hill Research Station
Bracknell, Berkshire, RG12 6EY, U.K.

[2]Institute of Catalysis and Surface Chemistry
Polish Academy of Sciences, 20-239 Cracow
Niezapominajek, Poland

The influence of addition of polymers eg.
poly(ethylene oxide) (PEO) or poly(vinyl alcohol) (PVA)
on the deposition of negative particles on a positive
substrate is discussed in terms of the balance between
steric repulsion and van der Waals and electrostatic
attraction. A theory for particle deposition is
derived based on the solution of the transport equation
for particles in a liquid jet in the vicinity of the
stagnent point. Both hydrodynamic and surfaces forces
are taken into account. Equations are derived for all
interaction forces involved. Numerical solutions of
these equations are produced to show the variation of
particle flux with polymer concentration for two cases.
In the first case steric interaction is neglected and
the effect of polymer is described in terms of the
presence of a hydrodynamic barrier that opposes
deposition. In the second case, the steric interaction
is superimposed on the hydrodynamic interactions. In
this case two regimes of reduction in particle flux are
observed as the polymer concentration is increased.
The first regime, at low polymer concentrations, shows
a slow reduction in flux since the steric repulsion is
relatively small compared to the van der Waals and
double layer attraction. In the high polymer
concentration regime, the reverse is true and the flux
decreases rapidly with increase in polymer
concentration.

Results were obtained for the deposition of
negative polystyrene latex on positive mica sheets in
the presence of various molecular weight PEO or PVA.
With PEO, the change in flux with increase in polymer
concentration (at low electrolyte concentration)
follows the trend predicted from theory. The relative
flux results could be used to obtain the hydrodynamic
thickness of the polymer layer. Using reasonable

Surface Phenomena and Fine Particles in Water-Based Coatings and Printing Technology
Edited by M.K. Sharma and F.J. Micale, Plenum Press, New York, 1991

309

values for the other parameters eg. the   (polymer
solvent) and   (segment distribution) parameters,
agreement between theory and experiment was
reasonable.

With PVA, on the other hand, the flux showed
a rapid reduction at low polymer concentration and
eventually the flux reached a low plateau value.
This different behaviour was attributed to the
formation of thick adsorbed layers which are
developed at low polymer concentrations. Under
these conditions, the reduction in flux is
simply due to the hydrodynamic drag produced by
the adsorbed polymer layer and steric repulsion
plays a minor role in this case.

INTRODUCTION

Particle deposition and adhesion onto solid surfaces (the substrate)
are of great importance in many industrial applications. In some cases,
deposition and adhesion is desirable, e.g. in paper coating, in paints
and in agrochemical application whereby high tenacity is required to
prevent rain-washing of the particles from the crop. In other cases,
deposition and adhesion of particles is undesirable, as for example is
the case with prevention of dental plaque, in detergency and in some
application in pharmaceuticals, e.g. targetted delivery of drugs where
particle adhesion has to be prevented before reaching the target.

It is perhaps useful to distinguish between deposition and adhesion
in terms of the range of interaction between the particles and the
collector. Deposition is the process of flow and attachment of particles
to a substrate and hence it may be considered as a limiting case of
heterocoagulation. The kinetics of this process is determined by the
rate of transport from the bulk ofthe suspension and by all types of
long-range interactions between particle and collector surfaces. These
long-range interaction forces include double layer, van der Waals and
steric interaction. Moreover, deposition is determined by the
hydrodynamic interaction between particle and surfaces. Adhesion, on
the other hand, is determined by the short-range interaction between
particle and collector that may involve particle and/or surface
deformation. The adhesion is measured by the force necessary to
separate the adherents after making intimate contact. In this respect,
adhesion is less understood phenomenon than deposition.

The general theory of convective transport from well defined flow
fields have been developed by Dabros, Adamzyck and collaborators (1-5).
This theory allows one to predict the effect of various parameters on
the deposition kinetics. Dabros et al (6-8) developed a convenient
method for studying particle deposition which gives the possibility of
direct in situ determination of deposition kinetics for transparent
colloids.   The stagnation point flow cell ensures uniform accessibility
of a collector surface in the area around the symmetry axis. This was
applied for determination of particle flux when deposition proceeded
without an energy barrier. This was achieved by making the collector
oppositely charged to the particles to be deposited. The dependence of
the particle flux on flow intensity, particle size and ionic strength
obtained from these measurements, was in quantitative agreement with
theoretical predictions.

Deposition in the presence of an energy barrier, e.g. under
conditions of double layer repulsion, is much more difficult to study
experimentally. This is due to the large changes that occur in
deposition rate with small changes in the parameters that determine the
energy barrier height. Results obtained under these repulsive
conditions only show a qualitative correlation between the deposition
rate and the physico-chemical variables influencing the energy barrier
height (9). The particle fluxes measured were always much higher than
predicted by the theory. This discrepancy was accounted for by the
roughness and heterogeneity of collector and particle surface. The
effect of heterogeneity can be described in terms of the energy barrier
fluctuation. This was clearly demonstrated by Adamczyk et al (10),
who showed that random fluctuations of barrier height (with low
frequency) can increase the deposition rate by a few orders of
magnitude.

When a polymer is adsorbed on both particle and collector
surfaces, the theory of particle deposition should be modified to take
into account the steric interactions that result from the presence of
adsorbed polymer layers. Moveover, one should consider the increase of
convective transport   due to the growth of the hydrodynamic radius of
a particle (intercept effect),increased solution viscosity and
modification of the hydrodynamic interaction between particle and
collector surface.

The steric interactions resulting from the presence of the
adsorbed layers start to play a role at a particle-surface distance
comparable to twice the adsorbed layer thickness. These steric
interactions are conveniently divided into two terms, namely mixing
(osmotic) and elastic (volume restriction) contributions (11). The
first terms results from the change infree energy of the system when
the polymer layers overlap which leads to an increase in segment
density in the overlap region. The second term, arises from the
decrease in configurational entropy of the chains when significant
overlap occurs. The steric interaction is determined by several
parameters, the most important of which are the segment density
distribution and the polymer - solvent interaction parameter χ (the
Flory-Huggins interaction parameter). It should be mentioned that
the presence of adsorbed polymer layer will also affect the other two
interactions, namely van der Waals and double layer. The effect of
adsorbed polymer layer on van der Waals attraction was first recognised
by Vold (12) and later analysed in more detail by Vincent and coworkers
(13,14). Polymer adsorption can modify the electrical double layer
repulsion in several ways, e.g. by modifying the distribution of ions
in the electrical double layer, by changing the ion mobility and by
the effect of the polymer on the dielectric permittivity of the medium
near an interface (15). All three effects are difficult to quantify
and in most cases, the effect of polymer layers is simply described
in terms of a simple shift in shear plane (16-18).

Recently, one of us (19) has developed a theory for particle
deposition in the presence of adsorbed polymer layer. A summary of the
theoretical treatment and the assumption made will be given in the next
Section. This is followed by a description of the technique used for
measuring particle deposition in the presence of adsorbed polymer
layers. A summary of the results obtained will be given and a
comparison of the experimental results with those predicted from theory
will be given. Full details of the results and their interpretation
were recently published (20).

THEORY OF PARTICLE DEPOSITION IN THE PRESENCE OF ADSORBED POLYMER LAYERS

The theory of particle deposition is based on the solution of the transport equation for particles in a liquid jet impinging in the vicinity of the stagnation point. Both hydrodynamic and surface forces are taken into account. The stationary transport equation takes the following dimensionless form,

$$\frac{d}{dH} F_1(H) \left[-\frac{d\bar{n}}{dH} - \frac{1}{2} Pe\, F_2(H)\ (H+1)^2 \bar{n} - \frac{d\bar{\phi}}{dH}\, \bar{n} \right] + Pe\, F_3(H)\ (H+1)\bar{n} = 0 \tag{1}$$

where H is the dimensionless surface-to-surface separation (i.e., $H = h/a$, where h is the surface-to-surface distance and a is the particle radius). $\bar{n}$ is the dimensionless particle number concentration that is equal to $n/\bar{n}_\infty$, where n_∞ is the bulk particle number concentration. Pe is the dimensionless shear rate or Peclet number, $F_1(H)$, $F_2(H)$ and $F_3(H)$ are universal hydrodynamic functions accounting for the Stokes' law correction due to the presence of the collector surface (5, 21, 22) $\bar{\phi} = \phi/kT$ is the energy that is the sum of the three contribution of van der Waals $\bar{\phi}_d$, double layer $\bar{\phi}_e$ and steric interaction $\bar{\phi}_s$ and any external field contribution eg. gravity or electric field,

$$\bar{\phi} = \bar{\phi}_d + \bar{\phi}_e + \bar{\phi}_s \tag{2}$$

$\bar{\phi}_d$ is given by the Hamaker equation for a sphere - flat plate interaction and introducing a retardation correction, i.e.

$$\bar{\phi}_d = -\ Ad\ \frac{\lambda'}{H(\lambda' + \bar{s}\ 11)} \tag{3}$$

where Ad = A/6 kT (A is the Hamaker constant), $\lambda' = \lambda/a$ is the dimensionless retardation parameter and $\bar{s}$ is equal to 11.116

$\bar{\phi}_e$ is given by the expression derived by Hogg, Healy and Furestenau (23) for a sphere - flat plate interaction, i.e.,

$$\bar{\phi}_e = D\ell \left\{ \ln\left[1 + \exp(-\tau H)\right] + \frac{Da}{2} \ln\left[1 - \exp(-2\tau H)\right] \right\} \tag{4}$$

where $\tau = a\kappa$ (κ is the Debye-Huckel parameter); $D\ell = \varepsilon\ \zeta_1\ \zeta_2\ a\ k/T$ is the dimensionless double layer number (ζ_1 and ζ_2 are the zeta potentials of particle and collector surface) and $Da = 0.5\ (\zeta_1 - \zeta_2)^2/\zeta_1 \zeta_2)$ is the double layer asymmetry parameter.

In the present theory, it was assumed that the presence of adsorbed polymer layer does not cause any change in distribution of ions in the double layer. The effect of the polymer layer was simply considered to cause a shift in the shear plane.

$\bar{\phi}_s$ is considered to consist of two contribution $\bar{\phi}_s^o$ (osmotic) and $\bar{\phi}_s^e$ (elastic)). Using Deryaguin approximation, expressions for $\bar{\phi}_s^o$ and $\bar{\phi}_s^e$ were derived (19) and expressed in terms of two dimensionless numbers, namely a dimensionless osmotic steric number St^o and a dimensionless elastic steric number St^e, i.e.

$$\bar{\phi}^{\circ}_{s} = St^{\circ} \int_{H}^{\infty} \left\{ \frac{\Gamma(2\alpha - 1)}{2^{2\alpha+1}\ \Gamma^{2}(\alpha)\ \beta^{'}} \left[\frac{\bar{\gamma}(2\alpha - 1, 2H/\beta^{'})}{\bar{\gamma}^{2}(\alpha, H/\beta^{'})} - 1 \right] \right.$$

$$\left. + \left[\frac{(\Gamma\Gamma^{'})^{2}}{\Gamma^{2} + (\Gamma^{'})^{2}} - 1 \right] \left(\frac{H}{\beta^{'}} \right)^{2\alpha-1} \frac{\exp(-H/\beta^{'})}{\beta^{'}\ \Gamma(2\alpha)\ \bar{\gamma}^{2}(\alpha, H/\beta^{'})} \right\} \partial H \qquad (5)$$

$$\bar{\phi}^{el}_{s} = St^{el} \int_{H}^{\infty} \left\{ \ln \frac{1}{\bar{\gamma}(\alpha, H/\beta^{'})} + (\alpha - 1) \frac{d}{d\alpha} \ln\bar{\gamma}(\alpha, H/\beta^{'}) \right.$$

$$\left. -\alpha \left[\frac{\bar{\gamma}(\alpha + 1, H/\beta^{'})}{\bar{\gamma}(\alpha, H/\beta^{'})} - 1 \right] \right\} dH \qquad (6)$$

where $St^{\circ} = 2\ \pi N_{A}\ (0.5 - \chi)\ a^{3}\ (\Gamma^{2} + (\Gamma^{'})^{2})\ /\ V_{1})$ and
$St^{el} = 2\ \pi \rho\ a^{3}\ (\Gamma + \Gamma^{'})/Mp$ where N_{A} is the Avogadros constant, Γ is
the polymer adsorption in gm cm^{-2}, $\Gamma^{'} = \Gamma/\rho_{p}$ a where ρ_{p} is the polymer
density, M_{p} is the polymer molecular weight and V_{1} is the molar volume
of the solvent. $\bar{\gamma}^{'} = \gamma/\Gamma$ where γ is the polymer distribution function.
α and β are parameters of the γ distribution.

Fig. 1 shows the dependence of the osmotic (A) and elastic
interaction energy on the dimensionless particle - collector surface
separation. Various curves A correspond to various values of the
$(1/2 - \chi)$ parameter and the same value of β (0.001) and Γ (8.7×10^{-4}).
The particle radius was taken to be 0.47 μm (which is the radius of the
particles used in the experimental work). It can be seen from Fig. 1
that the elastic term is relatively small compared to the osmotic term.

84

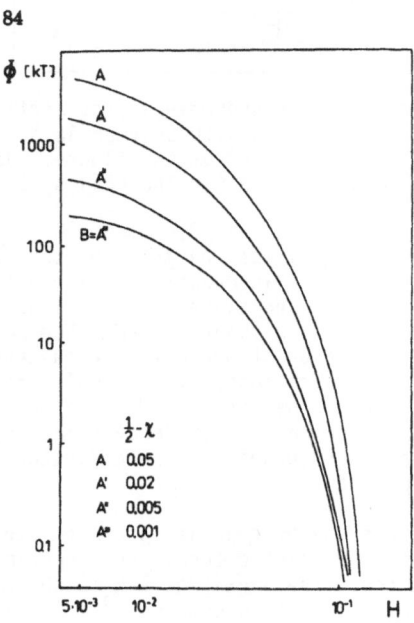

Fig. 1 - Dependence of osmotic (A) and elastic (B) terms of steric
interaction energy on the dimensionless particle-collector separation H.
Various curves A correspond to various values of the $(1/2 - \chi)$ parameter
and the same values of β (0.001) and Γ ($8.7\ 10^{-4}$). Particle radius
equal to 0.47 μm.

Fig. 2 shows the results of calculation of the hydrodynamic correction function F_1 for particles and collector both covered with an adsorbed polymer layer. Since the adsorbed polymer layer increases the effective hydrodynamic radius of the particle and hence the hydrodynamic drag exerted on it (24,25), the value of any hydrodynamic function $F_n(H)$ for a given distance H between particle and collector both bearing polymer layers is taken as equal to the values of $F_n(H)$ for a particle of radius a_h and bare surface at the distance $h' = h - 2\delta_h$ where $\delta_h = a_h - a$, is the hydrodyanmic thickness of the polymer layer, assumed to be equal for both particle and collector surfaces.

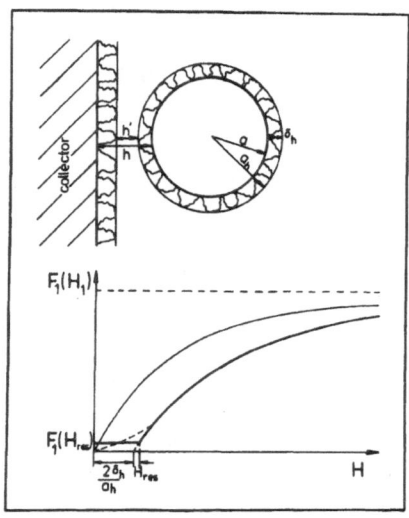

Fig. 2 - Postulated change of hydrodynamic correction function F_1 for particles and collector covered with polymer layers according to the model presented in the upper part of the figure. Dashed line represents a probable dependence for the real system.

The transport equation (1) was solved using the fourth order Kulter method and the integrals in equations (5) and (6) were solved numerically. Fig. 3 shows the relative particle flux (nomalised to the flux under the same conditions without any adsorbed polymer) to the collector surface as a function of hydrodynamic thickness δ_h at three different electrolyte concentrations. In the calculations, steric interactions were assumed to be absent and the adsorbed polymer layer was considered to simply increase the hydrodynamic drag exerted on a particle moving in the neighbourhood of the collector surface in the presence of adsorbed polymer.

Fig. 4 shows the results of calculation, when steric interaction (including both osmotic and elastic terms) are superimposed on hydrodynamic interactions. The particle flux is shown is a function of the dimensionless osmotic steric number St^o. It can be seen from Fig. 4 that at low adsorption and/or weak polymer-solvent interaction (small St^o) the particle flux is completely determined by the hydrodynamic thickness of the adsorbed layer since in this case, the weak steric interaction cannot counter balance the stronger van der Waals and electrostatic attraction. At high St^o values, on the other hand, an abrupt drop in particle flux appears at a certain value of adsorption. Under these conditions, the steric interaction becomes

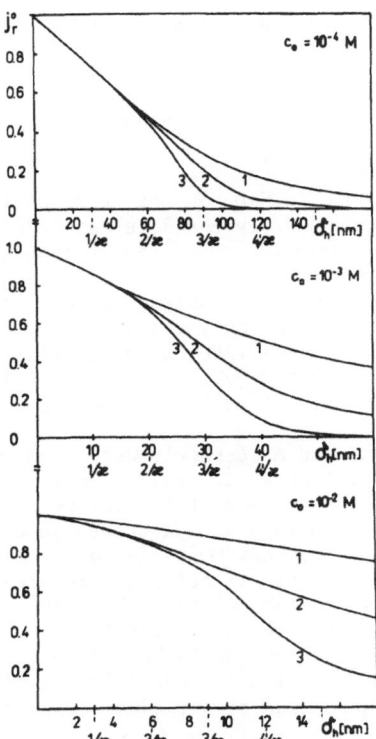

Fig. 3 - Relative particle flux as a function of hydrodynamic layer thickness (at three electrolyte concentrations) calculated assuming that only the hydrodynamic properties of the system were changed in the presence of polymer (with interaction energy the same as for bare particle and collector surfaces). Curves 1, 2 and 3 were calculated for values of H_{res} equal to 10^{-1}, 10^{-2} and 10^{-3}, respectively.

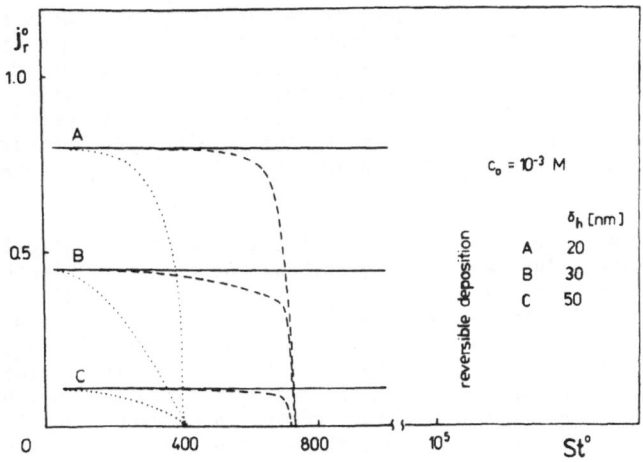

Fig. 4 - Relative particle flux as a function of dimensionless osmotic steric number, calculated when both steric and hydrodynamic effects were taken into account. Values of $\kappa\beta$ were: 0.33, 0.67 and 1.00 for full, dashed and dotted lines, respectively. The value of H_{res} is equal to 10^{-2}.

comparable in magnitude to the van der Waals and electrostatic
attraction.

INVESTIGATION OF PARTICLE-SURFACE DEPOSITION

For this purpose, the stagnent flow method was used. The
experimental set up is shown in Fig. 5. As model particles, negatively
changed fairly monodisperse polystyrene latex paricles with radius of
0.47 μ m were used. The collector consisted of thin mica shells that
were made positively charged by immersion in 1% aqueous solution of
γ -aminopropyltriethylsilane for 5 minutes, washed with water and
heated to 80°C for 24 hours. The polymers used were commercial
materials of poly(ethylene oxide) (PEO) with weight average molecular
weights of 90,000 and 300,000 and 600,000, and poly(vinyl alcohol) (PVA)
with average molecular weights of 45,000 and 88,000.

The latex suspensions and mica sheets were conditioned in polymer
solutions for about 20 hours prior to measurements. Deposition kinetics
was determined by direct observation of the collector surface exposed to
the jet of suspension. The method was described in detail by Dabros and
van de Ven (26) and applied for determination of particle fluxes under
barrierless conditions. All results refer to the area in the vicinity
of the stagnation point where the collector surface is uniformly
accessible for the particles. The measurements were performed at a
constant flow viscosity corresponding to Reynolds number $\sim$ 28 and at
constant particle number concentration equal to $\sim$ 5 x 10^7 cm^{-3}. In most
cases four measurements were made at any given experimental conditions.

The electrophoretic mobility of latex particles was measured using a
Pen Ken Electrokinetic Analyser system 3000 at the same conditions as the
deposition studies.

RESULTS AND DISCUSSION

Figs. 6 and 7 shows a comparison of the kinetics of deposition
studies in the absence and the presence of PEO respectively. The polymer
concentrations shown in Fig. 2 are the initial polymer amounts added,
which is close to the equilibrium concentration since the particle number
concentration was low (less than 1% of the polymer added was adsorbed).
The initial shape of these curves give the value of the initial particle
flux J^0 expressd as the number of particles per $\mu m^2 s$. In order to avoid
any uncertainty of the particle number, the results were normalised using
the bulk suspension concentration n and the data were represented as the
normalised flux J^0 ($\mu m\ s^{-1}$).

Fig. 8 shows the results for the effect of molecular weight of PEO
on the normalized flux, whereas Fig.9 shows the effect of electrolyte
concentration on deposition for one molecular weight (90,000). At low KCl
concentration (10^{-4} and 10^{-3} mol dm^{-3}) there is an initial slow decrease of
flux with increasing polymer concentration. At a critical polymer
concentration there is a sharp drop in particle flux with further increase
in polymer concentration. At high KCl concentration (10^{-3} mol dm^{-3}), the
flux decreases almost linearly with increase in polymer concentration.
At any given polymer concentration, the average flux decreases with
increase in molecular weight of the polymer.

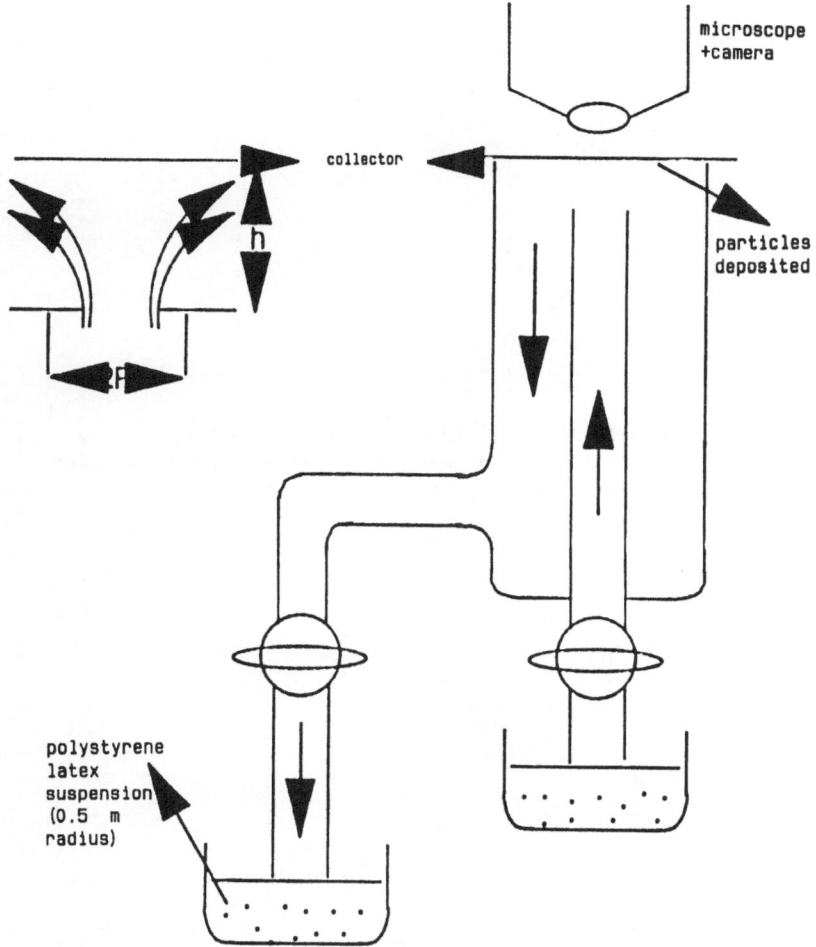

microscope
+camera

collector

particles
deposited

h

F

polystyrene
latex
suspension
(0.5 m
radius)

Fig. 5 - Schematic diagram of the experimental set up
to measure particle deposition

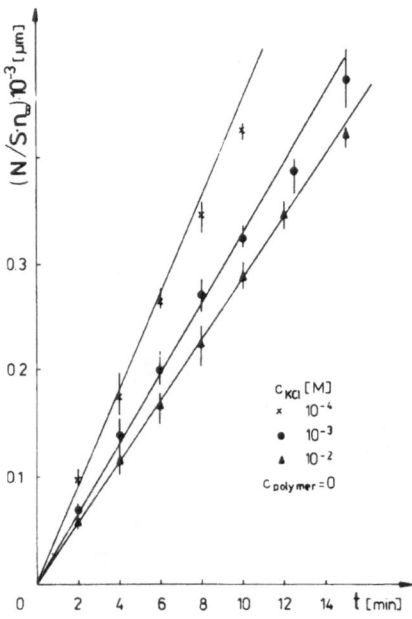

Fig. 6 - Number of particles deposited on unit collector area from suspension of unit particle concentration as a function of time (no polymer added).

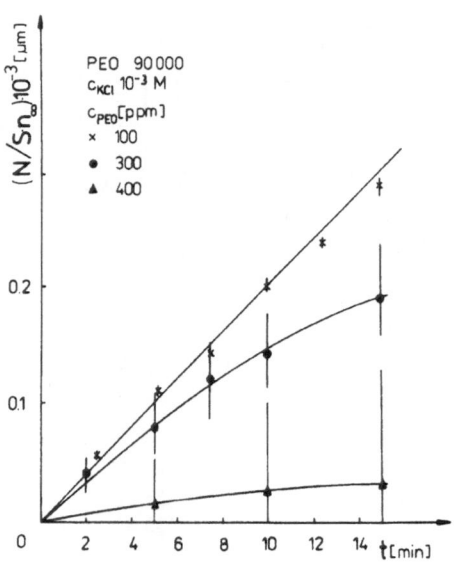

Fig. 7 - Number of particles deposited on unit collector area from suspension of unit particle concentration in the presence of PEO 90,000 as a function of time.

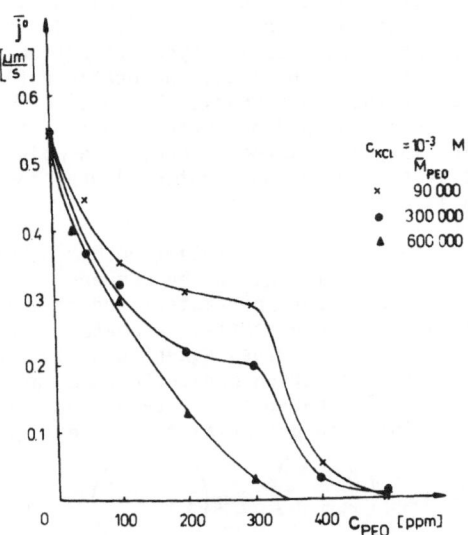

Fig. 8 - Dependence of the normalized particle flux on the polymer
concentration for PEO of various molecular weights.

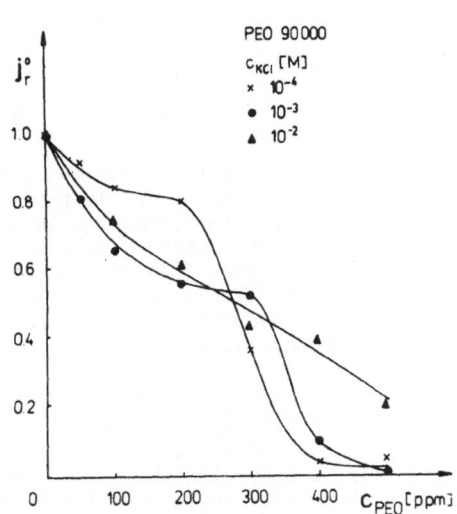

Fig. 9 - Normalized particle flux versus the PEO 90,000 concentration
for various ionic strengths.

The trend observed at low electrolyte concentration implies that there are at least two different mechanisms of particle deposition with increasing polymer concentration. Such behaviour is predicted theoretically as discussed in Section 2. The initial slow decrease in particle flux is consistant with the relatively small steric repulsion compared with van der Waals and double layer attraction, at such low polymer dosage. Under these conditions, the particle flux is completely determined by the hydrodynamic thickness of the adsorbed layer. However, at a critical polymer concentration, there is a sharp decrease in particle flux with further increase in polymer concentration. In this regime, the steric repulsion counterbalances any attraction and particle deposition is significantly reduced.

The experimental results obtained using 90,000 PEO at three different electrolyte concentrations can be compared with theoretical predictions provided the necessary parameters for calculating the steric interaction are available. The essential parameters are δ_h (the hydrodynamic layer thickness), β (the segment distribution parameter) and χ (the polymer solvent interaction parameter). δ_h was calculated from the deposition results and compared to δ_e, (electrokinetic adsorbed layer thickness) estimated using zeta potential measurement, i.e.

$$\tanh\left(\frac{Z\, e\, \zeta}{4kT}\right) = \tanh\left(\frac{Z\, e\, \zeta_o}{4kT}\right)\, \exp\left(1 - \kappa\, \delta_e\right) \tag{7}$$

where ζ is the electrokinetic potential of latex particles in electrolyte concentration containing polymer and ζ_o is the value in absence of added polymer. Values of δ_h, δ_e are given in Table 1 for PEO, $M_w = 90,000$.

Table 1

C_{KCl} mol dm^{-3}	δ_h	β	δ_e
10^{-4}	24	16	43
10^{-3}	28	6	22
10^{-2}	28	2	7

It can be seen that δ_h, estimated from the relative particle flux J^o and assuming $H_{res} = 10^{-2}$ is independent of electrolyte concentration. In contrast δ_e seems to decrease with increase in electrolyte. For the calculation of J^o the δ_h values were used. χ was taken to be equal 0.484. A comparison of theoretical calculated with the experimentally determined J^o value is shown in Fig. 10. Considering all the assumptions and approximations made, the agreement between theory and experiment is quite reasonable.

Figs. 11 and 12 show the results of J^o for two PVA molecular weights (45,000 and 88,000 respectively) at three different electrolyte concentrations. The trends obtained are quite different from these obtained using PEO. Firstly, the results shown considerable scatter, that may be attributed to the larger polydispersity of PVA compared to the PEO. Secondly, the results show a rapid reduction in flux reaching a near plateau low value with the PVA concentration studied. This difference in behaviour between the PEO and PVA may be attributed to the difference in structure of the adsorbed polymer layer. Neutron scattering experiment (27) showed that PVA forms long dangling tails giving much higher adsorbed layer thicknesses when compared with PEO with similar molecular weight. Such thick adsorbed layers are also reflected in electrophoretic mobility measurements which show a rapid

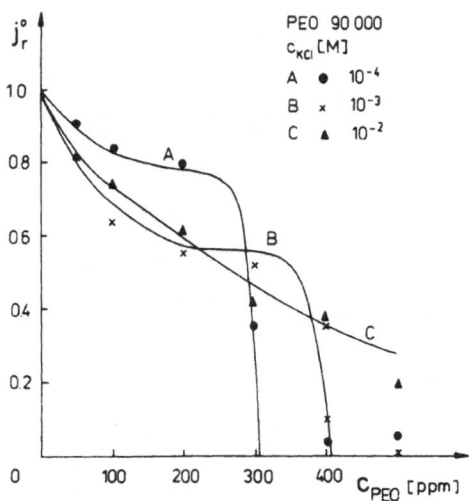

Fig. 10 - Relative particle flux versus polymer concentration for various values of ionic strength. Points represent experimentally obtained values. Curves were calculated theoretically using estimated values of δ_h and β (cf. Table 1) with $\chi = 0.484$, $H_{resd} = 10^{-2}$ and adsorption isotherm taken from (15).

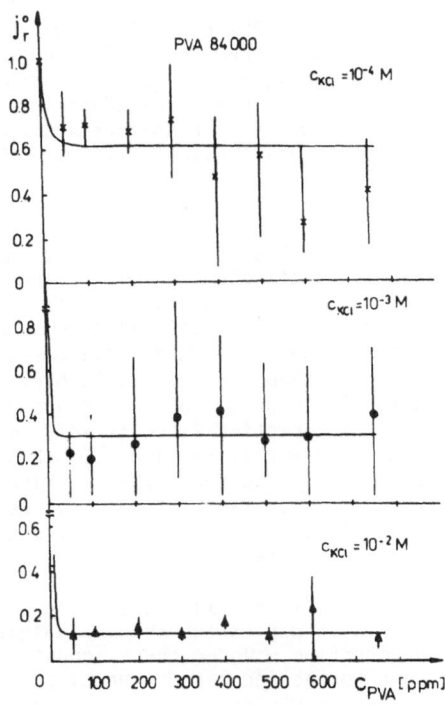

Fig. 11 - Relative particle flux versus the PVA 84,000 concentration for various ionic strengths.

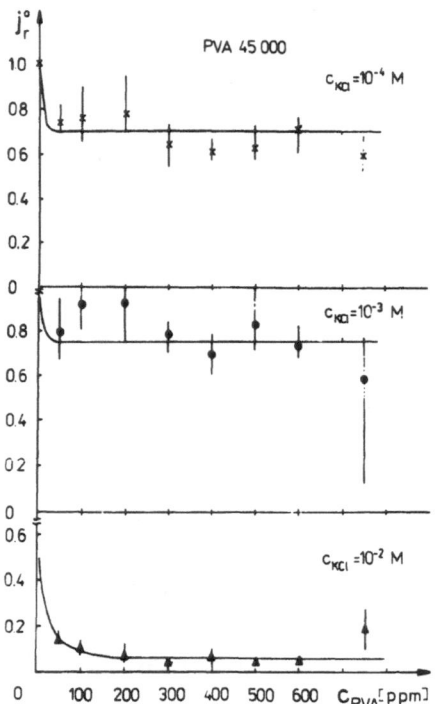

Fig. 12 - Relative particle flux versus the PVA 45,000 concentration for various ionic strengths.

decrease with initial increase in PVA concentration. When the polymer adsorbed layer becomes fully developed (and this occurs at relatively low PVA concentration), the particle flux becomes independent of polymer concentration. Under these conditions, the reduction in deposition rate is due to the hydrodynamic drag with the polymer adsorption layer and steric interaction plays a secondary role.

CONCLUSIONS

Adsorbed polymer layers of PEO or PVA cause a significant reduction in the initial deposition rate of negative polystyrene latex particles on positive mica sheets compared to the values obtained in pure electrolyte solutions. With PEO the flux shows an initial reduction with increase in polymer concentration followed by a rapid reduction above a critical polymer concentration. This trend agrees with theoretical predictions of particle deposition in the presence of adsorbed polymer. At low polymer concentrations, the steric interaction is relatively small compared to the van der Waals and double layer attraction. In this regime, the deposition rate is determined by the hydrodynamic thickness of the polymer layer. At higher polymer concentrations, the steric repulsion predominates over van der Waals and double layer attraction and this results in significant reduction in the flux.

With PVA, the particle flux initially decreases and rapidly attains a constant low level. This is due to the thicker polymer layers that develop at low PVA concentration. Under these conditions, the reduction in deposition rate is due to the increased hydrodynamic drag within the polymer adsorption layer and steric interaction plays a secondary role.

Deposition rate measurements can be used to estimate the adsorption.

REFERENCES

1. T. Dabros, Z. Adamczyk and J. Czarnecki, J. Colloid Interface Sci. 62, 529 (1977).
2. D. C. Prieve and M. J. Lin, J. Colloid Interface Sci., 76, 32 (1980).
3. Z. Adamczyk and T. G. M. van de Ven, J. Colloid Interface Sci. 54, 497 (1981).
4. T. Dabros and T. G. M. van de Ven, J. Colloid Interface Sci. 89, 232 (1982).
5. Z. Adamczyk, T. Dabros, J. Czarnecki and T. G. M. van De Ven, Adv. Colloid Interface Sci. 19, 183 (1983).
6. T. Dabros and T. G. M. van de Ven, Colloid Polymer Sci. 261, 694 (1983).
7. Z. Adamczyk, M. Zembala, B. Ziwek and J. Czarnecki, J. Colloid Interface Sci. 110, 188 (1986)
8. Z. Adamczyk, B. Ziwek, M. Zembala and P. Warszynski, J. Colloid Interface Sci., in press.
9. M. Hull and J. A. Kitchener, Trans. Faraday Soc. 65, 3093 (1969).
10. Z. Adamczyck, J. Czarnecki and P Warsynski, J. Colloid Interface Sci. 106, 299 (1985).
11. D. H. Napper "Polymeric Stabilisation etc.
12. M. J. Vold, J. Colloid Interface Sci.,
13. D. W. J. Osmond, B. Vincent and F. A. Waite, J. Colloid Interface Sci. 42, 262 (1977).
14. B. Vincent, J. Colloid Interface Sci. 42, 270 (1973).
15. B. V. Deryaguin, S. S. Du Khin and A. E. Yaroschuk, J. Colloid Interface Sci., 115, 234 (1987).
16. H. Maier, J. A. Baker and J. C. Berg, J. Colloid Interface Sci. 119, 512 (1987).
17. A. A. Baran, N. H. Soboleva and L. M. Du Khin, Kolloid Zh. 46, 840 (1984).
18. M. J. Garvey, Th. F. Tadros and B. Vincent, J. Colloid Interface Sci. 55, 440 (1976).
19. P. Warzynski, Colloids and Surfaces, 39, 79 (1989).
20. Th. F. Tadros and M. Zembala, Colloids and Surfaces, 39, 93 (1989).
21. L. A. Spielman and J. A. Fitzpatrick, J. Colloid Interface Sci. 42, 607 (1973).
22. L. A. Spielman and P. M. Cukor, J. Colloid Interface Sci. 43, 51 (1973).
23. R. Hogg, T. W. Healy and D. W. Fuerstenau, Trans. Faraday Soc. 62, 1638 (1966).
24. J. H. Masliyah, G. Neale, K. Malysa and T. G. M. van de Ven, Chem. Eng. Sci. 42, 245 (1987).
25. S. Saaki, Colloid Polymer Sci. 263, 935 (1985).
26. T. Dabros and T. G. M. van de Ven, Colloid Polym. Sci. 261, 694 (1983).
27. K. Barnet, T. Cosgrove, T. L. Crowley, Th. F. Tadros and B. Vincent, in "The Effect of Polymers on Dispersion Stability", Editor Th. F. Tadros, Academic Press, London (1982) pp.183-197).

ORGANIC PIGMENTS AND THEIR RELATIONSHIPS TO AQUEOUS INK

George Sonn

MAGRUDER COLOR CO INC
1029 Newark Ave
Elizabeth NJ   07208

INTRODUCTION

Aqueous printing ink is essentially an exercise in colloidal chemistry.
Organic pigments are the basic colorants used to pigment aqueous printing
ink systems. As such organic pigments present vast physical surface areas
compared to inorganic pigments. An acrylic color chip represents a new
concept in the manufacture of aqueous printing inks. The basic premise is
the use of water soluble acrylic carrier to produce an acrylic color
concentrate in chip form. This chip is then solubilized in a basic aqueous
solution to yield a homogeneous color base which can be combined with a
vehicle to produce a finished ink or coating.

The printing ink industry is currently on a stampede course to replace
solvent coatings with aqueous coatings. A prime concern in the non-
commodity coatings area is the quality of the aqueous film versus the
solvent film. This concern manifests itself in the area of packaging where
in many cases the colored package actually sells the product.

An acrylic chip with locked in colorant is an approach whereby the advantageous properties of a quality solvent coatings can be realized and
environmental concerns can be minimized. The Clean Air Act is shifting its
impact. The staffs of the nine regional EPA groups are forcefully exerting
pressure on their state counterparts to enforce the amounts to dealing with
both state and federal enforcers.

The Clean Air Act specifies the amount of VOC or volatile organic compound
emissions limit that a plant is permitted to operate within. The major VOC
sources (Solvent and oil users) have already been identified in the United
States by region and are pin-pointed within the region. The ability of
generators of VOC emissions to comply with the governmental regulations
will be examined on a case by case basis but the key point is that the
specific plants must make good faith attempts to comply. Consent orders
are used to implement compliance. These orders, together in come cases
with fines will allow the environmental authorities to enforce the VOC
emission limits. In metropolitan New york City Area there are 20 to 25
major convertors which can emit 100 tons or more of VOC per year. In
California area, there are 35 major convertor sources of VOC emission.

Compliance via the use of very low solvent inks must be implemented. Water inks and coatings have taken tremendous strides in film and print quality.

A key parameter to be noted for acrylic color concentrates is that a whole new type of pigment application must be utilized to produce pigments that have increased aqueous stability. Aqueous pigment stability means a stabilized pigment system which does not gel, flocculate, or separate over time. Pigment selection within the same color index is the key premise for aqueous colloidal system stability.

The specific system being discussed in this paper contains 60% organic pigment and 40% of a water soluble acrylic resin. The acrylic resin is compatible with other resins and has excellent pigment wetting characteristics that further promote dispersion. The resin is water white and does not degrade on processing. Common pigments used include Azo and Naphthol Red, Diarylide Yellow, Phthalocyanine Blue and Green, Methyl and Carbazole Violet and Dianisidine and DNA Orange. No additives, dispersants or defoamers have been added to this system due to potential deleterious effects in the convertor process stream. The pigment and acrylic resin system is combined via pre-mixing in a change can mixer. This dry mix is dispersed on a high horse power heavy duty two roll mill with power knives. The two roll mill affords theoretical optimum dispersion as a mechanical dispersion means. The two roll mill sheets are cooled and chipped via a mechanical chipper to a uniform non-dusting physical particle size. The final color concentrate is blended to conform to SPC color control system.

The advantages of a solid system versus a liquid system are obvious. A base that will have an extremely long shelf life, no settling, bodying freeze thaw stability or colloidal adverse phenomena are present. The dry non-dusting base is the most environmentally desirous as spills can be swept or vacuumed up without color loss to the surrounding environment. A color chip represents a pigment that is locked into a vehicle system that will only be released by the color user in the amount needed. The dry color concentrate may be cut at a customer site in the amount and pigment percents desired. The key to the use of the acrylic chip is the pH of the cutting solution which must always be above the 8.5 region, while 9 gives a better margin for any plant errors. Common bases to impart basic pH include morpholine, sodium hydroxide and various other amine and ammonium derivatives.

The color chips produced have maximum transparency and greater gloss than color concentrates prepared by conventional dispersing equipment. In side by side testing of dry color milling in an aqueous acrylic system the color chips produced 20 to 25% greater color strength. Favorable strength, gloss and transparency also exist (versus fluidized presscake systems). In a high performance pigment such as Carbazole Violet this can relate to a saving of $8.00 per pound on an equivalent color strength basis. In addition the optimum dispersion of the color chips assures a better and more continuous trouble free quality and more homogeneous film coating. SPC programs show better consistency due to improved product quality. Table 1 compares a high speed dry pigment with a solubilized acrylic chip. As indicated on spectrophotometric adjusted readings the gloss and strength are much imrpoved for the acrylic chip. Gloss is evaluated via gloss meter.

EXPERIMENTAL

The acrylic color concentrate chip can be used in all aqueous ink and print systems. The acrylic resin chosen is a styrene acrylic type which has

TABLE 1 SPECTROPHOTOMETRIC DATA

	GLOSS	ADJUSTED COLOR STRENGTH
Carbazole Violet Acrylic Chip	62	+ 20%
Carbazole Violet Dry Pigment	28	
BON Rubine Acrylic Chip	57	+ 18%
BON Rubine Dry Pigment	23	
HR Yellow Acrylic Chip	60	+ 17%
HR Yellow Dry Pigment	19	
Calcuim Lithol Acrylic Chip	57	+ 18%
Calcium Lithol Dry Pigment	21	

excellent compatibility. This acrylic resin also has excellent pigment wetting characteristics which help to promote ultimate dispersion level. The resin may also be used as a pigment additive treatment. The resin is water white and does not degrade on processing so that no deleterious properties are imparted to the color concentrate.

The acrylic color concentrate is prepared in base form from 25 to 40% solids content. Hot or cold water may be utilized according to local plant procedures. The cutting equipment utilized spans a variety of equipment geared for solubilization. Equipment included high speed mixers, Kady mills and other high speed equipment. An especially efficient piece of equipment is the Hill or Eppenbach mixer which provides multi-directional cross currents in the liquid system. The basic cutting system utilized a mixture of water and basic compound to produce a pH in the range of 8.5 to 9.0. The color concentrate acrylic chip is added slowly to this system with adequate agitation. Complete solution should be accomplished in one hour.

The finished solution may be blended with other color components and may be utilized as a finished aqueous colorant base. This base may be added to various resin vehicles as long as the overall pH of the system is maintained above 8.5. The acrylic chip color concentrate system affords a systems which is stable until used whereas the use of fluidized aqueous colorants allows all the problems of flocculation, settling, freezing and general pigment solution problems to be present.

The use of color concentrate acrylic chips affords a convenient methodology to comply with the new wave of aqueous printing inks.

RESULTS AND DISCUSSION

Aqueous systems employ entirely different theoretical concepts from those upon which solvent systems are based. Aqueous systems are extremely polar and colloidal. Aqueous systems are dynamic and, therefore, much more vulnerable than solvent systems to the effects of double-layer phenomena and ionic charges. Pigments for aqueous systems must, therefore, be formulated with their stabilization requirements constantly in mind.

The development of aqueous colors is based on the study of interfacial phenomena; when pigments are "struck" and are consequently given final surface treatment or particle-size alteration, such treatment is governed by interfacial phenomena. In colloidal, the surface is enormous compared to the mass of substance. Consequently, the magnitude of surface forces such as absorption, capillarity, and surface tension becomes extremely large and their effect is critical.

The primary pigment particle is built up by crystal growth and accretion. In actual use, the ultimate pigment particle represents a formation of clusters of particles. This phenomenon is termed flocculation and its occurrence is determined by the degree to which the primary particles are kept separate. These phenomena affect both transparency and gloss as well as the color yield of the pigment.

Classical colloidal solutions are stabilized by a electrical double layer around the colloidal particles. Deflocculation is determined by the variation in potential energy between two approaching particles. Attraction and repulsion are in effect at the same time: London-Vander Waals forces attract the particles; the double electrical layer repel them.

The inner of the two layers is caused by the preferential absorption of one species of ion at the surface of the particle. When two particles approach each other, the two negatively charged clouds begin to interact and give rise to a repulsive force, which is dependent upon the distance between them.

The double layer is one of the stabilizing factors in colloidal systems. In some cases, stabilization may be the result of the stearic nature of the absorbed layer. One way of improving colloidal stability is to chemically modify the pigment surface so as to take advantage of the stearic nature of the absorbed layer.

A surface parameter profile can be used to predict surface activity of pigments in the aqueous phase. Each pigment has its own electrical-charge profile, which can be correlated to the clear polymer system. Pigments are produced either by the grinding or mechanical attrition of crude material or by actual chemical synthesis. Both methods produce salts that create high electrical charges. once the pigment is in its finished fluid form, it can be left as is or treated with surface-active agents or various resins that alter its surface charges. The pigment is then washed and filtered; the presscake is dried and micropulverized and is ready to be converted to an acrylic chip.

The zeta potential is closely correlated to colloidal stability. Salts present in many phases of pigment manufacture interact proportionally with the power of the zeta potential. Positively charged particles adhere to negative particles.

Again, the closer the pigment surface can be made to fit the colloidal charge profile of the resin, the better the color value and dispersion of the final color concentrate. Aquaflo was designed with these concepts in mind.

CONCLUSION

Formulation and application of water-based systems can be more accurately qualified by the use of the theoretical models. A theoretical statistical model of pigments can be developed in which the constant is water solvent with basic pH level. The design of pigments for aqueous systems must take

account of colloidal science. the polar characteristics of the aqueous
system are of critical importance. Table 2 indicates various typical
Aquaflo compositions of matter in varius color ranges.

TABLE 2 AQUAFLO TYPICAL COMPOSITION

	% Pigment		Color Index
Black	45	Black 7	
Phthalocyanine Blue		60	Blue 15:3
Phthalocyanine Green		60	Grren 7
Dianisidine Orange		60	Orange 16
Naphthol Red BS		60	Unassigned
Rhodamine Y	60	Red 81	
Carbazole Violet		60	Violet 23
Yellow AAoT	60	Yellow 14	
Yellow HR		60	Yellow 83
BON Rubine	60	Red 52:1	
Red Lake C	60	Red 53:1	
Calcium Lithol		60	Red 49:2
Methyl Violet		60	Violet 3
Red 2b YS		60	Red 48:1

Aquaflo was designed with these factors in mind. When compared with
conventional pastes, Aquaflo offers better print quality, brilliance,
and color strength.

REFERENCES

1. Adam NK Physical Chemistry Oxford, 1956.
2. J.T. & Rideal E.K. Interfacial Phenomena, Academic Press,
    1963.
3. Shinoda K, Tamamushi B, Nakagawa T & Isemura T, Collodial
    Surfactants, Academic Press, 1983.
4. Gibbs, Collected Works, University Press, 1948.
5. Hartley & Brunskill, Surface Phenomena in Chemistry and
    Biology, Press 1958.

AUTHOR INDEX

SUBJECT INDEX